# Texts in Theoretical Computer Science
## An EATCS Series

T0190222

Editors: W. Brauer G. Rozenberg A. Salomaa

On behalf of the European Association
for Theoretical Computer Science (EATCS)

---

Advisory Board: G. Ausiello M. Broy C.S. Calude
A. Condon D. Harel J. Hartmanis T. Henzinger
J. Hromkovič N. Jones T. Leighton M. Nivat
C. Papadimitriou D. Scott

Marcus Hutter

# Universal Artificial Intelligence

## Sequential Decisions Based on Algorithmic Probability

Springer

*Author*

Dr. Marcus Hutter
Istituto Dalle Molle
di Studi sull'Intelligenza
Artificiale (IDSIA)
Galleria 2
CH-6928 Manno-Lugano
Switzerland
marcus@idsia.ch
www.idsia.ch/~marcus

*Series Editors*

Prof. Dr. Wilfried Brauer
Institut für Informatik der TUM
Boltzmannstr. 3, 85748 Garching, Germany
Brauer@informatik.tu-muenchen.de

Prof. Dr. Grzegorz Rozenberg
Leiden Institute of Advanced Computer Science
University of Leiden
Niels Bohrweg 1, 2333 CA Leiden, The Netherlands
rozenber@liacs.nl

Prof. Dr. Arto Salomaa
Turku Centre for Computer Science
Lemminkäisenkatu 14 A, 20520 Turku, Finland
asalomaa@utu.fi

ACM Computing Classification (1998): I.3, I.2.6, F.0, F.1.3, F.4.1, E.4, G.3

ISBN 978-3-642-06052-6 e-ISBN 978-3-540-26877-2

This work is subject to copyright. All rights are reserved, whether the whole or part of the material is concerned, specifically the rights of translation, reprinting, reuse of illustrations, recitation, broadcasting, reproduction on microfilm or in any other way, and storage in data banks. Duplication of this publication or parts thereof is permitted only under the provisions of the German Copyright Law of September 9, 1965, in its current version, and permission for use must always be obtained from Springer. Violations are liable for prosecution under the German Copyright Law.

Springer is a part of Springer Science+Business Media

springeronline.com

© Springer-Verlag Berlin Heidelberg 2010
Printed in Germany

The use of general descriptive names, registered names, trademarks, etc. in this publication does not imply, even in the absence of a specific statement, that such names are exempt from the relevant protective laws and regulations and therefore free for general use.

Cover design: KünkelLopka, Heidelberg

Printed on acid-free paper 45/3142/YL - 5 4 3 2 1 0

# Preface

**Personal motivation.** The dream of creating artificial devices that reach or outperform human intelligence is an old one. It is also one of the dreams of my youth, which have never left me. What makes this challenge so interesting? A solution would have enormous implications on our society, and there are reasons to believe that the AI problem can be solved in my expected lifetime. So, it's worth sticking to it for a lifetime, even if it takes 30 years or so to reap the benefits.

**The AI problem.** The science of artificial intelligence (AI) may be defined as the construction of intelligent systems and their analysis. A natural definition of a *system* is anything that has an input and an output stream. Intelligence is more complicated. It can have many faces like creativity, solving problems, pattern recognition, classification, learning, induction, deduction, building analogies, optimization, surviving in an environment, language processing, and knowledge. A formal definition incorporating every aspect of intelligence, however, seems difficult. Most, if not all known facets of intelligence can be formulated as goal driven or, more precisely, as maximizing some utility function. It is, therefore, sufficient to study goal-driven AI; e.g. the (biological) goal of animals and humans is to survive and spread. The goal of AI systems should be to be useful to humans. The problem is that, except for special cases, we know neither the utility function nor the environment in which the agent will operate in advance. The major goal of this book is to develop a theory that solves these problems.

**The nature of this book.** The book is theoretical in nature. For most parts we assume availability of unlimited computational resources. The first important observation is that this does not make the AI problem trivial. Playing chess optimally or solving NP-complete problems become trivial, but driving a car or surviving in nature do not. This is because it is a challenge itself to well-define the latter problems, not to mention presenting an algorithm. In other words: The AI problem has not yet been well defined. One may view the book as a suggestion and discussion of such a mathematical definition of AI.

**Extended abstract.** The *goal* of this book is to develop a universal theory of sequential decision making akin to Solomonoff's celebrated universal theory of induction. Solomonoff derived an optimal way of predicting future data, given previous observations, provided the data is sampled from a computable probability distribution. Solomonoff's unique predictor is universal in the sense that it applies to every prediction task and is the output of a universal Turing machine with random input. We extend this approach to derive an optimal rational reinforcement learning agent, called AIXI, embedded in an unknown environment. The *main idea* is to replace the unknown environmental distribution $\mu$ in the Bellman equations by a suitably generalized universal distribution $\xi$. The state space is the space of complete histories. AIXI is a universal theory without adjustable parameters, making no assumptions about the environment except that it is sampled from a computable distribution. From an algorithmic complexity perspective, the AIXI model generalizes optimal passive universal induction to the case of active agents. From a decision-theoretic perspective, AIXI is a suggestion of a new (implicit) "learning" algorithm, which may overcome all (except computational) problems of previous reinforcement learning algorithms.

*Chapter 1.* We start with a survey of the contents and main results in this book.

*Chapter 2.* How and in which sense induction is possible at all has been subject to long philosophical controversies. Highlights are Epicurus' principle of multiple explanations, Occam's razor, and Bayes' rule for conditional probabilities. Solomonoff elegantly unified all these aspects into one formal theory of inductive inference based on a universal probability distribution $\xi$, which is closely related to Kolmogorov complexity $K(x)$, the length of the shortest program computing $x$. We classify the (non)existence of universal priors for several generalized computability concepts.

*Chapter 3.* We prove rapid convergence of $\xi$ to the unknown true environmental distribution $\mu$ and tight loss bounds for arbitrary bounded loss functions and finite alphabet. We show Pareto optimality of $\xi$ in the sense that there is no other predictor that performs better or equal in all environments and strictly better in at least one. Finally, we give an Occam's razor argument showing that predictors based on $\xi$ are optimal. We apply the results to games of chance and compare them to predictions with expert advice. All together this shows that Solomonoff's induction scheme represents a universal (formal, but incomputable) solution to all *passive* prediction problems.

*Chapter 4.* Sequential decision theory provides a framework for finding optimal reward-maximizing strategies in *reactive* environments (e.g. chess playing as opposed to weather forecasting), assuming the environmental probability distribution $\mu$ is known. We present this theory in a very general form (called AI$\mu$ model) in which actions and observations may depend on arbitrary past events. We clarify the connection to the Bellman equations and discuss minor parameters including (the size of) the I/O spaces and the lifetime of the agent and their universal choice which we have in mind. Optimality of AI$\mu$ is obvious by construction.

*Chapter 5.* Reinforcement learning algorithms are usually used in the case of unknown $\mu$. They can succeed if the state space is either small or has effectively been made small by generalization techniques. The algorithms work only in restricted, (e.g. Markovian) domains, have problems with optimally trading off exploration versus exploitation, have nonoptimal learning rate, are prone to diverge, or are otherwise ad hoc. The formal solution proposed in this book is to generalize the universal prior $\xi$ to include actions as conditions and replace $\mu$ by $\xi$ in the AI$\mu$ model, resulting in the AIXI model, which we claim to be universally optimal. We investigate what we can expect from a universally optimal agent and clarify the meanings of *universal*, *optimal*, etc. We show that a variant of AIXI is self-optimizing and Pareto optimal.

*Chapter 6.* We show how a number of AI problem classes fit into the general AIXI model. They include sequence prediction, strategic games, function minimization, and supervised learning. We first formulate each problem class in its natural way for known $\mu$, and then construct a formulation within the AI$\mu$ model and show their equivalence. We then consider the consequences of replacing $\mu$ by $\xi$. The main goal is to understand in which sense the problems are solved by AIXI.

*Chapter 7.* The major drawback of AIXI is that it is incomputable, or more precisely, only asymptotically computable, which makes an implementation impossible. To overcome this problem, we construct a modified model AIXI$tl$, which is still superior to any other time $t$ and length $l$ bounded algorithm. The computation time of AIXI$tl$ is of the order $t \cdot 2^l$. A way of overcoming the large multiplicative constant $2^l$ is presented at the expense of an (unfortunately even larger) additive constant. The constructed algorithm $M_p^\varepsilon$ is capable of solving all well-defined problems $p$ as quickly as the fastest algorithm computing a solution to $p$, save for a factor of $1+\varepsilon$ and lower-order additive terms. The solution requires an implementation of first-order logic, the definition of a universal Turing machine within it and a proof theory system.

*Chapter 8.* Finally we discuss and remark on some otherwise unmentioned topics of general interest. We also critically review what has been achieved in this book, including assumptions, problems, limitations, performance, and generality of AIXI in comparison to other approaches to AI. We conclude the book with some less technical remarks on various philosophical issues.

**Prerequisites.** I have tried to make the book as self-contained as possible. In particular, I provide all necessary background knowledge on algorithmic information theory in Chapter 2 and sequential decision theory in Chapter 4. Nevertheless, some prior knowledge in these areas could be of some help. The chapters have been designed to be readable independently of one another (after having read Chapter 1). This necessarily implies minor repetitions. Additional information to the book (FAQs, errata, prizes, ...) is available at http://www.idsia.ch/~marcus/ai/uaibook.htm.

**Problem classification.** Problems are included at the end of each chapter of different motivation and difficulty. We use Knuth's rating scheme for exercises [Knu73] in slightly adapted form (applicable if the material in the corresponding chapter has been understood). In-between values are possible.

C00 *Very easy.* Solvable from the top of your head.
C10 *Easy.* Needs 15 minutes to think, possibly pencil and paper.
C20 *Average.* May take 1–2 hours to answer completely.
C30 *Moderately difficult or lengthy.* May take several hours to a day.
C40 *Quite difficult or lengthy.* Often a significant research result.
C50 *Open research problem.* An obtained solution should be published.

The rating is possibly supplemented by the following qualifier(s):

*i* Especially *interesting* or *instructive* problem.
*m* Requires more or higher *math* than used or developed here.
*o* *Open* problem; could be worth publishing; see web for prizes.
*s* *Solved* problem with published solution.
*u* *Unpublished* result by the author.

The problems represent an important part of this book. They have been placed at the end of each chapter in order to keep the main text better focused.

**Acknowledgements.** I would like to thank all those people who in one way or another have contributed to the success of this book. For interesting discussions I am indebted to Jürgen Schmidhuber, Ray Solomonoff, Paul Vitányi, Peter van Emde Boas, Richard Sutton, Leslie Kaelbling, Leonid Levin, Peter Gács, Wilfried Brauer, and many others. Shane Legg, Jan Poland, Viktor Zhumatiy, Alexey Chernov, Douglas Eck, Ivo Kwee, Philippa Hutter, Paul Vitányi, and Jürgen Schmidhuber gave valuable feedback on drafts of the book. Thanks also collectively to all other IDSIAnies and to the Springer team for the pleasant working atmosphere and their support. This book would not have been possible without the financial support of the SNF (grant no. 2000-61847.00). Thanks also to my father, who taught me to think sharply and to my mother who taught me to do what one enjoys. Finally, I would like to thank my wife and children who patiently supported my decision to write this book.

Lugano, Switzerland, August 2004 *Marcus Hutter*

# Contents

# Tables, Figures, Theorems, ...

# Notation

The following is a list of commonly used notation. The first entry is the symbol itself, followed by its meaning or name (if any) and the page number where the definition appears. Some standard symbols like $\mathbb{R}$ are not defined in the text. There appears a * in place of the page number for these symbols.

| Symbol | Explanation | Page |
|---|---|---|
| [C35s] | classification of problems | viii |
| [Hut04b] | paper, book or other reference | * |
| (5.3) | label/reference for a formula/theorem/definition/... | * |
| $\infty$ | infinity | * |
| $\{a,...,z\}$ | set containing elements $a,b,...,y,z$. $\{\}$ is the empty set | * |
| $[a,b)$ | interval on the real line, closed at $a$ and open at $b$ | * |
| $\cap,\cup,\setminus,\in$ | set intersection, union, difference, membership | * |
| $\wedge,\vee,\neg$ | Boolean conjunction (and), disjunction (or), negation (not) | * |
| $\subseteq,\subset$ | subset, proper subset | * |
| $\Rightarrow$ | implies | * |
| $\Leftrightarrow$ | equivalence, if and only if, iff | * |
| $\Box$ | q.e.d. (Latin), which was to be demonstrated | * |
| $\forall,\exists$ | for all, there exists | * |
| $\approx,\lesssim,\gtrsim$ | approximately equal, less equal, greater equal | 33 |
| $\ll,\gg$ | much smaller/greater than | * |
| $\equiv$ | equivalent, identical, equal by definition | * |
| $\cong$ | isomorphic | * |
| $:=$ | define as | * |
| $\widehat{=}$ | corresponds to, informal equality | * |
| $\sim$ | asymptotically proportional to | 33 |
| $\propto$ | proportional to | * |
| $=,\neq$ | equal to, not equal to | * |
| $+,-,\cdot,/$ | standard arithmetic operations: sum, difference, product, ratio | * |
| $\sqrt{\ }$ | square root | * |
| $\leq,\geq,<,>$ | standard inequalities | * |
| $\|S\|,\|a\|$ | size/cardinality of set $S$, absolute value of $a$ | * |

| | | |
|---|---|---|
| $\rightarrow$ | mapping, approaches, Boolean implication | * |
| $\rightarrow$ | converge to each other | 33 |
| $\lim_{n\to\infty}$ | limiting value of argument for $n$ tending to infinity | * |
| $\rightsquigarrow$ | replace with | * |
| $\lceil x\rceil$ | ceiling of $x$: smallest integer larger or equal than $x$ | * |
| $\lfloor x\rfloor$ | floor of $x$: largest integer smaller or equal than $x$ | * |
| $\delta_{ab}$ | Kronecker symbol, $\delta_{ab}=1$ if $a=b$ and 0 otherwise | * |
| $\sum_{k=1}^n$ | summation from $k=1$ to $n$ | * |
| $\sum_x'$ | summation over $x$ for which $\mu(x)\neq 0$ | 69 |
| $\prod_{k=1}^n$ | product from $k=1$ to $n$ | * |
| $\int,\int_a^b dx$ | Lebesgue integral, integral from $a$ to $b$ over $x$ | * |
| $i,k,n,t$ | natural numbers | 33 |
| $x,y,z$ | finite strings | 33 |
| min/max | min-/maximal element of set: $\min_{x\in\mathcal{X}} f(x)=\min\{f(x):x\in\mathcal{X}\}$ | * |
| argmin | $\text{argmin}_x f(x)$ is the $x$ minimizing $f(x)$ (ties broken arbitrarily) | * |
| l.h.s. | left-hand side | * |
| r.h.s. | right-hand side | * |
| w.r.t. | with respect to | * |
| e.g. | exempli gratia (Latin), for example | * |
| i.e. | id est (Latin), that is | * |
| etc. | et cetera (Latin), and so forth | * |
| cf. | confer (Latin, imperative of conferre), compare with | * |
| et al. | et alii (Latin), and others | * |
| q.e.d. | quod erat demonstrandum (Latin), which was to be shown | * |
| i.i.d. | independent identically distributed (random variables) | * |
| iff | if and only if | * |
| w.p.1/i.p. | with probability 1 / in probability | 71 |
| i.m./i.m.s. | in the mean / in mean sum | 71 |
| log | logarithm to some basis | * |
| $\log_b$ | logarithm to basis $b$ | * |
| ln | natural logarithm to basis e=2.71828... | * |
| e | base of natural logarithm e=2.71828... | * |
| $\mathbb{R}$ | set of real numbers | * |
| $\mathbb{R}^+$ | set of nonnegative real numbers | * |
| $\mathbb{N}$ | set of natural numbers $\{1,2,3,...\}$ | 33 |
| $\mathbb{N}_0$ | set of natural numbers including zero $\{0,1,2,3,...\}$ | 33 |
| $\mathbb{Z}$ | set of integers $\{...,-2,-1,0,1,2,3,...\}$ | * |
| $\mathbb{Q}$ | set of rational numbers $\{\frac{n}{d}\}$ | * |

| | | |
|---|---|---|
| $\mathbb{B}=\{0,1\}$ | binary alphabet | * |
| $y_t\in\mathcal{Y}$ | action (output of agent) in cycle $t$, followed by ... | 128 |
| $x_t\in\mathcal{X}$ | perception (feedback/input to agent) in cycle $t$ | 45, 128 |
| $o_t\in\mathcal{O}$ | informative input/observation in cycle $t$ | 128 |
| $r_t\in\mathcal{R}\subset\mathbb{R}$ | reward in cycle $t$ | 128 |
| $\varepsilon$ | some small positive real number | * |
| $\epsilon$ | empty string | 33 |
| $*$ | wildcard for some string (prefix, finite, or infinite) | 33 |
| $x_{1:n}$ | $=x_1...x_n=$ string of length $n$ | 45, 68, 128 |
| $x_{<t}$ | $=x_1...x_{t-1}=$ string of length $t-1$ | 45, 68, 128 |
| $y\!x_{k:n}$ | action-perception sequence $y_k x_k ... y_n x_n$ | 128 |
| $\dot{y}\!\dot{x}_{<k}$ | actually realized action-perception sequence $\dot{y}_1\dot{x}_1...\dot{y}_{k-1}\dot{x}_{k-1}$ | 130 |
| $\omega$ | infinite sequence, elementary event | 33 |
| $\Omega$ | sample space | 42, 68 |
| $\Gamma_{x_{1:n}}$ | $=\{\omega:\omega_{1:n}=x_{1:n}\}=$ cylinder set | 46, 68 |
| $\ell(x)$ | length of string $x$ | 33 |
| $\langle o\rangle$ | coding of object $o$ | 33 |
| $\langle x,y\rangle$ | uniquely decodable pairing of $x$ and $y$ | 33 |
| $x'$ | prefix coding of $x$ | 33 |
| $O(),o()$ | big and small oh-notation | 33 |
| $a\overset{+}{\leq}b$ | less within an additive const., i.e. $a\leq b+O(1)$. Similarly $\overset{+}{=}$ | 33 |
| $a\overset{\times}{\leq}b$ | less within a multiplicative const., i.e. $a=O(b)$. Similarly $\overset{\times}{=}$ | 33 |
| $K(x)$ | prefix Kolmogorov complexity of string $x$ | 37 |
| $Km(x_{1:n})$ | monotone (Kolmogorov) complexity of string $x_{1:n}$ | 47, 190 |
| $K(o_1\|o_2)$ | Kolmogorov complexity of object $o_1$, given object $o_2$ | 37 |
| $M\overset{\times}{=}\xi_U$ | Solomonoff-Levin's universal semimeasure | 46, 48 |
| $\mathcal{M}=\{\nu\}$ | (usually countable) set of (semi)measures | 48, 81 |
| EC | $\in\{$AI, SP, FM, EX, SG, ...$\}$ is an environmental class | * |
| AI | artificial or algorithmic intelligence, | 2 |
| | most general computational environmental class | 130, 154 |
| SP | sequence prediction | 187 |
| CF | classification | 108 |
| SG | strategic two-player informed zero-sum games | 192 |
| FM | function minimization | 197 |
| EX | supervised learning (by examples) | 204 |
| pd | probability density function / distribution / measure | * |
| $\rho(x_{1:n})$ | probability of string/sequence starting with $x_{1:n}$ | 46, 68 |
| $\mu\in\mathcal{M}$ | true generating environmental pd | 68 |

| | | |
|---|---|---|
| $\mathbf{E}$ | expectation value, usually w.r.t. the true distribution $\mu$ | 68 |
| $\mathbf{P}$ | probability, usually w.r.t. the true distribution $\mu$ | 68 |
| $\mu(x_1\underline{x}_2x_3\underline{x}_4)$ | $\mu$ probability that the $2^{nd}$ and $4^{th}$ symbols of a string are $x_2$ and $x_4$, given the $1^{st}$ and $3^{rd}$ symbols are $x_1$ and $x_3$ | 132 |
| $\nu\in\mathcal{M}$ | any pd in $\mathcal{M}$ | 70 |
| $\rho$ | any pd not necessarily in $\mathcal{M}$ usually specifying a policy | 68 |
| $\xi$ | $=\sum_{\nu\in\mathcal{M}}w_\nu\nu=$ mixture (belief) pd | 48, 70 |
| $w_\nu$ | prior degree of belief in $\nu$ –or– weight of $\nu$ | 48, 70 |
| $\rho^{\text{EC}}$ | pd of environmental argument type EC | 185 |
| $\xi^{\text{EC}}$ | mixture distribution of type EC for class EC | 185 |
| $\ell_{x_ty_t}$ | incurred loss when predicting $y_t$ and $x_t$ is next symbol | 86 |
| $l_{t\nu}^{\Lambda}$ | $\nu$-expected instantaneous loss in step $t$ of predictor $\Lambda$ | 99, 87 |
| $L_{n\nu}^{\Lambda}$ | $\nu$-expected cumulative loss of steps $1...n$ of predictor $\Lambda$ | 100 |
| $\Theta_\rho$ | predictor with minimal number of $\rho$-expected errors | 82 |
| $\Lambda_\rho$ | predictor that minimizes the $\rho$-expected loss | 87 |
| $e_{t\nu}^{\Theta}$ | $\nu$-probability that $\Theta$-predictor errs in step $t$ | 83 |
| $E_{n\nu}^{\Theta}$ | $\nu$-expected number of errors in steps $1...n$ of predictor $\Theta$ | 83 |
| $L_n^{\Lambda}\equiv L_{n\mu}^{\Lambda}$ | abbreviation for true $\mu$-expected loss | 86 |
| $V_{km}^{p\nu}(\dot{y}\dot{x}_{<k})$ | value of policy $p$ in environment $\nu$ given history $\dot{y}\dot{x}_{<k}$ | 153 |
| $y_t^{\Lambda}$ | prediction/decision/action of predictor $\Lambda$ in step $t$ | 87 |
| $y_k^p$ | action of policy $p$ in cycle $k$ | * |
| $\gamma_k$ | discounting sequence | 159 |
| $\Gamma_k$ | value function normalization $(\sum_{i=k}^{\infty}\gamma_k)$ | 159 |
| $m$,$h$ | agent's lifespan, horizon | 129, 169 |
| $p$ | agent's policy | 126 |
| $q$ | deterministic environment | 126 |
| $p^\nu$ | policy that maximizes value $V_\nu^p$ | 130 |
| $V_\mu^*\equiv V_{1m}^{p^\mu\mu}$ | true or generating value | 130 |
| $V_\xi^*\equiv V_{1m}^{p^\xi\xi}$ | universal value | 146 |
| $D_n\equiv D_{n\mu}^{\xi}$ | relative entropy between $\mu$ and $\xi$ for the first $n$ cycles | 73 |

Albert Einstein
(1879–1955)

*I have no particular talent. I am merely inquisitive.*
— Albert Einstein

# 1 A Short Tour Through the Book

This Chapter represents a short tour through the book. It is not meant as a gentle introduction for novices, but as a condensed presentation of the most important concepts and results of the book. The price for this brevity is that in this chapter we mostly forgo mathematical rigor, subtleties, proofs, discussions, references and comparisons to other work. More seriously, some sections demand high background knowledge. Readers unfamiliar with algorithmic information theory should first read Chapter 2 or consult the textbooks [LV97, Cal02]. Readers unfamiliar with sequential decision theory should first read Chapter 4 or consult the textbooks [BT96, SB98]. Before becoming discouraged by the complexity of some of the sections, it is better to skip them completely.

## 1.1 Introduction

**Artificial Intelligence.** The science of artificial intelligence (AI) might be defined as the construction of intelligent systems and their analysis. A natural definition of a *system* is anything that has an input and an output stream. Intelligence is more complicated. It can have many faces like creativity, solving problems, pattern recognition, classification, learning, induction, deduction, building analogies, optimization, surviving in an environment, language processing, knowledge and many more. A formal definition incorporating every aspect of intelligence, however, seems difficult. Further, intelligence is graded: There is a smooth transition between systems, which everyone would agree to be not intelligent, and truly intelligent systems. One simply has to look in nature, starting with, for instance, inanimate crystals, then amino acids, then some RNA fragments, then viruses, bacteria, plants, animals, apes, followed by the truly intelligent homo sapiens, and possibly continued by AI systems or ETs. So, the best we can expect to find is a partial or total order relation on the set of systems, which orders them w.r.t. their degree of intelligence (like intelligence tests do for human systems, but for a limited class of problems). Having this order we are, of course, interested in large elements, i.e. highly intelligent systems. If a largest element exists, it would correspond to the most intelligent system which could exist.

Most, if not all, known facets of intelligence can be formulated as goal driven or, more precisely, as maximizing some utility function. It is therefore sufficient to study goal-driven AI. For example, the (biological) goal of animals and humans is to survive and spread. The goal of AI systems should be to be useful to humans. The problem is that, except for special cases, we know neither the utility function nor the environment in which the agent will operate in advance.

**Main idea.** This book presents a theory that formally[1] solves the problem of unknown goal and environment. It might be viewed as a unification of the ideas of universal induction, probabilistic planning and reinforcement learning, or as a unification of sequential decision theory with algorithmic information theory. We apply this model to some of the facets of intelligence, including induction, game playing, optimization, reinforcement and supervised learning, and show how it solves these problem classes. This, together with general convergence theorems, supports the belief that the constructed universal AI system is the best one in a sense to be clarified in the following, i.e. that it is the most intelligent environment-independent system possible. The intention of this book is to introduce the universal AI model and give an extensive analysis.

## 1.2 Simplicity & Uncertainty

This section introduces Occam's razor principle, Kolmogorov complexity, and objective/subjective probabilities. We finally arrive at the problem of universal prediction, and its solution by Solomonoff.

### 1.2.1 Introduction

An important and nontrivial aspect of intelligence is inductive inference. Simply speaking, induction is the process of predicting the future from the past, or, more precisely, it is the process of finding rules in (past) data and using these rules to guess future data. Weather or stock-market forecasting or continuing number series in an IQ test are nontrivial examples. Making good predictions plays a central role in natural and artificial intelligence in general, and in machine learning in particular. All induction problems can be phrased

[1] With a formal solution we mean a rigorous mathematically definition, uniquely specifying the solution. In the following, a solution is always meant in this formal sense.

as sequence prediction tasks. This is, for instance, obvious for time-series prediction, but also includes classification tasks. Having observed data $x_t$ at times $t<n$, the task is to predict the $n^{th}$ symbol $x_n$ from sequence $x_1...x_{n-1}$. This *prequential approach* [Daw84] skips over the intermediate step of learning a model based on observed data $x_1...x_{n-1}$ and then using this model to predict $x_n$. The prequential approach avoids problems of model consistency, how to separate noise from useful data, and many other issues. The goal is to make "good" predictions, where the prediction quality is usually measured by a loss function, which shall be minimized. The key concept to well-defining and solving induction problems is *Occam's razor* (simplicity) principle, which says that "*Entities should not be multiplied beyond necessity.*" This may be interpreted as keeping the simplest theory consistent with the observations $x_1...x_{n-1}$ and using this theory to predict $x_n$. Before we can present Solomonoff's formal solution, we have to quantify Occam's razor in terms of Kolmogorov complexity, and introduce the notions of subjective and objective probabilities.

### 1.2.2 Algorithmic Information Theory

Intuitively, a string is simple if it can be described in a few words, like "the string of one million ones", and is complex if there is no such short description, like for a random string whose shortest description is specifying it bit by bit. We can restrict the discussion to binary strings, since for other (non-stringy mathematical) objects we may assume some default coding as binary strings. Furthermore, we are only interested in effective descriptions, and hence restrict decoders to be Turing machines. Let us choose some universal (so-called prefix) *Turing machine* $U$ with unidirectional binary input and output tapes and a bidirectional work tape. We can then define the *prefix Kolmogorov complexity* [Cha75, Gác74, Kol65, Lev74] of a binary string $x$ as the length $\ell$ of the shortest program $p$ for which $U$ outputs the binary string $x$

$$K(x) \;:=\; \min_p\{\ell(p) : U(p) = x\}.$$

Simple strings like 000...0 can be generated by short programs, and, hence have low Kolmogorov complexity, but irregular (e.g. random) strings are their own shortest description, and hence have high Kolmogorov complexity. An important property of $K$ is that it is nearly independent of the choice of $U$. Furthermore, it shares many properties with Shannon's entropy (information measure) $S$, but $K$ is superior to $S$ in many respects. Figure 2.11 on page 38 contains a schematic graph of $K$. To be brief, $K$ is an excellent universal complexity measure, suitable for quantifying Occam's razor. There is (only) one severe disadvantage: $K$ is not finitely computable. More precisely, a function $f$ is said to be *finitely computable* (or *recursive*) if there exists a Turing machine which, given $x$, computes $f(x)$ and then halts. Some functions are not finitely computable but still *approximable* in the sense that there is a nonhalting Turing machine with an infinite output sequence $y_1,y_2,y_3,...$ with $\lim_{t\to\infty}y_t=f(x)$.

If additionally the output sequence is monotone increasing/decreasing, then $f$ is said to be *lower/upper semicomputable* (or *enumerable/co-enumerable*). Finally, we call $f$ *estimable* if some Turing machine, given $x$ and a precision $\varepsilon$, finitely computes an $\varepsilon$-approximation of $x$. The major algorithmic property of $K$ is that it is co-enumerable, but not finitely computable.

### 1.2.3 Uncertainty & Probabilities

For the *objectivist*, probabilities are real aspects of the world.[2] The outcome of an observation or an experiment is not deterministic, but involves physical random processes. Kolmogorov's axioms of probability theory formalize the properties which probabilities should have. In the case of independent and identically distributed (i.i.d.) experiments the probabilities assigned to events can be interpreted as limiting frequencies (*frequentist* view), but applications are not limited to this case. Conditionalizing probabilities and Bayes' rule are the major tools in computing posterior probabilities from prior ones. For instance, given the initial binary sequence $x_1...x_{n-1}$, what is the probability of the next bit being 1? The probability of observing $x_n$ at time $n$, given past observations $x_1...x_{n-1}$ can be computed with multiplication or the chain rule[3] if the true generating distribution $\mu$ of the sequences $x_1x_2x_3...$ is known: $\mu(x_n|x_{<n}) = \mu(x_{1:n})/\mu(x_{<n})$, where we introduced the abbreviations $x_{1:n} \equiv x_1x_2...x_n$ and $x_{<n} \equiv x_1x_2...x_{n-1}$. The problem, however, is that one often does not know the true distribution $\mu$ (e.g. in the cases of weather and stock-market forecasting).

The *subjectivist* uses probabilities to characterize an agent's degree of belief in (or plausibility of) something, rather than to characterize physical random processes. This is the most relevant interpretation of probabilities in AI. It is somewhat surprising that plausibilities can be shown to also respect Kolmogorov's axioms of probability and the chain rule by assuming only a few plausible qualitative rules they should follow [Cox46]. Hence, if the plausibility of $x_{1:n}$ is $\rho(x_{1:n})$, the degree of belief in $x_n$ given $x_{<n}$ is, again, given by the chain rule: $\rho(x_n|x_{<n}) = \rho(x_{1:n})/\rho(x_{<n})$.

The chain rule allows the computation of posterior probabilities/plausibilities from prior ones, but leaves open the question of how to determine the priors themselves. In statistical physics, the principle of indifference (symmetry principle) and the maximum entropy principle can often be exploited to determine prior probabilities, but only Occam's razor is general enough to assign prior probabilities in *every* situation, especially to cope with complex domains typical for AI.

---

[2] Readers not believing in objective and/or subjective probabilities should read the remark at the beginning of Section 2.3.

[3] Strictly speaking, it is just the definition of conditional probabilities.

### 1.2.4 Algorithmic Probability & Universal Induction

Occam's razor (appropriately interpreted and in compromise with Epicurus' principle of indifference) tells us to assign high/low a priori plausibility to simple/complex strings $x$. Using $K$ as complexity measure, any monotone decreasing function of $K$, e.g. $\rho(x)=2^{-K(x)}$, would satisfy this criterion. But $\rho$ also has to satisfy the probability axioms, so we have to be a bit more careful. Solomonoff [Sol64, Sol78] defined the *universal prior* $M(x)$ as the probability that the output of a universal Turing machine $U$ starts with $x$ when provided with fair coin flips on the input tape. Formally, $M$ can be defined as

$$M(x) \;:=\; \sum_{p\,:\,U(p)=x*} 2^{-\ell(p)} \qquad (1.1)$$

where the sum is over all (so-called minimal) programs $p$ for which $U$ outputs a string starting with $x$. Strictly speaking $M$ is only a *semimeasure* since it is not normalized to 1, but this is acceptable/correctable. We derive the following bound:

$$\sum_{t=1}^{\infty}(1-M(x_t|x_{<t}))^2 \leq -\tfrac{1}{2}\sum_{t=1}^{\infty}\ln M(x_t|x_{<t}) = -\tfrac{1}{2}\ln M(x_{1:\infty}) \leq \tfrac{1}{2}\ln 2\cdot Km(x_{1:\infty})$$

where $Km(x_{1:\infty})$ is the length of the shortest (nonhalting) program computing $x_{1:\infty}$. In the first inequality we have used $(1-a)^2 \leq -\frac{1}{2}\ln a$ for $0 \leq a \leq 1$. In the equality we exchanged the sum with the logarithm and eliminated the resulting product by the chain rule. In the last inequality we used $M(x) \geq 2^{-Km(x)}$, which follows from definition (1.1) by dropping all terms in $\sum_p$ except for the shortest $p$ computing $x$. If $x_{1:\infty}$ is a computable sequence, then $Km(x_{1:\infty})$ is finite, which implies $M(x_t|x_{<t}) \to 1$ ($\sum_{t=1}^{\infty}(1-a_t)^2 < \infty \Rightarrow a_t \to 1$). This means that if the environment is a computable sequence (whichever, e.g. the digits of $\pi$ or $e$ in binary representation), after having seen the first few digits, $M$ correctly predicts the next digit with high probability, i.e. it recognizes the structure of the sequence.

Assume now that the true sequence is drawn from the distribution $\mu$, i.e. the true (objective) probability of $x_{1:n}$ is $\mu(x_{1:n})$, but $\mu$ is unknown. How is the posterior (subjective) belief $M(x_n|x_{<n}) = M(x_n)/M(x_{<n})$ related to the true (objective) posterior probability $\mu(x_n|x_{<n})$? Solomonoff's [Sol78] central result is that the posterior (subjective) beliefs converge to the true (objective) posterior probabilities, if the latter are computable. More precisely, he showed that

$$\sum_{t=1}^{\infty}\sum_{x_{<t}\in\{0,1\}^{t-1}} \mu(x_{<t})\Big(M(0|x_{<t}) - \mu(0|x_{<t})\Big)^2 \;\leq\; \tfrac{1}{2}\ln 2\cdot K(\mu) + O(1). \qquad (1.2)$$

The complexity $K(\mu)$ is finite if $\mu$ is a computable function, but the infinite sum on the l.h.s. can only be finite if the difference $M(0|x_{<t}) - \mu(0|x_{<t})$ tends to zero for $t \to \infty$ with $\mu$-probability 1 (w.$\mu$.p.1). This shows that using $M$ as an estimate for $\mu$ may be a reasonable thing to do.

### 1.2.5 Generalized Universal (Semi)Measures

One can derive a universal prior in a different way: Solomonoff [Sol64, Eq.(13)] defines a somewhat problematic mixture over all computable probability distributions. Levin [ZL70] considers the larger class $\mathcal{M}_U := \{\nu_1,\nu_2,...\}$ of all so-called enumerable semimeasures. Let $\mu \in \mathcal{M}_U$, and assign (consistent with Occam's razor) a prior plausibility of $2^{-K(\nu_a)}$ to $\nu_a$. Then the prior plausibility of $x_{1:n}$ is, by elementary probability theory,

$$\xi_U(x_{1:n}) := \sum_{\nu \in \mathcal{M}_U} 2^{-K(\nu)} \nu(x_{1:n}). \tag{1.3}$$

One can show that $\xi_U$ coincides with $M$ within an (irrelevant) multiplicative constant, i.e. $M(x) \stackrel{\times}{=} \xi_U(x)$, where $f(x) \stackrel{\times}{\leq} g(x)$ abbreviates $f(x) = O(g(x))$, and $\stackrel{\times}{=}$ denotes $\stackrel{\times}{\leq}$ *and* $\stackrel{\times}{\geq}$. Both $\xi_U$ and $M$ can be shown to be lower semicomputable. The dominance $M(x) \stackrel{\times}{=} \xi_U(x) \geq 2^{-K(\mu)} \mu(x)$ is the central ingredient in the proof of (1.2). The advantage of $\xi_U$ over $M$ is that the definition immediately generalizes to arbitrary weighted sums of (semi)measures in $\mathcal{M}$ for arbitrary countable $\mathcal{M}$. Most proofs in this book go through for generic $\mathcal{M}$ and weights.

So, what is so special about the class of all enumerable semimeasures $\mathcal{M}_U$? The larger we choose $\mathcal{M}$, the less restrictive is the assumption that $\mathcal{M}$ should contain the true distribution $\mu$, which will be essential throughout the book. Why not restrict to the still rather general class of estimable or finitely computable (semi)measures? For *every* countable class $\mathcal{M}$, the mixture $\xi(x) := \xi_{\mathcal{M}}(x) := \sum_{\nu \in \mathcal{M}} w_\nu \nu(x)$ with $w_\nu > 0$, the important dominance $\xi(x) \geq w_\nu \nu(x)$ is satisfied. The question is, what properties does $\xi$ possess. The distinguishing property of $\mathcal{M}_U$ is that $\xi_U$ is itself an element of $\mathcal{M}_U$. On the other hand, in this book $\xi_{\mathcal{M}} \in \mathcal{M}$ is not by itself an important property. What matters is whether $\xi$ is computable in one of the senses we defined above. There is an enumerable semimeasure ($M$) that dominates all enumerable semimeasures in $\mathcal{M}_U$. As we will see, there is *no* estimable semimeasure that dominates all computable measures, and there is *no* approximable semimeasure that dominates all approximable measures. From this it follows that for a universal (semi)measure which at least satisfies the weakest form of computability, namely being approximable, the largest dominated class among the classes considered in this book is the class of enumerable semimeasures, but there are even larger classes [Sch02a]. This is the reason why $\mathcal{M}_U$ and $M$ play a special role in this (and other) works. In practice though, one has to restrict to a finite subset of finitely computable environments $\nu$ to get a finitely computable $\xi$.

## 1.3 Universal Sequence Prediction

In the following we more closely investigate sequence prediction (SP) schemes based on Solomonoff's universal prior $M \stackrel{\times}{=} \xi_U$ and on more general Bayes mixtures $\xi$, mainly from a decision-theoretic perspective. In particular, we show that they are optimal w.r.t. various optimality criteria.

### 1.3.1 Setup & Convergence

Let $\mathcal{M}:=\{\nu_1,\nu_2,...\}$ be a countable set of candidate probability distributions on strings over the finite alphabet $\mathcal{X}$. We define a weighted average on $\mathcal{M}$:

$$\xi(x_{1:n}) := \sum_{\nu\in\mathcal{M}} w_\nu\cdot\nu(x_{1:n}), \quad \sum_{\nu\in\mathcal{M}} w_\nu = 1, \quad w_\nu > 0. \tag{1.4}$$

It is easy to see that $\xi$ is a probability distribution as the weights $w_\nu$ are positive and normalized to 1 and the $\nu\in\mathcal{M}$ are probabilities. We call $\xi$ universal relative to $\mathcal{M}$, as it multiplicatively dominates all distributions in $\mathcal{M}$ in the sense that $\xi(x_{1:n})\geq w_\nu\cdot\nu(x_{1:n})$ for all $\nu\in\mathcal{M}$. In the following, we assume that $\mathcal{M}$ is known and contains the true but unknown distribution $\mu$, i.e. $\mu\in\mathcal{M}$, and $x_{1:\infty}$ is sampled from $\mu$. We abbreviate expectations w.r.t. $\mu$ by $\mathbf{E}[..]$; for instance, $\mathbf{E}[f(x_{1:n})]=\sum_{x_{1:n}\in\mathcal{X}^n}\mu(x_{1:n})f(x_{1:n})$. We use the (total) relative entropy $D_n$ and squared Euclidian distance $S_n$ to measure the distance between $\mu$ and $\xi$:

$$D_n := \mathbf{E}\left[\ln\frac{\mu(x_{1:n})}{\xi(x_{1:n})}\right], \quad S_n := \sum_{t=1}^n \mathbf{E}\left[\sum_{x'_t\in\mathcal{X}}\Big(\mu(x'_t|x_{<t})-\xi(x'_t|x_{<t})\Big)^2\right]. \tag{1.5}$$

The following sequence of inequalities can be shown, which generalize Solomonoff's result (1.2): $S_n\leq D_n\leq\ln w_\mu^{-1}<\infty$. The finiteness of $S_\infty$ implies $\xi(x'_t|x_{<t})-\mu(x'_t|x_{<t})\to 0$ for $t\to\infty$ w.$\mu$.p.1 for any $x'_t$ ($\sum_{t=1}^\infty s_t^2<\infty\Rightarrow s_t\to 0$). We also show that $\sum_{t=1}^n\mathbf{E}[(\sqrt{\xi(x_t|x_{<t})/\mu(x_t|x_{<t})}-1)^2]\leq D_n\leq\ln w_\mu^{-1}<\infty$, which implies $\xi(x_t|x_{<t})/\mu(x_t|x_{<t})\to 1$ for $t\to\infty$ w.$\mu$.p.1. This convergence motivates the belief that predictions based on (the known) $\xi$ are asymptotically as good as predictions based on (the unknown) $\mu$, with rapid convergence.

### 1.3.2 Loss Bounds

Most predictions are eventually used as a basis for some decision or action, which itself leads to some reward or loss. Let $\ell_{x_t y_t}\in[0,1]\subset I\!R$ be the received loss when performing prediction/decision/action $y_t\in\mathcal{Y}$, and $x_t\in\mathcal{X}$ is the $t^{th}$ symbol of the sequence. Let $y_t^\Lambda\in\mathcal{Y}$ be the prediction of a (causal) prediction scheme $\Lambda$. The true probability of the next symbol being $x_t$, given $x_{<t}$, is $\mu(x_t|x_{<t})$. The expected loss when predicting $y_t$ is $\mathbf{E}[\ell_{x_t y_t}]$. The total $\mu$-expected loss suffered by the $\Lambda$ scheme in the first $n$ predictions is

$$L_n^\Lambda := \sum_{t=1}^n \mathbf{E}[\ell_{x_t y_t^\Lambda}].$$

The goal is to minimize the expected loss. More generally, we define the $\Lambda_\rho$ sequence prediction scheme (later also called SP$\rho$) $y_t^{\Lambda_\rho} := \text{argmin}_{y_t\in\mathcal{Y}}\sum_{x_t}\rho(x_t|x_{<t})\ell_{x_t y_t}$, which minimizes the $\rho$-expected loss. If $\mu$ is known, $\Lambda_\mu$ is obviously the best prediction scheme in the sense of achieving minimal expected loss ($L_n^{\Lambda_\mu} \leq L_n^{\Lambda}$ for any $\Lambda$). We prove the following loss bound for the universal $\Lambda_\xi$ predictor

$$0 \leq L_n^{\Lambda_\xi} - L_n^{\Lambda_\mu} \leq D_n + \sqrt{4L_n^{\Lambda_\mu} D_n + D_n^2} \leq 2D_n + 2\sqrt{L_n^{\Lambda_\mu} D_n}. \quad (1.6)$$

Together with $L_n \leq n$ and $D_\infty \leq \ln w_\mu^{-1} < \infty$, this shows that $\frac{1}{n}L_n^{\Lambda_\xi} - \frac{1}{n}L_n^{\Lambda_\mu} = O(n^{-1/2})$, i.e. asymptotically $\Lambda_\xi$ achieves the optimal average loss of $\Lambda_\mu$ with rapid convergence. Moreover, $L_\infty^{\Lambda_\xi}$ is finite if $L_\infty^{\Lambda_\mu}$ is finite, and $L_n^{\Lambda_\xi}/L_n^{\Lambda_\mu} \to 1$ if $L_\infty^{\Lambda_\mu}$ is not finite. Bound (1.6) also implies $L_n^{\Lambda} \geq L_n^{\Lambda_\xi} - 2\sqrt{L_n^{\Lambda_\xi} D_n}$, which shows that *no* (causal) predictor $\Lambda$ whatsoever achieves significantly less (expected) loss than $\Lambda_\xi$. Note that for $w_\nu = 2^{-K(\nu)}$, $D_n \leq \ln 2 \cdot K(\mu)$ is of "reasonable" size. Instantaneous loss bounds can also be proven.

### 1.3.3 Optimality Properties

For any predictor $\Lambda$, a worst-case lower bound that asymptotically matches the upper bound (1.6) can be derived. More precisely, let $\Lambda$ be any deterministic predictor not knowing from which distribution $\mu \in \mathcal{M}$ the observed sequence $x_1 x_2...$ is sampled. Predictor $\Lambda$ knows (depends on) $\mathcal{M}$, $w_\nu$, and $\ell$, and has at time $t$ access to the previous outcomes $x_{<t}$. Then for every $n$ there is an $\mathcal{M}$ and $\mu \in \mathcal{M}$ and $\ell$ and weights $w_\nu$ such that

$$L_n^{\Lambda} - L_n^{\Lambda_\mu} \geq \tfrac{1}{2}[S_n + \sqrt{4L_n^{\Lambda_\mu} S_n + S_n^2}], \qquad \text{and} \quad D_n/S_n \to 1 \quad \text{for} \quad n \to \infty.$$

For the universal predictor $\Lambda = \Lambda_\xi$, the lower bound holds even without the factor $\frac{1}{2}$. This shows that bound (1.6) is quite tight in the sense that no other predictor can lead to significantly smaller bounds without making extra assumptions on $\mathcal{M}$, $w_\nu$, or $\ell$. For instance, for logarithmic and quadratic loss functions the regret $L_\infty^{\Lambda_\xi} - L_\infty^{\Lambda_\mu}$ is finite and bounded by $\ln w_\mu^{-1}$.

A different kind of optimality is *Pareto optimality*. Let $\mathcal{F}(\mu,\rho)$ be any performance measure of $\rho$ relative to $\mu$. The universal prior $\xi$ is called Pareto optimal w.r.t. $\mathcal{F}$ if there is no $\rho$ with $\mathcal{F}(\nu,\rho) \leq \mathcal{F}(\nu,\xi)$ for all $\nu \in \mathcal{M}$ and strict inequality for at least one $\nu$. We show that the universal prior $\xi$ is Pareto optimal w.r.t. the squared distance $S_n$, the relative entropy $D_n$, and the losses $L_n$. That is, for all performance measures that are relevant from a decision-theoretic point of view (i.e. for all loss functions $\ell$) any improvement achieved by some predictor $\Lambda_\rho$ over $\Lambda_\xi$ in some environments $\nu$ is balanced by a deterioration in other environments. There are non-decision-theoretic performance measures w.r.t. which $\xi$ is *not* Pareto optimal. Pareto optimality is a rather weak notion of optimality, but it emphasizes the distinctiveness of Bayes mixture strategies.

Pareto optimality of $\xi$ still leaves open the question of how to choose the class $\mathcal{M}$ and the weights $w_\nu$. We have argued that $\mathcal{M}_U$ is the largest $\mathcal{M}$ suitable from a computational point of view. $\mathcal{M}_U$ is also sufficiently large if we make the mild assumption that strings are sampled from a computable probability distribution. We show that within the class of enumerable weight functions with short program, the universal weights $w_\nu = 2^{-K(\nu)}$ lead to the smallest performance bounds within an additive (to $\ln w_\nu^{-1}$) constant in all enumerable environments. This argument justifies the selection of Solomonoff-Levin's prior (1.3) among all possible Bayes mixtures.[4]

### 1.3.4 Miscellaneous

**Games of chance.** The general loss bound (1.6) can, for instance, be used to estimate the time needed to reach the winning threshold in a game of chance (defined as a sequence of bets, observations and rewards). At time $t$ we bet, depending on the history $x_{<t}$, a certain amount of money $s_t$, take some action $y_t$, observe outcome $x_t$, and receive reward $r_t$. Our net profit, which we want to maximize, is $p_t = r_t - s_t \in [p_{max} - p_\Delta, p_{max}]$. The loss, which we want to minimize, can be identified with the negative (scaled) profit, $\ell_{x_k y_t} = (p_{max} - p_t)/p_\Delta \in [0,1]$. The $\Lambda_\rho$-system acts as to maximize the $\rho$-expected profit. Let $\bar{p}_n^{\Lambda_\rho}$ be the average expected profit of the first $n$ rounds. Bound (1.6) shows that the average profit of the $\Lambda_\xi$ system converges to the best possible average profit $\bar{p}_n^{\Lambda_\mu}$ achieved by the $\Lambda_\mu$ scheme ($\bar{p}_n^{\Lambda_\xi} - \bar{p}_n^{\Lambda_\mu} = O(n^{-1/2}) \to 0$ for $n \to \infty$). If there is a profitable scheme at all, then asymptotically the universal $\Lambda_\xi$ scheme will also become profitable with the same average profit. We further show using $\xi_U$ that $(2p_\Delta / \bar{p}_n^{\Lambda_\mu})^2 \cdot \ln 2 \cdot K(\mu)$ is an upper bound on the number of bets $n$ needed to reach the winning zone. The bound is proportional to the complexity of the environment $\mu$.

**Continuous probability classes $\mathcal{M}$.** We have considered thus far countable probability classes $\mathcal{M}$, which makes sense from a computational point of view. On the other hand, in statistical parameter estimation one often has a continuous hypothesis class (e.g. a Bernoulli($\theta$) process with unknown $\theta \in [0,1]$). Let $\mathcal{M} := \{\mu_\theta : \theta \in \Theta \subseteq I\!R^d\}$ be a family of probability distributions parameterized by a $d$-dimensional continuous parameter $\theta$. Let $\mu \equiv \mu_{\theta_0} \in \mathcal{M}$ be the true generating distribution. For a continuous weight density $w(\theta) > 0$ the sums in (1.4) are naturally replaced by integrals: $\xi(x_{1:n}) := \int_\Theta w(\theta) \cdot \mu_\theta(x_{1:n}) d\theta$ with $\int_\Theta w(\theta) d\theta = 1$. The most important property of $\xi$ in the discrete case was the dominance $\xi(x_{1:n}) \geq w_\nu \cdot \nu(x_{1:n})$, which was obtained from (1.4) by dropping the sum over $\nu$. The analogous construction here is to restrict the integral over $\Theta$ to a small vicinity $N_\delta$ of $\theta$. For sufficiently smooth $\mu_\theta$ and $w(\theta)$ we expect $\xi(x_{1:n}) \gtrsim |N_{\delta_n}| \cdot w(\theta) \cdot \mu_\theta(x_{1:n})$, where $|N_{\delta_n}|$ is the volume of

[4] Readers who smell some free lunch here [WM97] should appease their hunger with Section 3.6.5.

$N_{\delta_n}$. This in turn leads to $D_n \lesssim \ln w_\mu^{-1} + \ln|N_{\delta_n}|^{-1}$, where $w_\mu := w(\theta_0)$. $N_{\delta_n}$ should be the largest possible region in which $\ln\mu_\theta$ is approximately flat on average. More precisely, generalizing [CB90] to the non-i.i.d. case, we show $D_n \le \ln w_\mu^{-1} + \frac{d}{2}\ln\frac{n}{2\pi} + O(1)$, where the $O(1)$ term depends on the smoothness of $\mu_\theta$, measured by the Fisher information. $D_n$ is no longer bounded by a constant, but still grows only logarithmically with $n$, the intuitive reason being the necessity to describe $\theta$ to an accuracy $O(n^{-1/2})$. So, bound (1.6) is also applicable to the case of continuously parameterized probability classes.

## 1.4 Rational Agents in Known Probabilistic Environments

### 1.4.1 The Agent Model

A very general framework for intelligent systems is that of rational agents [RN95]. In cycle $k$, an agent performs *action* $y_k \in \mathcal{Y}$ (output), which results in a *perception* $x_k \in \mathcal{X}$ (input), followed by cycle $k+1$, and so on. We assume that the action and perception spaces $\mathcal{X}$ and $\mathcal{Y}$ are finite. We write $p(x_{<k}) = y_{1:k}$ to denote the output $y_{1:k}$ of the agent's policy $p$ on input $x_{<k}$, and similarly $q(y_{1:k}) = x_{1:k}$ for the environment $q$ in the case of deterministic environments. We call policy $p$ and environment $q$ behaving in this way *chronological.* The figure on the book cover and on page 128 depicts this interaction in the case where $p$ and $q$ are modeled by Turing machines. Note that policy and environment are allowed to depend on the complete history. We do not make any MDP or POMDP assumption here, and we do not talk about states of the environment, only about observations. In the more general case of a *probabilistic environment*, given the history $y\!x_{<k}y_k \equiv y\!x_1...y\!x_{k-1}y_k \equiv y_1x_1...y_{k-1}x_{k-1}y_k$, the probability that the environment leads to perception $x_k$ in cycle $k$ is (by definition) $\mu(y\!x_{<k}y\underline{x}_k)$. The underlined argument $\underline{x}_k$ in $\mu$ is a random variable, and the other non-underlined arguments $y\!x_{<k}y_k$ represent conditions.[5] We call probability distributions like $\mu$ *chronological.* Since value-optimizing policies (see below) can always be chosen deterministic, there is no real need to generalize the setting to probabilistic policies.

### 1.4.2 Value Functions & Optimal Policies

The goal of the agent is to maximize future *rewards*, which are provided by the environment through the inputs $x_k$. The inputs $x_k \equiv r_k o_k$ are divided into a regular part $o_k$ and some (possibly empty or delayed) reward $r_k \in [0, r_{max}]$.[6] We use the abbreviation

[5] The standard notation $\mu(x_k|y\!x_{<k}y_k)$ for conditional probabilities destroys the chronological order and would become confusing in later expressions.

[6] In the reinforcement learning literature when dealing with (PO)MDPs the reward is usually considered to be a function of the environmental state. The zero-

$$\mu(yx_{<k}\underline{yx}_{k:m}) \;=\; \mu(yx_{<k}\underline{yx}_k)\cdot\mu(yx_{1:k}\underline{yx}_{k+1})\cdot\ ...\ \cdot\mu(yx_{<m}\underline{yx}_m),$$

which is essentially the chain rule, and $\epsilon = yx_{<1}$ for the empty string. We define the (total) *value* of policy $p$ in environment $\mu$, or shorter, the $\mu$-value of $p$, as the $\mu$-expected reward sum

$$V_\mu^p \;:=\; \sum_{x_{1:m}}(r_1+\ ...\ +r_m)\mu(\underline{yx}_{1:m})_{|y_{1:m}=p(x_{<m})}, \tag{1.7}$$

where $m$ is the *lifespan* or initial *horizon* of the agent. The optimal policy $p^\mu$ that maximizes the value $V_\mu^p$ is

$$p^\mu := \arg\max_p V_\mu^p, \qquad V_\mu^* := V_\mu^{p^\mu} = \max_p V_\mu^p \geq V_\mu^p\ \forall p.$$

The policy $p^\mu$, which we call *AI$\mu$ model*, is optimal in the sense that no other policy for an agent leads to higher $\mu$-expected reward. Explicit expressions for the action $y_k$ in cycle $k$ of the $\mu$-optimal policy $p^\mu$ and their value $V_\mu^*$ are

$$y_k \;=\; y_k^\mu \;:=\; \arg\max_{y_k}\sum_{x_k}\max_{y_{k+1}}\sum_{x_{k+1}}\ ...\ \max_{y_m}\sum_{x_m}(r_k+\ ...\ +r_m)\cdot\mu(yx_{<k}\underline{yx}_{k:m}), \tag{1.8}$$

$$V_\mu^* \;=\; \max_{y_1}\sum_{x_1}\max_{y_2}\sum_{x_2}\ ...\ \max_{y_m}\sum_{x_m}(r_1+\ ...\ +r_m)\cdot\mu(\underline{yx}_{1:m}), \tag{1.9}$$

where $yx_{<k}$ is the actual history. We show that these definitions are consistent and correctly capture our intention. For instance, consider the expectimax expression (1.9): The best expected reward is obtained by averaging over possible perceptions $x_i$ and by maximizing over the possible actions $y_i$. This has to be done in chronological order $y_1x_1...y_mx_m$ to correctly incorporate the dependencies of $x_i$ and $y_i$ on the history. This is the origin of the alternating *expectimax* sequence, which is similar to the well-known minimax sequence/tree/algorithm in game theory.

### 1.4.3 Sequential Decision Theory & Reinforcement Learning

One can relate (1.9) to the Bellman equations [Bel57] of sequential decision theory by identifying complete histories $yx_{<k}$ with states, $\mu(yx_{<k}\underline{yx}_k)$ with the state transition matrix, $V_\mu^*$ with the value function, and $y_k$ with the action in cycle $k$ [BT96, RN95]. Due to the use of complete histories as state space, the AI$\mu$ model assumes neither stationarity nor the Markov property nor complete accessibility of the environment. Every state occurs at most once in the lifetime of the system. For this and other reasons the explicit formulation (1.8) is more

assumption analogue here is that the reward $r_k$ is some probabilistic function $\mu'$ depending on the complete history. It is very convenient to integrate $r_k$ into $x_k$ and $\mu'$ into $\mu$.

natural and useful here than to enforce a pseudo-recursive Bellman equation form.

As we have in mind a universal system with complex interactions, the action and perception spaces $\mathcal{Y}$ and $\mathcal{X}$ are huge (e.g. video images), and every action or perception itself occurs usually only once in the lifespan $m$ of the agent. As there is no (obvious) universal similarity relation on the state space, an effective reduction of its size is impossible, but there is no principle problem in determining $y_k$ from (1.8) as long as $\mu$ is known and computable and $\mathcal{X}$, $\mathcal{Y}$ and $m$ are finite.

Things drastically change if $\mu$ is unknown. Reinforcement learning algorithms [BT96, KLM96, SB98] are commonly used in this case to learn the unknown $\mu$ or directly its value. They succeed if the state space is either small or has effectively been made small by generalization or function approximation techniques. In any case, the solutions are either ad hoc, work in restricted domains only, have serious problems with state space exploration versus exploitation, are prone to diverge, or have nonoptimal learning rates. There is no universal and optimal solution to this problem so far. The central theme of this book is to present a new model and to argue that it formally solves all these problems in an optimal way. The true probability distribution $\mu$ will not be learned directly, but will be replaced by some generalized universal prior $\xi_U$, which converges to $\mu$, similarly to the induction (SP) case.

## 1.5 The Universal Algorithmic Agent AIXI

### 1.5.1 The Universal AIXI Model

We have developed enough formalism to present the universal AIXI model. All we have to do is to suitably generalize Solomonoff's universal prior $M$ and to replace the true but unknown probability $\mu$ in the AI$\mu$ model by this generalized $M$. Similarly to (1.1), we define $M$ as the $2^{-\ell(q)}$ weighted sum over all chronological programs (environments) $q$ that output $x_{1:k}$, but with $y_{1:k}$ provided on the input tape. This also generalizes $\xi_U$ (within an irrelevant multiplicative constant):

$$\xi(y\!x_{1:k}) \;=\; \xi_U(y\!x_{1:k}) \;\stackrel{\times}{=}\; M(y\!x_{1:k}) \;:=\; \sum_{q:q(y_{1:k})=x_{1:k}} 2^{-\ell(q)}. \tag{1.10}$$

If not clear from context, we add superscripts SP and AI to $\xi$, to resolve ambiguities between (1.3) and (1.10). Replacing $\mu$ by $\xi$ in (1.8) the *AIXI system* outputs

$$y_k \;=\; y_k^\xi \;:=\; \arg\max_{y_k}\sum_{x_k}\ldots\max_{y_m}\sum_{x_m}(r_k+\,\ldots\,+r_m)\cdot\xi(y\!x_{<k}y\!x_{k:m}) \tag{1.11}$$

in cycle $k$ given the history $y\!x_{<k}$. The $\xi$-value $V_\xi^p$ and the universal value $V_\xi^*$ are defined as in (1.7) and (1.9), with $\mu$ replaced by $\xi$. The AIXI model and

its behavior is completely defined by (1.10) and (1.11). It (slightly) depends on the choice of the universal Turing machine, because $K()$ and $\ell()$ depend on $U$ and hence are defined only up to terms of order one. The AIXI model also depends on the choice of $\mathcal{X}$ and $\mathcal{Y}$, but we do not expect any bias when the spaces are chosen sufficiently large and simple, e.g. all strings of length $2^{16}$. Choosing $I\!N$ as the I/O spaces would be ideal, but whether the maxima (or suprema) exist in this case has to be shown beforehand. The only nontrivial dependence is on the horizon $m$. Ideally, we would like to chose $m=\infty$, but there are several subtleties to be unraveled later, which prevent at least a naive limit $m\to\infty$. So apart from $m$ and unimportant details, *the AIXI system is uniquely defined by (1.10) and (1.11) without adjustable parameters.*

### 1.5.2 On the Optimality of AIXI

**Universality and convergence of $\xi$.** One can show that also $\xi$ defined in (1.10) is universal and rapidly converges to $\mu$ analogous to the induction (SP) case. If we take a finite product of conditional $\xi$'s and use the chain rule, we see that also $\xi(y\!x_{<k}\underline{y\!x}_{k:k+h})$ converges to $\mu(y\!x_{<k}\underline{y\!x}_{k:k+h})$ for $k\to\infty$. This gives confidence that the outputs $y_k^\xi$ of the AIXI model (1.11) could converge to the outputs $y_k^\mu$ of the AI$\mu$ model (1.8), at least for a bounded moving horizon $h$. The problems with a fixed horizon $m$ and especially $m\to\infty$ will be discussed at the end of this section.

**Universally optimal AI systems.** We call an AI model *universal* if it is independent of the true environment $\mu$ (unbiased, model-free) and is able to solve any solvable problem and learn any learnable task. Further, we call a universal model *universally optimal* if there is no program that can solve or learn significantly faster (in terms of interaction cycles). As the AIXI model is parameter-free, $\xi$ converges to $\mu$, the AI$\mu$ model is itself optimal, and we expect no other model to converge faster to AI$\mu$ by analogy to the SP case,

*we expect AIXI to be universally optimal.*

This is our main claim. Further support is given below.

**Intelligence order relation.** We want to call a policy $p$ *more or equally intelligent* than a policy $p'$ and write $p\succeq p'$ if $p$ yields in every cycle $k$ and for every fixed history $y\!x_{<k}$ higher (future) $\xi$-expected reward sum than $p'$. It is a formal exercise to show that $p^\xi\succeq p$ for all $p$. The AIXI model is hence the most intelligent agent w.r.t. $\succeq$. Relation $\succeq$ is a universal order relation in the sense that it is free of any parameters (except $m$) or specific assumptions about the environment. A proof that $\succeq$ is a reasonable intelligence order (which we believe to be true) would prove that AIXI is universally optimal.

**Value bounds.** The values $V_\rho^*$ associated with the AI$\rho$ systems correspond roughly to the negative total loss $-L_n^{\Lambda_\rho}$ (with $n=m$) of the SP$\rho$ ($=\Lambda_\rho$) systems.

In the SP case we were interested in small bounds for the regret $L_n^{\Lambda_\xi} - L_n^{\Lambda_\mu}$. Unfortunately, simple value bounds for AIXI or any other AI system in terms of $V_\nu^*$ analogous to the loss bound (1.6) cannot hold. We even have difficulties in specifying what we can expect to hold for AIXI or any AI system that claims to be universally optimal. In SP, the only important property of $\mu$ for proving loss bounds was its complexity $K(\mu)$. In the AI case, there are no useful bounds in terms of $K(\mu)$ only. We either have to study restricted problem or environmental classes or consider bounds depending on other properties of $\mu$, rather than on its complexity only.

### 1.5.3 Value-Related Optimality Results

**The mixture distribution $\xi$.** In the following, we consider general Bayes mixtures $\xi$ over classes $\mathcal{M}$ of chronological probability distributions $\nu$:

$$\xi(y\!x_{1:m}) = \sum_{\nu\in\mathcal{M}} w_\nu \nu(y\!x_{1:m}) \quad \text{with} \quad \sum_{\nu\in\mathcal{M}} w_\nu = 1 \quad \text{and} \quad w_\nu > 0 \quad \forall\nu\in\mathcal{M}.$$

We define $V_\xi^p$, $p^\xi$, and $V_\xi^*$ as in (1.7)–(1.9) with $\mu$ replaced by $\xi$. Policy $p^\xi$ is called the AI$\xi$ model. For $\xi = \xi_U$ the AIXI$\equiv$AI$\xi_U$ model is recovered. If $\mu$ is unknown, but known to belong to the known class $\mathcal{M}$, it is natural to follow policy $p^\xi$, which maximizes $V_\xi^p$. The (true $\mu$-)expected reward when following policy $p^\xi$ is $V_\mu^{p^\xi}$. The optimal (but infeasible) policy $p^\mu$ yields reward $V_\mu^{p^\mu} \equiv V_\mu^*$. It is now of interest ($a$) whether there are policies with uniformly larger value than $V_\mu^{p^\xi}$ and ($b$) how close $V_\mu^{p^\xi}$ is to $V_\mu^*$.

**Linearity and convexity of $V_\rho$ in $\rho$.** The following properties of $V_\rho$ are crucial. $V_\rho^p$ is a linear function in $\rho$, and $V_\rho^*$ is a convex function in $\rho$ in the sense that

$$V_\xi^p = \sum_{\nu\in\mathcal{M}} w_\nu V_\nu^p \qquad \text{and} \qquad V_\xi^* \le \sum_{\nu\in\mathcal{M}} w_\nu V_\nu^*.$$

Linearity is obvious from the definition of $V_\rho^p$, and convexity follows easily from the convexity of $\max_p$ and nonnegativity of the weights $w_\nu$. One loose interpretation of the convexity is that a mixture can never increase performance.

**Pareto optimality of AI$\xi$.** Similarly to the SP case, one can show that $p^\xi$ is *Pareto optimal* in the sense that there is no other policy $p$ with $V_\nu^p \ge V_\nu^{p^\xi}$ for all $\nu\in\mathcal{M}$ and strict inequality for at least one $\nu$. In particular, AIXI is Pareto optimal.

**Self-optimizing policy $p^\xi$ w.r.t. average value.** Since we do not know the true environment $\mu$ in advance, we are interested under which circumstances[7]

[7] Here and elsewhere we interpret $a_m \to b_m$ as an abbreviation for $a_m - b_m \to 0$. $\lim_{m\to\infty} b_m$ may not exist.

$$\tfrac{1}{m}V_\nu^{p^\xi} \to \tfrac{1}{m}V_\nu^* \quad \text{for horizon} \quad m\to\infty \quad \text{for } all \quad \nu\in\mathcal{M}. \tag{1.12}$$

Note that $V_\nu$ as well as $p^\xi = p^\xi_m$ depend on $m$. The least we must demand from $\mathcal{M}$ to have a chance that (1.12) is true is that there exists a policy (sequence) $\tilde p = \tilde p_m$ at all with this property, i.e.

$$\exists\tilde p: \tfrac{1}{m}V_\nu^{\tilde p} \to \tfrac{1}{m}V_\nu^* \quad \text{for horizon} \quad m\to\infty \quad \text{for } all \quad \nu\in\mathcal{M}. \tag{1.13}$$

We show that this necessary condition is also sufficient, i.e. (1.13) implies (1.12). This is another (asymptotic) optimality property of policy $p^\xi$. *If* universal convergence in the sense of (1.13) is possible at all in a class of environments $\mathcal{M}$, then policy $p^\xi$ converges in the same sense (1.12). We call policies $\tilde p$ with a property like (1.13) *self-optimizing* [KV86].

Unfortunately, the result is not an asymptotic convergence statement of a single policy $p^\xi$, since $p^\xi$ depends on $m$. The result merely says that under the stated conditions the average value of $p^\xi_m$ is arbitrarily close to optimum for sufficiently large (pre-chosen) horizon $m$. This weakness will be resolved in the following.

**Discounted future value function.** We now shift our focus from the total value to future values (value-to-go). First, we have to get rid of the horizon parameter $m$. We eliminate the horizon by discounting the rewards $r_k \rightsquigarrow \gamma_k r_k$ with $\gamma_k \geq 0$ and $\sum_{i=1}^\infty \gamma_i < \infty$ and taking $m\to\infty$. The analogue of $m$ is now an effective horizon $h_k^{eff}$, which may be defined by $\sum_{i=k}^{k+h_k^{eff}}\gamma_i \approx \sum_{i=k+h_k^{eff}}^\infty \gamma_i$. Furthermore, we renormalize the value $V$ by $\sum_{i=k}^\infty \gamma_i$ and denote it by $V_{k\gamma}$. Finally, we extend the definition to probabilistic policies $\pi$ (which is not essential). We define the $\gamma$-discounted weighted-average future *value* of (probabilistic) policy $\pi$ in environment $\rho$ given history $y\!x_{<k}$, or shorter, the $\rho$-value of $\pi$ given $y\!x_{<k}$, as

$$V_{k\gamma}^{\pi\rho}(y\!x_{<k}) \;:=\; \frac{1}{\Gamma_k}\lim_{m\to\infty}\sum_{y\!x_{k:m}}(\gamma_k r_k + ... + \gamma_m r_m)\rho(y\!x_{<k}y\!\underline{x}_{k:m})\pi(y\!x_{<k}\underline{y}x_{k:m}),$$

with $\Gamma_k := \sum_{i=k}^\infty \gamma_i$. The policy $p^\rho$ is defined as to maximize the future value $V_{k\gamma}^{\pi\rho}$:

$$p^\rho := \arg\max_\pi V_{k\gamma}^{\pi\rho}, \qquad V_{k\gamma}^{*\rho} := V_{k\gamma}^{p^\rho\rho} = \max_\pi V_{k\gamma}^{\pi\rho} \geq V_{k\gamma}^{\pi\rho}\,\forall\pi.$$

Setting $\gamma_k = 1$ for $k \leq m$ and $\gamma_k = 0$ for $k > m$ gives back the old undiscounted model with horizon $m$ and $V_{1\gamma}^{p\rho} = \frac{1}{m}V_\rho^p$. Note that $V_{k\gamma}$ depends on the realized history $y\!x_{<k}$. More important, $p^\rho$ can be shown to be independent of $k$. Similarly to the undiscounted case, one can prove that for every $k$ and history $y\!x_{<k}$, $V_{k\gamma}^{\pi\rho}$ is a linear function in $\rho$, $V_{k\gamma}^{*\rho}$ is a convex function in $\rho$, and $p^\xi$ is Pareto optimal in the sense that there is no other policy $\pi$ with $V_{k\gamma}^{\pi\nu} \geq V_{k\gamma}^{p^\xi\nu}$ for all $\nu\in\mathcal{M}$ and strict inequality for at least one $\nu$. Finally, $p^\xi$ is self-optimizing (w.r.t. discounted value) if $\mathcal{M}$ admits self-optimizing policies:

$$\text{If } \exists\tilde{\pi}\forall\nu : V_{k\gamma}^{\tilde{\pi}\nu} \stackrel{k\to\infty}{\longrightarrow} V_{k\gamma}^{*\nu} \quad \text{w.}\nu\text{.p.1} \quad \Longrightarrow \quad V_{k\gamma}^{p^\xi\mu} \stackrel{k\to\infty}{\longrightarrow} V_{k\gamma}^{*\mu} \quad \text{w.}\mu\text{.p.1.}$$

The probability qualifier refers to the historic perceptions $x_{<k}$. The historic actions $y_{<k}$ are arbitrary. Note that $k$ is a real running value, namely the current cycle number, whereas $m$ was a pre-chosen fixed horizon.

### 1.5.4 Markov Decision Processes

From all possible environments, Markov (decision) processes are probably the most intensively studied ones. $\mu$ is called a (completely observable stationary) *Markov decision process* (MDP) if the probability of perceiving $x_k \in \mathcal{X}$, given history $yx_{<k}y_k$ only depends on the last action $y_k \in \mathcal{Y}$ and the last perception $x_{k-1}$, i.e. if $\mu(yx_{<k}y_k\underline{x}_k) = \mu(x_{k-1}y_k\underline{x}_k)$. In this case $x_k$ is called a *state*, $\mathcal{X}$ the *state space*, and $\mu(x_{k-1}y_k\underline{x}_k)$ the *transition matrix*. An MDP $\mu$ is called *ergodic* if there exists a policy under which every state is visited infinitely often with probability 1. If an MDP $\mu(x_{k-1}y_k\underline{x}_k)$ is independent of the action $y_k$ it is a *Markov process*; if it is independent of the last perception $x_{k-1}$ it is an *i.i.d.* process.

Stationary MDPs $\mu$ with geometric discounting $\gamma_k = \gamma^k$ have stationary optimal policies $p^\mu$ mapping the same state/perception $x_k$ always to the same action $y_k$. On the other hand, a mixture $\xi$ of MDPs is itself not an MDP, i.e. $\xi \notin \mathcal{M}_{\text{MDP}}$, which implies that $p^\xi$ is, in general, not a stationary policy.

One can construct self-optimizing policies for the class of ergodic MDPs w.r.t. the average value $\frac{1}{m}V_\rho^p$ and if $\frac{\gamma_{k+1}}{\gamma_k} \to 1$ also w.r.t. to the discounted future value $V_{k\gamma}^{\pi\rho}$. The necessary condition $\frac{\gamma_{k+1}}{\gamma_k} \to 1$ ensures unboundedly increasing effective horizon $h_k^{eff}$. The existence of self-optimizing policies for ergodic MDPs implies that for a countable class $\mathcal{M}$ of ergodic MDPs, the policies $p_m^\xi$ maximizing $V_\xi^p$ and $p^\xi$ maximizing $V_{k\gamma}^{\pi\xi}$ are self-optimizing in the sense that

$$\forall\nu\in\mathcal{M} : \frac{1}{m}V_{1m}^{p_m^\xi\nu} \stackrel{m\to\infty}{\longrightarrow} \frac{1}{m}V_{1m}^{*\nu} \quad \text{and} \quad V_{k\gamma}^{p^\xi\nu} \stackrel{k\to\infty}{\longrightarrow} V_{k\gamma}^{*\nu} \quad \text{if } \frac{\gamma_{k+1}}{\gamma_k} \to 1. \tag{1.14}$$

We also show that if $\mathcal{M}$ is finite, then the speed of the first convergence is at least $O(m^{-1/3})$. The conditions $\Gamma_k < \infty$ and $\frac{\gamma_{k+1}}{\gamma_k} \to 1$ on the discount sequence are, for instance, satisfied for $\gamma_k = 1/k^2$, but *not* for the popular geometric discount $\gamma_k = \gamma^k$, which has finite effective horizon.

Limits (1.14) show that $p^\xi$ is self-optimizing for bandits, i.i.d. processes, and classification tasks, since they are special (degenerate) cases of ergodic MDPs. The existence of self-optimizing policies is not limited to (subclasses of ergodic) MDPs. Certain classes of POMDPs, $k^{th}$-order ergodic MDPs, factorizable environments, repeated games, and prediction problems are not MDPs, but nevertheless admit self-optimizing policies. Hence the corresponding Bayes optimal mixture policy $p^\xi$ is self-optimizing.

### 1.5.5 The Choice of the Horizon

The only significant arbitrariness in the AIXI model lies in the choice of the lifespan $m$ or in the discounted case in the discount sequence $\gamma_k$. We will not discuss ad hoc choices for specific problems. We are interested in universal choices. In many cases the time we are willing to run a system depends on the quality of its actions. Hence, the lifetime, if finite at all, is not known in advance. Geometric discounting $r_k \rightsquigarrow r_k \cdot \gamma^k$ solves the mathematical problem of $m \to \infty$ but is not a real solution, since an effective horizon $h^{eff} \sim \ln\gamma^{-1} < \infty$ has been introduced. The scale-invariant discounting $r_k \rightsquigarrow r_k \cdot k^{-\alpha}$ with $\alpha > 1$ has a dynamic horizon $h \sim k$. This choice has some appeal, as it seems that humans of age $k$ years also usually do not plan their lives for more than the next $\sim k$ years. It also satisfies the condition $\frac{\gamma_{k+1}}{\gamma_k} \to 1$, necessary for AI$\xi$ being self-optimizing in ergodic MDPs. The largest lower semicomputable horizon with guaranteed finite reward sum $\Gamma_1 < \infty$ is obtained by the discount $r_k \rightsquigarrow r_k \cdot 2^{-K(k)}$, where $K(k)$ is the Kolmogorov complexity of $k$. This is maybe the most attractive universal discount. It is similar to a near-harmonic discount $r_k \rightsquigarrow r_k \cdot k^{-(1+\varepsilon)}$, since $2^{-K(k)} \leq 1/k$ for most $k$ and $2^{-K(k)} \geq c/(k \log^2 k)$ for some constant $c$. We are not sure whether the choice of the horizon is of marginal importance, as long as it is chosen sufficiently large, or whether the choice will turn out to be a central topic for the AIXI model or for the planning aspect of any universal AI system in general. Most, if not all, problems in agent design of balancing exploration and exploitation vanish by a sufficiently large choice of the (effective) horizon and a sufficiently general prior.

## 1.6 Important Environmental Classes

In this and the next section we define $\xi = \xi_U \stackrel{\times}{=} M$ be Solomonoff's prior, i.e. AI$\xi$=AIXI. Each subsection represents an abstract on what will be done in the corresponding section of Chapter 6.

### 1.6.1 Introduction

In order to give further support for the universality and optimality of the AI$\xi$ theory, we apply AI$\xi$ to a number of problem classes. They include sequence prediction, strategic games, function minimization and, especially, how AI$\xi$ learns to learn supervised. For some classes we give concrete examples to illuminate the scope of the problem class. We first formulate each problem class in its natural way (when $\mu^{\text{problem}}$ is known) and then construct a formulation within the AI$\mu$ model and prove its equivalence. We then consider the consequences of replacing $\mu$ by $\xi$. The main goal is to understand why and how the problems are solved by AI$\xi$. We only highlight special aspects of each problem class. The goal is to give a better picture of the flexibility of the AI$\xi$ model.

### 1.6.2 Sequence Prediction (SP)

Using the AI$\mu$ model for sequence prediction (SP) is identical to Bayesian sequence prediction SP$\mu$. One might expect, when using the AI$\xi$ model for sequence prediction, one would recover exactly the universal sequence prediction scheme SP$\xi$, as AI$\xi$ was a unification of the AI$\mu$ model and the idea of universal probability $\xi$. Unfortunately, this is not the case. One reason is that $\xi$ is only a probability distribution in the inputs $x$ and not in the outputs $y$. This is also one of the origins of the difficulty of proving loss/value bounds for AI$\xi$. Nevertheless, we argue that AI$\xi$ is as well suited for sequence prediction as SP$\xi$. In a very limited setting we prove a (weak) error bound for AI$\xi$, which gives hope that a general proof is attainable.

### 1.6.3 Strategic Games (SG)

A very important class of problems are strategic games (SG). We restrict ourselves to deterministic strictly competitive strategic games like chess. If the environment is a minimax player, the AI$\mu$ model itself reduces to a minimax strategy. Repeated games of fixed lengths are a special case of factorizable $\mu$. The consequences of variable game lengths are sketched. The AI$\xi$ model has to learn the rules of the game under consideration, as it has no prior information about these rules. We describe how AI$\xi$ actually learns these rules.

### 1.6.4 Function Minimization (FM)

Many problems fall into the category 'resource-bounded function minimization' (FM). They include the traveling salesman problem, minimizing production costs, inventing new materials or even producing, e.g. nice paintings, which are (subjectively) judged by a human. The task is to (approximately) minimize some function $f:\mathcal{Y}\to Z$ within a minimal number of function calls. We will see that a greedy model trying to minimize $f$ in every cycle fails. Although the greedy model has nothing to do with downhill or gradient techniques (there is nothing like a gradient or direction for functions over $\mathcal{Y}$), which are known to fail, we discover the same difficulties. FM has already nearly the full complexity of general AI. The reason being that FM can actively influence the information gathering process by its trials $y_k$ (whereas SP and CF=classification cannot). We discuss in detail the optimal FM$\mu$ model and its inventiveness in choosing the $y\in\mathcal{Y}$. A discussion of the subtleties when using AI$\xi$ for function minimization follows.

### 1.6.5 Supervised Learning from Examples (EX)

Reinforcement learning, as the AI$\xi$ model does, is an important learning technique, but not the only one. To improve the speed of learning, supervised learning, i.e. learning by acquiring knowledge, or learning from a constructive teacher, is necessary. We show how AI$\xi$ learns to learn supervised. It actually establishes supervised learning very quickly within $O(1)$ cycles.

### 1.6.6 Other Aspects of Intelligence

Finally, we give a brief survey of other general aspects, ideas and methods in AI, and their connection to the AI$\xi$ model. Some aspects are directly included in the AI$\xi$ model, while others are or should be emergent.

## 1.7 Computational Aspects

Up to now we have shown the universal character of the AIXI model but have completely ignored computational aspects. We start by developing an algorithm $M_{p^*}^{\varepsilon}$ that is capable of solving any well-defined problem $p$ as quickly as the fastest algorithm computing a solution to $p$, save for a factor of $1+\varepsilon$ and lower-order additive terms. Based on a similar idea we then construct a computable version of the AIXI model.

### 1.7.1 The Fastest & Shortest Algorithm for All Well-Defined Problems

**Introduction.** A wide class of problems can be phrased in the following way. Given a formal specification $f:\mathcal{X}\to\mathcal{Y}$ of a problem depending on some parameter $x\in\mathcal{X}$, we are interested in a fast algorithm computing solution $y\in\mathcal{Y}$.

Levin search is (within a large constant factor) the fastest algorithm to invert a function $g:\mathcal{Y}\to\mathcal{X}$, *if* $g$ can be evaluated quickly [Lev73b, Lev84]. Levin search can also handle time-limited optimization problems [Sol86]. Prime factorization, graph coloring, and truth assignments are example problems suitable for Levin search, if we want to find a solution, since verification is quick. Levin search cannot decide the corresponding decision problems. It is also not applicable to, e.g. matrix multiplication and reinforcement learning, since the verification task $g$ is as hard as the computation task. Blum's speed-up theorem [Blu67, Blu71] shows that there are types of problems $f$ for which an (incomputable) sequence of speed-improving algorithms (of increasing size) exists, but no fastest algorithm.

In the approach presented here, we consider only those algorithms that *provably* solve a given problem and have a fast (i.e. quickly computable) time bound. Neither the programs themselves nor the proofs need to be known in advance. Under these constraints we construct the asymptotically fastest algorithm save a factor of $1+\varepsilon$ that solves any well-defined problem $f$.

**The fast algorithm $M_{p^*}^{\varepsilon}$.** Let $p^*$ be a given algorithm computing $p^*(x)$ from $x$, or, more generally, a specification of a function $f$. One ingredient to our fastest algorithm $M^\varepsilon_{p^*}$ to compute $p^*(x)$ is an enumeration of proofs of increasing length in some formal axiomatic system. If a proof actually proves that some $p$ is functionally equivalent to $p^*$, and $p$ has time bound $t_p$, the tuple $(p,t_p)$ is added to a list $L$. The program $p$ in $L$ with the currently smallest time bound $t_p(x)$ is executed. By construction, the result $p(x)$ is identical to $p^*(x)$. The trick to achieve a small runtime is to schedule everything in a proper way, in order not to lose too much performance by computing slow $p$'s and $t_p$'s before *the* $p$ has been found.

More formally, we say that a program "$p$ computes function $f$", when a universal reference Turing machine $U$ on input $(p,x)$ computes $f(x)$ for all $x$. This is denoted by $U(p,x)=f(x)$. To be able to talk about proofs, we need a formal logic system $(\forall,\lambda,y_i,c_i,f_i,R_i,\rightarrow,\wedge,=,...)$ and axioms and inference rules. A proof is a sequence of formulas, where each formula is either an axiom or inferred from previous formulas in the sequence by applying the inference rules. We only need to know that *provability*, *Turing Machines*, and *computation time* can be formalized, and that the set of (correct) proofs is enumerable. We say that $p$ is provably equivalent to $p^*$ if the formula $[\forall y : U(p,y)=U(p^*,y)]$ can be proven. Let us fix $\varepsilon\in(0,\frac{1}{2})$. $M^\varepsilon_{p^*}$ runs three algorithms $A$, $B$, and $C$ in parallel:

$M^\varepsilon_{p^*}(x)$

Initialize the shared variables
$L:=\{\}$, $t_{fast}:=\infty$, $p_{fast}:=p^*$.
Start algorithms $A$, $B$, and $C$
in parallel with relative computational
resources $\varepsilon$, $\varepsilon$, and $1-2\varepsilon$, respectively.

$A$

Run through all proofs.
**if** a proof proves for some $(p,t)$ that
$p(\cdot)$ is equivalent to (computes) $p^*(\cdot)$
and has time bound $t(\cdot)$
**then** add $(p,t)$ to $L$.

$B$

Compute all $t(x)$ in parallel
for all $(p,t)\in L$ with
relative computation time $2^{-\ell(p)-\ell(t)}$.
**if** for some $t$, $t(x)<t_{fast}$,
**then** $t_{fast}:=t(x)$ and $p_{fast}:=p$.
**continue**

$C$

run $U$ on $(p_{fast},x)$.
For each time step decrease $t_{fast}$ by 1.
**if** $U$ halts **then** print result $U(p_{fast},x)$
and abort computation of $A$, $B$ and $C$.

Note that $A$ and $B$ only terminate when aborted by $C$. It is obvious that $M^\varepsilon_{p^*}$ is equivalent to (computes) $p^*$. We show that the computation time of $M^\varepsilon_{p^*}$ is bounded by

$$time_{M^\varepsilon_{p^*}}(x) \;\leq\; (1+\varepsilon)\cdot t_p(x) + \tfrac{d_p}{\varepsilon}\cdot time_{t_p}(x) + \tfrac{c_p}{\varepsilon},$$

$$d_p \;=\; 3\cdot 2^{\ell(p)+\ell(t_p)}, \quad c_p \;=\; 3\cdot 2^{\ell(proof p))+1}\cdot O(\ell(proof p)^2),$$

where $p$ is any algorithm, provably computing the same function as $p^*$ with computation time provably bounded by the function $t_p(x)$ for all $x$, and $time_{t_p}(x)$ is the time needed to compute the time bound $t_p(x)$. Known time bounds for practical problems can often be computed quickly, i.e. $time_{t_p}(x)/time_p(x)$ often converges very quickly to zero. Furthermore, from a practical point of view, the provability restrictions are often rather weak. Hence, we have constructed for all those problems a solution that is asymptotically only a factor $1+\varepsilon$ slower than the (provably) fastest algorithm. On the flip side, for realistically sized problems, the lower-order terms usually dominate, which limits the practical use of $M_{p^*}^\varepsilon$.

**Algorithmic complexity and the shortest algorithm.** A natural definition for the (Kolmogorov) complexity of a function $f$ is the length of the shortest program computing $f$: $K'(f) := \min_p\{\ell(p) : U(p,x) = f(x)\,\forall x\}$. Unfortunately, $K'$ suffers from not even being approximable, since functional equality of programs is in general undecidable. Let $p^*$ be a formal specification or a program for $f$. Using $K(p^*)$ is also not a suitable alternative, since it essentially depends on the choice of $p^*$ because, e.g. "dead code" in $p^*$ contributes to $K(p^*)$. A satisfactory solution is to take the length of the shortest program *provably* equivalent to $p^*$:

$$K''(p^*) \;:=\; \min_p\{\ell(p) : \text{a proof of } [\forall y\!:\!U(p,y) = U(p^*,y)] \text{ exists}\}.$$

$K''$ (like $K$) is upper semicomputable. Let $p'$ be some short description of $p^*$. We are now concerned with the computation time of $p'$. Could we get slower and slower algorithms by compressing $p^*$ more and more? Interestingly, this is not the case. Inventing complex (long) programs is *not* necessary to construct asymptotically fast algorithms, under the stated provability assumptions, in contrast to Blum's theorem [Blu67, Blu71]. We show that there exists a program $\tilde p$, equivalent to $p^*$ with

$$\begin{array}{lll} (i) & \ell(\tilde p) & \leq\; K''(p^*) + O(1), \\ (ii) & time_{\tilde p}(x) & \leq\; (1+\varepsilon)\!\cdot\! t_p(x) + \frac{d_p}{\varepsilon}\!\cdot\! time_{t_p}(x) + \frac{c_p}{\varepsilon}, \end{array}$$

where $p$ is any program provably equivalent to $p^*$ with computation time provably less than $t_p(x)$. That is, $\tilde p$ is simultaneously among the shortest *and* fastest programs.

**Generalizations.** Algorithm $M_{p^*}^\varepsilon$ can be modified to handle I/O streams, definable by a Turing machine with unidirectional input and output tapes (and bidirectional work tapes) receiving an input stream and producing an output stream, as is the case in the agent setup.

### 1.7.2 Time-Bounded AIXI Model

The major drawback of the AIXI model is that it is uncomputable. To overcome this problem, we construct a modified algorithm AIXI*tl*, which is still superior to any other time $t$ and length $l$ bounded agent. The computation time of AIXI$tl$ is of the order $t \cdot 2^l$. Reducing the large factor $2^l$ along the lines of the previous subsection is possible, but will not be presented here.

**Non-effectiveness of AIXI.** $\xi^{\text{AI}} = \xi_U^{\text{AI}}$ is not a computable but only an enumerable semimeasure. Hence, the output $\dot{y}_k$ of the AIXI model is only asymptotically computable (approximable). AIXI yields an algorithm that produces a sequence of trial outputs eventually converging to the correct output $\dot{y}_k$, but one can never be sure whether one has already reached it. Besides this, convergence is extremely slow, so this type of asymptotic computability is of no direct practical use. Furthermore, the replacement of $\xi^{\text{AI}}$ by time-limited versions [LV91, LV97], which is suitable for sequence prediction, fails for the AIXI model. This leads to the issues addressed next.

**Time bounds and effectiveness.** Let $\tilde{p}$ be a policy that calculates an acceptable output within a reasonable time $\tilde{t}$ per interaction cycle. This sort of computability assumption, namely, that a general-purpose computer of sufficient power and appropriate program is able to behave in an intelligent way, is the very basis of AI research. Here it is not necessary to discuss what exactly is meant by 'reasonable time/intelligence' and 'sufficient power'. What we are interested in is whether there is a computable version of the AIXI system that is superior or equal to any policy $p$ with computation time per cycle of at most $\tilde{t}$.

What one can realistically hope to construct is an AIXI$\tilde{t}\tilde{l}$ system of computation time $c \cdot \tilde{t}$ per cycle for some constant $c$. The idea is to run all programs $p$ of length $\leq \tilde{l} := \ell(\tilde{p})$ and time $\leq \tilde{t}$ per cycle and pick the best output in the sense of maximizing the *universal value* $V_\xi^*$. The total computation time is $c \cdot \tilde{t}$ with $c \approx 2^{\tilde{l}}$. Unfortunately, $V_\xi^*$ cannot be used directly since this measure is itself only semicomputable and the approximation quality by using computable versions of $\xi^{\text{AI}}$ given a time of order $c \cdot \tilde{t}$ is crude [LV97]. On the other hand, we *have* to use a measure that converges to $V_\xi^*$ for $\tilde{t}, \tilde{l} \to \infty$, since we want the AIXI$\tilde{t}\tilde{l}$ model to converge to the AIXI model in that case.

**Valid approximations.** We suggest the following solution satisfying the above conditions: The main idea is to consider *extended chronological incremental policies* $p$, which in addition to the regular output $y_k^p$ *rate* their own output with $w_k^p$. The AIXI$\tilde{t}\tilde{l}$ model selects the output $\dot{y}_k = y_k^p$ of the policy $p$ with highest rating $w_k^p$. Policy $p$ might suggest any output $y_k^p$, but it is not allowed to rate itself with an arbitrarily high $w_k^p$ if one wants $w_k^p$ to be a reliable criterion for selecting the best $p$. One must demand that no policy $p$ is allowed to claim that it is better than it actually is. We define a logical predicate VA($p$), called *valid approximation*, which is true if and only if $p$ *always* satisfies $w_k^p \leq V_\xi^p(y\!x_{<k})$, i.e. never overrates itself. $V_\xi^p(y\!x_{<k})$ is the $\xi^{\text{AI}}$-expected future reward under policy $p$. Valid policies $p$ can then be (partially) ordered w.r.t. their rating $w_k^p$.

**The universal time-bounded AIXI$tl$ system.** In the following, we describe the algorithm $p^*$ underlying the AIXI$\tilde{t}\tilde{l}$ system. It is essentially based on the selection of the best algorithms $p_k^*$ out of the time $\tilde{t}$ and length $\tilde{l}$ bounded policies $p$, for which there exists a proof $P$ of VA($p$) with length $\leq l_P$.

1. Create all binary strings of length $l_P$ and interpret each as a coding of a mathematical proof in the same formal logic system in which VA($\cdot$) has been formulated. Take those strings that are proofs of VA($p$) for some $p$ and keep the corresponding programs $p$.
2. Eliminate all $p$ of length $>\tilde{l}$.
3. Modify the behavior of all remaining $p$ in each cycle $k$ as follows: Nothing is changed if $p$ outputs some $w_k^p y_k^p$ within $\tilde{t}$ time steps. Otherwise stop $p$ and write $w_k=0$ and some arbitrary $y_k$ to the output tape of $p$. Let $P$ be the set of all those modified programs.
4. Start first cycle: $k:=1$.
5. Run every $p\in\mathcal{P}$ on extended input $\dot{y}\dot{x}_{<k}$, where all outputs are redirected to some auxiliary tape: $p(\dot{y}\dot{x}_{<k}) = w_1^p y_1^p ... w_k^p y_k^p$. This step is performed incrementally by adding $\dot{y}\dot{x}_{k-1}$ for $k>1$ to the input tape and continuing the computation of the previous cycle.
6. Select the program $p$ with highest rating $w_k^p$: $p_k^*:=\text{argmax}_p w_k^p$.
7. Write $\dot{y}_k:=y_k^{p_k^*}$ to the output tape.
8. Receive input $\dot{x}_k$ from the environment.
9. Begin next cycle: $k:=k+1$, goto step 5.

**Properties of the $p^*$ algorithm.** Let $p$ be any extended chronological (incremental) policy of length $\ell(p)\leq\tilde{l}$ and computation time per cycle $t(p)\leq\tilde{t}$, for which there exists a proof of VA($p$) of length $\leq l_P$. The algorithm $p^*$, depending on $\tilde{l}$, $\tilde{t}$ and $l_P$ but not on $p$, has always higher rating than any such $p$. The setup time of $p^*$ is $t_{setup}(p^*)=O(l_P^2\cdot 2^{l_P})$, and the computation time per cycle is $t_{cycle}(p^*)=O(2^{\tilde{l}}\cdot\tilde{t})$. Furthermore, for $\tilde{t},\tilde{l},l_P\to\infty$, policy $p^*$ converges to the behavior of the AIXI model.

Roughly speaking, this means that if there exists a computable solution to some AI problem at all, then the explicitly constructed algorithm $p^*$ is such a solution. This claim is quite general, but there are some limitations and open questions regarding the setup time, regarding the necessity that the policies must rate their own output, regarding true but not (efficiently) provable VA($p$), and regarding "inconsistent" policies.

## 1.8 Discussion

**What has been achieved.** We suggested an elegant mathematical foundation of artificial intelligence. More specifically, we developed a theory for rational agents acting optimally in any environment. Thereby we touched various scientific areas, including reinforcement learning, algorithmic information theory, computational complexity theory, probability theory, sequential decision theory, and many more. We presented sequential decision theory in a very general form and unified it with Solomonoff's theory of universal induction, both shown to be optimal in their own domain. The resulting parameter-free AIXI model constitutes an agent for which we gave strong arguments that it behaves optimally in any environment. For restricted environmental classes and Bayes mixtures $\xi$ we showed that AI$\xi$ is self-optimizing and Pareto optimal. We discussed the choice of the horizon and motivated the use of non-geometric discounting of rewards. We also discussed a number of important problem classes, including sequence prediction, strategic games, function minimization, and supervised learning. All in all, this shows that artificial intelligence *can* be framed by an elegant mathematical theory. Some progress has also been made toward an elegant *computational* theory of intelligence. AIXI$tl$ has optimal order of computation time, apart from a large multiplicative constant, which we could get rid of at the expense of an (unfortunately even larger) additive constant.

**Comparison to other approaches.** There are many other approaches to AI, too many to mention them all. Among the models that can learn from experience are the "classical" reinforcement learning algorithms like temporal difference learning [SB98], adaptive variants of Levin search [SZW97, Sch04], prediction with expert advice [CB97], market/economy-based reinforcement learning [Bau99, KHS01b], etc. All these models have been implemented and are applicable in limited domains, often with reasonable performance. In contrast, AIXI($tl$) behaves optimally in every (completely general) environment, is data efficient, has generalization capabilities, addresses the exploration versus exploitation problem, etc., but is computationally not feasible without further approximations.

**Outlook & open questions.** The major theoretical challenge is to derive good non-asymptotic bounds on the value or related quantities for AI$\xi$ and AIXI, ideally as strong as in the sequence prediction case. The major practical challenge is to scale the AI$\xi$ model down, e.g. by using more restricted forms of $\xi$, like the minimum description length principle does for universal induction. The AIXI$tl$ model is a different, very general approach toward a computational model. Unfortunately, it suffers from the same large factor $2^{\tilde{l}}$ in computation time as Levin search for inversion problems [Lev73b, Lev84]. On the other hand, Levin search has been implemented and successfully adapted and applied to a variety of problems [Sch97, Sch04, SZW97], and the multiplicative constant can be eliminated as in $M_{p^*}^{\varepsilon}$ or reduced by the Gödel machine [Sch03b]. Inspecting existing approaches suggests that, while AIXI is an elegant mathematical theory that seems to serve all formal needs, computational AI may be messy. For instance, special-purpose algorithms for preprocessing inputs and postprocessing outputs are likely to be necessary in any efficient AI system. Another issue is that of incorporating extra knowledge. In principal there is no need to modify AIXI, since any prior knowledge can simply be presented as first input $x_1$ in any format. As long as the algorithm to interpret the data is of size $O(1)$, AIXI will "understand" the data after a few cycles. Another important issue is that of the training process itself. By a training process we mean a sequence of simple-to-complex tasks to solve, with the simpler ones helping in learning the more complex ones. These and many other conceptual, practical, and philosophical issues, including concurrent actions and perceptions, the choice of the I/O spaces, treatment of encrypted information, peculiarities of mortal embodied agents, the free will paradox, the existence of objective probabilities, the Turing test, the existence of efficient and elegant universal theories of intelligence related to Penrose's non-computable environments and Chaitin's 'number of wisdom' $\Omega$ will be addressed later in the book.

## 1.9 History & References

**Introductory textbooks.** The book of Hopcroft and Ullman, and in the new revision coauthored by Motwani [HMU01], is a very readable elementary introduction to automata theory, formal languages, and computation theory. The artificial intelligence book [RN95] by Russell and Norvig gives a comprehensive overview over AI approaches in general. For an excellent introduction to algorithmic information theory, Kolmogorov complexity, and Solomonoff induction one should consult the book of Li and Vitányi [LV97], or the book of Calude [Cal02] which focuses more on algorithmic randomness. The reinforcement learning book by Sutton and Barto [SB98] requires no background knowledge, describes the key ideas, open problems, and great applications of this field. A tougher and more rigorous book by Bertsekas and Tsitsiklis on sequential decision theory provides all (convergence) proofs [BT96].

**Algorithmic information theory.** Kolmogorov [Kol65] suggested to define the information content of an object as the length of the shortest program computing a representation of it. Solomonoff [Sol64] invented the closely related universal prior probability distribution and used it for binary sequence prediction [Sol64, Sol78] and function inversion and minimization [Sol86]. Together with Chaitin [Cha66, Cha75], this was the invention of what is now called algorithmic information theory. For further literature and many applications see [LV97, Cal02]. Other interesting applications can be found in [Cha91, Sch99, VW98, CV03]. Related topics are the weighted majority algorithm invented by Littlestone and Warmuth [LW94], universal forecasting by Vovk [Vov92], Levin search [Lev73b], PAC-learning introduced by Valiant [Val84] and minimum description length [LV92a, Ris89]. Resource-bounded complexity is discussed in [Dal73, Dal77, FMG92, Ko86, PF97], resource-bounded universal probability in [LV91, LV97, Sch02b]. Implementations are rare and mainly due to Schmidhuber [Sch95, WS96, Sch97, SZW97, Sch03a,

Sch04]. Good reviews with a philosophical touch are [LV92b, Sol97]. For an older general review of inductive inference see Angluin [AS83].

**Sequential decision theory.** The other ingredient in our AIXI model is sequential decision theory. We do not need much more than the maximum expected utility principle and the expectimax algorithm [Mic66, RN95]. The book of von Neumann and Morgenstern [NM44] might be seen as the initiation of game theory, which already contains the expectimax algorithm as a special case. If the true environmental $\mu$ is unknown, it needs to be learned with, e.g. the help of reinforcement learning algorithms. Existing reinforcement learning algorithms are [Sam59, BSA83, Sut88, Wat89, WD92, MA93, Tes94, BT96, KLM96, KLC98, WS98, KS98], but they are rather limited in view of AIXI. The literature on reinforcement learning and sequential decision theory is so vast that we refer to the textbooks [SB98, BT96, KV86] for further references.

**The author's contributions.** Many of the issues addressed in this book can already be found scattered in various reports and publications by the author: The AIXI model was first introduced and discussed in March 2000 in [Hut00] in a 62-page-long report. More succinct descriptions were published in [Hut01d, Hut01e]. The AIXI model has been argued to formally solve a number of problem classes, including sequence prediction, strategic games, function minimization, reinforcement and supervised learning [Hut00]. The generalization AI$\xi$ has recently been shown to be self-optimizing and Pareto optimal [Hut02b]. The construction of a general fastest algorithm (within a factor of 5) for all well-defined problems [Hut02a] arose from the construction of the time-bounded AIXI$tl$ model [Hut00, Hut01d]. Convergence [Hut03b] and tight [Hut03c] error [Hut01c, Hut01a] and loss [Hut01b, Hut03a] bounds for Solomonoff's universal sequence prediction scheme have been proven. These and other papers are available at http://www.idsia.ch/~marcus/ai.

*Nulla pluralitas est ponenda nisi per rationem vel experientiam vel auctoritatem illius, qui non potest falli nec errare, potest convinci.*

*A plurality should only be postulated if there is some good reason, experience, or infallible authority for it.*

— William of Ockham

William of Ockham (1285–1349)

# 2 Simplicity & Uncertainty



This chapter deals with the question of how to make predictions in unknown environments. Following a brief description of important philosophical attitudes regarding inductive reasoning and inference, we describe more accurately what we mean by induction, and motivate why we can focus on sequence prediction tasks. The most important concept is Occam's razor (simplicity) principle. Indeed, one can show that the best way to make predictions is based on the shortest ($\widehat{=}$ simplest) description of the data sequence seen so far. The most general effective descriptions can be obtained with the help of general recursive functions, or equivalently by using programs on Turing machines, especially on the universal Turing machine. The length of the shortest program describing the data is called the Kolmogorov complexity of the data. Unfortunately, the Kolmogorov complexity is not finitely computable, which makes it necessary to introduce several weaker computability concepts. Probability theory is needed to deal with uncertainty. The environment may be a stochastic process (e.g. gambling houses or quantum physics), which can be described by "objective" probabilities. But also uncertain knowledge about the environment, which leads to beliefs about it, can be modeled by "subjective" probabilities. The old question left open by subjectivists of how to choose the a priori probabilities is solved by Solomonoff's universal prior, which is closely related to Kolmogorov complexity. Solomonoff's major result is that the universal (subjective) posterior converges to the true (objective) environment (probability) $\mu$. The only assumption on $\mu$ is that $\mu$ (which needs not be known!) is computable. The problem of the unknown environment $\mu$ is hence solved for all problems of inductive type, like sequence prediction and classification. Finally, we show the (non)existence of universal priors for the other introduced computability concepts.

For a slower and more detailed introduction into Kolmogorov complexity and Solomonoff induction and most proofs one should consult the excellent book of Li and Vitányi [LV97].

## 2.1 Introduction

One very important and nontrivial aspect of intelligence is inductive inference. After discussing some examples we present the philosophical foundations, and thereafter the sequential setup we are interested in.

### 2.1.1 Examples of Induction Problems

What is the probability that the sun will rise tomorrow? Several answers come into mind: The probability is undefined, because there has never been an experiment that tested the existence of the sun *tomorrow* (reference class problem). The probability is 1, because in all experiments in the past the sun rose. The probability is $1-\epsilon$, where $\epsilon \ll 1$ is the proportion of stars in the universe that explode in a supernova per day. The probability can be derived from the type, age, size and temperature of the sun, even though we never have observed another star with exactly these properties. The probability is $\frac{d+1}{d+2}$, where $d$ is the number of past days the sun rose (Laplace' rule, see Problem 2.11).

Another example is extending binary sequences, like 1100100100001111-1101101010100.... The sequence looks random, so likely also its continuation. A closer look reveals that the sequence is the binary expansion of $\pi$, so we are probably better off predicting its continuation 010001.... We prefer answer 010001..., since we see more structure in the sequence than just random digits.

As another example, consider number sequences $x_1$,$x_2$,$x_3$,$x_4$,..., like 1,2,3,4,... of IQ tests. Virtually everybody predicts $x_5 = 5$ as the next number since $x_i = i$ for $i = 1...4$, but $x_5 = 29$ could also be argued for since $x_i = i^4 - 10i^3 + 35i^2 - 49i + 24$. We prefer answer 5 since a linear relation involves less arbitrary parameters than a $4^{th}$-order polynomial. More difficult is 2,3,5,7,11,13,17,19,23,29,31,37,41,43,47,53,59,?. The next number may be 61 since this is the next prime, or 60 since this is the order of the next simple group. Most will answer 61, since primes are a more familiar concept than simple groups (see [Slo04] for an encyclopedia of integer sequences).

The examples above demonstrate that finding prediction rules for every particular (new) problem is cumbersome and prone to disagreement or contradiction. What we need is a formal general theory for prediction.

### 2.1.2 Ockham, Epicurus, Hume, Bayes, Solomonoff

Generally speaking, induction is the process of predicting the future from the past, or more precisely, it is the process of finding rules in (past) data and using these rules to guess future data. Weather prediction, stock-market forecasting, or continuing number series in an IQ test are nontrivial examples. Making good predictions plays a central role in natural and artificial intelligence in general, and in machine learning in particular.

On the one hand, induction seems to happen in every day life by finding regularities in past observations and using them to predict the future. On the other hand, this procedure seems to add knowledge about the future from past observations. But how can we know something about the future? This dilemma and the induction principle in general have a long philosophical history:

- *Epicurus' principle of multiple explanations* (342?–270? B.C.)
  If more than one theory is consistent with the observations, keep all theories.
- *Occam's razor*[1] *(simplicity) principle* (1290?–1349?)
  Entities should not be multiplied beyond necessity – or – keep the simplest theory consistent with the observations.
- *Hume's negation of induction* (1711–1776) [Hum39]
  The belief in the possibility of true induction cannot be justified rationally.
- *Bayes' rule for conditional probabilities* (1702–1761) [Bay63]
  It tells us how to update our beliefs/probabilities when acquiring new data.

[1] Whereas *William of Ockham* is spelled with *ckh*, for some reason *Occam's razor* is usually spelled with *cc*.

Solomonoff [Sol64] cleverly unified the principles of Epicurus, Occam, and Bayes into one formal universal theory of inductive inference. Among all possible induction schemes it is the optimal method for making predictions.

### 2.1.3 Problem Setup

Every induction problem can be phrased as a sequence prediction task. This is most clearly illustrated in the domain of time-series prediction. Having observed data $x_t$ at times $t<n$, the task is to predict the $n^{th}$ symbol $x_n$ from sequence $x_1...x_{n-1}$. Classification can also be seen as a sequence prediction task. The task of classifying a new instance $z_n$ after having seen (instance,class) pairs $(z_1,c_1),...,(z_{n-1},c_{n-1})$ can be phrased as to predict the continuation of the sequence $z_1c_1...z_{n-1}c_{n-1}z_n$.[2] Machine learning is often concerned with finding the *true* or a *predictive* or a *causal model* based on observed data. This step is important for *understanding* the domain under consideration. Understanding is often a goal in itself, but finally the goal is to apply the model to make predictions. In this view, model learning is only an intermediate step. The direct study of predictions based on past observations without discussing models has been coined *prequential approach* by Dawid [Daw84] for sequence predictions and *transductive inference* by Vapnik [Vap99, Sec.9.1] for classification and regression. Several difficult issues are avoided by abandoning models. This includes questions about model consistency, i.e. whether the true model can be learned, and how to separate noise from useful data [GTV01, VV02]. One may even go one step further and ask why we want to make predictions. Usually the goal of prediction is to maximize one's profit/value, or equivalently to minimize one's loss. In considering only profits or losses one avoids questions on whether prediction algorithms converge to the best possible prediction algorithm (i.e. whether they are *self-tuning* [KV86, p232,p272]). Algorithms for which the loss converges to the minimal possible loss are called *self-optimizing* [KV86, p234]. This is a weaker demand than the ability to be self-tuning, but is often all we really care about. The main purpose of this book is to study algorithms that minimize loss. Convergence of posterior probability distributions or algorithms themselves or models are only considered if this is useful for the ultimate goal of minimizing loss. To summarize our setup:

- Every induction problem can be phrased as a sequence prediction task.
- Classification is a special case of sequence prediction.
  (With some tricks the other direction is also true)
- We are interested in maximizing profit or minimizing loss.
  We are not primarily interested in finding (true/predictive/causal) models or even in convergence of the predictor itself.

[2] Sequence prediction may also be phrased as a classification task by adding time tags and if one does not assume a random generation of instances. Predicting the next symbol of sequence $x_1x_2...x_{n-1}$ is the same as trying to find the class label of $n$ after having seen (instance,class) pairs $(1,x_1),...,(n-1,x_{n-1})$.

- Separating noise from data is *not* necessary in this setting.

After having clarified the setup we now must delve into math before we can present Solomonoff's induction scheme.

## 2.2 Algorithmic Information Theory

In this section we give a very brief introduction to Kolmogorov complexity. For a slower, more thorough and comprehensive introduction see [LV97].

### 2.2.1 Definitions and Notation

We write $I\!N=\{1,2,3,...\}$ for the set of natural numbers, $I\!B^*$ for the set of finite binary strings, and $I\!B^\infty$ for the set of infinite binary sequences. We use letters $i,k,n$ for natural numbers, $x,y,z$ for finite strings, $\epsilon$ for the empty string, $1^n$ the string of $n$ ones, $\ell(x)$ for the length of string $x$, and $\omega$ for infinite strings. We write $xy$ for the concatenation of string $x$ with $y$.

Every countable set may be identified with $I\!N$ by means of a bijection. We can interpret a string as a binary representation of a natural number. Unfortunately, a naive identification will not be unique since, for instance, strings 00101 and 101 both represent the number 5. We get a bijection if we map $x$ to the natural number that has binary representation $1x$ ($x$ prefixed with 1). We subtract 1 from this number, since we need a bijection between $I\!B^*$ and $I\!N_0:=\{0,1,2,3,...\}$ (see Table 2.2). With this identification $\log_2(x+1)-1<\ell(x)\leq\log_2(x+1)$. String $x$ is called a (proper) prefix of $y$ if there is a $z(\neq\epsilon)$ such that $xz=y$. A set of strings is called prefix-free if no element is a proper prefix of another. A prefix-free set $\mathcal{P}$ is also called a prefix code. Prefix codes have the important property of satisfying Kraft's inequality

$$\sum_{x\in\mathcal{P}} 2^{-\ell(x)} \leq 1. \tag{2.1}$$

This can be shown by assigning to each $x\in\mathcal{P}$ an interval $\Gamma_x:=[0.x\,,0.x+2^{-\ell(x)})\subseteq[0,1)$, where $0.x\equiv x\cdot 2^{-\ell(x)}$ is the real number with binary expansion $x$ after the comma. The length of interval $\Gamma_x$ is $2^{-\ell(x)}$. The intervals are disjoint, since $\mathcal{P}$ is prefix free, hence $\sum_{x\in\mathcal{P}}2^{-\ell(x)}=\sum_{x\in\mathcal{P}}\text{length}(\Gamma_x)\leq\text{length}([0,1])=1$. A converse of (2.1) can also be shown.

For $\bar{x}:=1^{\ell(x)}0x$ the set $\{\bar{x}:x\in I\!B^*\}$ forms a prefix code with $\ell(\bar{x})=2\ell(x)+1$. For $x':=\overline{\ell(x)}x=1^{\ell(\ell(x))}0\ell(x)x$ the set $\{x':x\in I\!B^*\}$ forms an asymptotically shorter prefix code with $\ell(x')=\ell(x)+2\ell(\ell(x))+1$ (see Table 2.2). We pair strings $x$ and $y$ (and $z$) by $\langle x,y\rangle:=x'y$ (and $\langle x,y,z\rangle:=x'y'z$), which are uniquely decodable, since $x'$ and $y'$ are prefix. Since ' serves as a separator, we also write $f(x,y)$ instead of $f(x'y)$ for functions $f$.

We abbreviate $\lim_{n\to\infty}[f(n)-g(n)]=0$ by $f(n)\stackrel{n\to\infty}{\longrightarrow}g(n)$ and say $f$ converges to $g$, without implying that $\lim_{n\to\infty}g(n)$ itself exists. We write

**Table 2.2 ((Prefix) coding of natural numbers and strings)** Bijection between natural numbers $I\!N$ and strings $I\!B^*$. Further, the length $\ell(x)$, and first and second-order prefix coding $\bar{x}:=1^{\ell(x)}0x$ and $x' := \overline{\ell(x)}x$. For illustrational purpose we separated the first part $\overline{\ell(x)}$ from the second part $x$ by a small space. $x'$ is longer or equal than $\bar{x}$ only for $x<15$, but shorter for all $x>30$.

| $x\in I\!N_0$ | 0 | 1 | 2 | 3 | 4 | 5 | 6 | 7 | ... |
|---|---|---|---|---|---|---|---|---|---|
| $x\in I\!B^*$ | $\epsilon$ | 0 | 1 | 00 | 01 | 10 | 11 | 000 | ... |
| $\ell(x)$ | 0 | 1 | 1 | 2 | 2 | 2 | 2 | 3 | ... |
| $\bar{x}$ | 0 | 100 | 101 | 11000 | 11001 | 11010 | 11011 | 1110000 | ... |
| $x'$ | 0 | 100 0 | 100 1 | 101 00 | 101 01 | 101 10 | 101 11 | 11000 000 | ... |

$f(n)\sim g(n)$ and say that $f$ is asymptotically proportional to $g$ if $\exists 0<c<\infty$: $\lim_{n\to\infty}f(n)/g(n)=c$. We write $a\lesssim b$ if $a$ is not much larger than $b$, with precision left unspecified. The big-$O$ notation $f(x)=O(g(x))$ means that there are constants $c$ and $x_0>0$ such that $|f(x)|\leq c|g(x)|\,\forall x>x_0$. The small-$o$ notation $f(x)=o(g(x))$ abbreviates $\lim_{x\to\infty}f(x)/g(x)=0$. We write $f(x)\stackrel{\times}{\leq}g(x)$ for $f(x)=O(g(x))$ and $f(x)\stackrel{+}{\leq}g(x)$ for $f(x)\leq g(x)+O(1)$. Corresponding equalities can be defined similarly. They hold if the corresponding inequalities hold in both directions.

### 2.2.2 Turing Machines

A Turing machine can be considered as an idealized form of a computer. It consists of tapes (memory), read/write heads, a table of rules (program), and an internal state (instruction pointer). A formal definition can be found in any textbook on computability theory, e.g. [HMU01]. The set of partial recursive functions coincides with the set of functions computable with a Turing machine. We say that a set of objects $S=\{o_1,o_2,o_3,...\}$ can be (effectively) enumerated if there is a Turing machine mapping $i$ to $\langle o_i\rangle$, where $\langle\rangle$ is some default coding of the elements in $S$.

The importance of partial recursive functions and Turing machines stems from the following theses:

**Thesis 2.3 (Turing)** Everything that can be reasonably said to be computable by a human using a fixed procedure can also be computed by a Turing machine.

**Thesis 2.4 (Church)** The class of algorithmically computable numerical functions (in the intuitive sense) coincides with the class of partial recursive functions.

We need to supplement Turing's and Church's theses in the following way:

**Assumption 2.5 (Short compiler)** Given two *natural* Turing-equivalent formal systems $F1$ and $F2$, then there always exists a single *short* program on $F2$ that is capable of interpreting all $F1$ programs.

This means that the difference of the size of the shortest $F1$ description and the shortest $F2$ description (of something) is not only bounded by a universal constant, but that this constant is also *reasonably small* for *natural* formal systems. It is easy to formally convert the interpreter into a compiler by attaching the interpreter to the program to be interpreted and by "selling" the result as a compiled version.

This extends Church's and Turing's theses in two respects. First, it says that the equivalence is effective, i.e. that there exists *one* program (interpreter/compiler) that effectively converts $F1$ programs to $F2$ programs. Church's & Turing's theses only state that the classes of computable functions coincide, leaving open the possibility that there is no effective way of transformation. Second, and more important, the extended thesis states that the compiler is short if both formal systems are natural.

The above theses cannot be proven true or false, since *human*, *reasonable*, *intuitive*, and *natural* have not been defined rigorously. One may *define intuitively computable* as Turing computable and a *natural Turing-equivalent system* as one which has a small (say $< 10^5$ bits) interpreter/compiler on a once and for all agreed-upon fixed reference universal Turing machine. The theses would then be that these definitions are reasonable.

For technical reasons we need the following variants of a Turing machine.

**Definition 2.6 (Prefix/Monotone Turing machine)** A prefix/monotone Turing machine is defined as a Turing machine with one unidirectional input tape, one unidirectional output tape, and some bidirectional work tapes. Input tapes are read only, output tapes are write only, unidirectional tapes are those where the head can only move from left to right. All tapes are binary (no blank symbol), work tapes initially filled with zeros.
**Prefix TM.** We say $T$ halts on input $p$ with output $x$, and write $T(p)=x$ if $p$ is to the left of the input head and $x$ is to the left of the output head after $T$ halts. The set of $p$ on which $T$ halts forms a prefix code. We call such codes $p$ *self-delimiting* programs.
**Monotone TM.** We say $T$ outputs/computes a string starting with $x$ (or a sequence $\omega$) on input $p$, and write $T(p)=x*$ (or $T(p)=\omega$) if $p$ is to the left of the input head when the last bit of $x$ is output ($T$ reads all of $p$ but no more). $T$ may continue operation and need not halt. For given $x$, the set of such $p$ forms a prefix code. We call such codes $p$ *minimal* programs.

The table of rules of a Turing machine $T$ can be encoded in a canonical way as a binary string, which we denote by $\langle T\rangle$. Hence, the set of Turing machines $\{T_1,T_2,...\}$ can be effectively enumerated. There are so-called universal

Turing machines that can "simulate" all other Turing machines. We define a particular one below, which also allows for side information $y$.

**Theorem 2.7 (Universal prefix/monotone Turing machine)**
There exists a universal prefix/monotone Turing machine $U$ which simulates prefix/monotone Turing machine $T_i$ with input $y'q$ if fed with input $y'i'q$, i.e.

$$U(y'i'q) = T_i(y'q)\,\forall i, q.$$

We call this particular $U$ the *reference* universal Turing machine. Note that for $p$ not of the form $y'i'q$, $U(p)$ does not halt. In case of no side information $y=\epsilon$, we suppress in the following the initial $y'=\epsilon'=0$ in the codes. We also drop the adjunct 'prefix/monotone' if clear from the context and identify objects with their coding $\langle\rangle$, i.e. we omit the $\langle\rangle$. The price we have to pay for the existence of a universal Turing machine is the undecidability of the halting problem [Tur36]: There is no TM $T$ with $\forall i,p$ $[T(i'p)=1 \Leftrightarrow T_i(p)$ does not halt$]$. Assume such a TM exists, then $R(i):=T(i'i)$ is computable, hence $\exists j: T_j \equiv R$, hence $R(j)=T(j'j)=1 \Leftrightarrow T_j(j)=R(j)$ does not halt, which is a contradiction.

### 2.2.3 Kolmogorov Complexity

In order to exploit Occam's razor beyond intuition we need to formalize the concept of simplicity and/or complexity. We first discuss the case of zero background knowledge $y=\epsilon$. Intuitively, a string is simple if it can be described in a few words, like "the string of one million ones", and is complex if there is no such short description, like for a random string whose shortest description is specifying it bit by bit. We are only interested in descriptions or *codes* that are effective and hence restrict the decoders to Turing machines. We say that (program) $p$ is a description of string $x$ relative to the prefix Turing machine $T$ if $T(p)=x$. The length of the shortest description is denoted by $K_T(x):=\min_p\{\ell(p):T(p)=x\}$. This complexity measure depends on $T$, and one may ask whether there exists a Turing machine which leads to shortest codes among *all* Turing machines for *all* $x$. Remarkably, there exists a Turing machine (the universal one) which "nearly" has this property. If $p$ is the shortest description of $x$ under $T=T_i$, then $i'p$ is a description of $x$ under $U$, hence

$$K_U(x) \leq K_T(x) + c_{TU} \tag{2.8}$$

with $c_{TU}=\ell(i')$, and similarly for other choices of universal Turing machines. The length of the shortest description of $x$ under $U$ is at most a constant number of bits longer than the shortest description under $T$. The statement and proof of this invariance theorem in [Sol64, Kol65, Cha69] is often regarded as the birth of algorithmic information theory. Furthermore, for each pair of universal Turing machines $U'$ and $U''$ satisfying the invariance theorem the complexities coincide up to an additive constant ($|K_{U'}(x)-K_{U''}(x)|\leq c_{U'U''}$).

Since $c_{U'U''}$ is essentially a compiler/interpreter constant, we recall Assumption 2.5 and interpret the assumption as $c_{U'U''}$ being small for natural universal Turing machines $U'$ and $U''$. Henceforth we write $O(1)$ for terms like $c_{U'U''}$ that only depend on the choice of universal Turing machines, but which are independent of the strings under consideration. We extend the definition of complexity to allow for side information $y$.

**Definition 2.9 (Kolmogorov complexity)** Let $U$ be the reference universal prefix Turing machine $U$ of Theorem 2.7. The (conditional) prefix Kolmogorov complexity is defined as the shortest program $p$, for which $U$ outputs $x$ (given $y$):

$$K(x) := \min_p\{\ell(p) : U(p) = x\}, \quad K(x|y) := \min_p\{\ell(p) : U(y`p) = x\}$$

For general (non-string) objects (like computable functions) one can specify some default coding and define $K(\text{object}):=K(\langle\text{object}\rangle)$, especially for numbers and pairs, e.g. we abbreviate $K(x,y) := K(\langle x,y\rangle) = K(x`y)$. The most important information-theoretic properties of $K$ are listed below.

**Theorem 2.10 (Properties of Kolmogorov complexity)**

(*i*) $K(x) \overset{+}{\leq} \ell(x)+2\log_2\ell(x), \quad K(n) \overset{+}{\leq} \log_2 n+2\log_2\log n$

(*ii*) $\sum_x 2^{-K(x)} \leq 1, \quad K(x) \geq \ell(x)$ for 'most' $x$, $\quad K(n)\to\infty$ for $n\to\infty$

(*iii*) $K(x|y) \overset{+}{\leq} K(x) \overset{+}{\leq} K(x,y)$

(*iv*) $K(xy) \overset{+}{\leq} K(x,y) \overset{+}{\leq} K(x)+K(y|x) \overset{+}{\leq} K(x)+K(y)$

(*v*) $K(x|y,K(y))+K(y) \overset{+}{=} K(x,y) \overset{+}{=} K(y,x) \overset{+}{=} K(y|x,K(x))+K(x)$

(*vi*) $K(f(x)) \overset{+}{\leq} K(x)+K(f)$ for recursive $f:\mathbb{B}^*\to\mathbb{B}^*$

(*vii*) $K(x) \overset{+}{\leq} -\log_2 P(x)+K(P)$ if $P:\mathbb{B}^*\to[0,1]$ is enum. and $\sum_x P(x)\leq 1$

All (in)equalities remain valid if $K$ is (further) conditioned under some $z$, i.e. $K(...)\rightsquigarrow K(...|z)$ and $K(...|y)\rightsquigarrow K(...|y,z)$. Those stated are all valid within an additive constant of size $O(1)$, but there are others that are only valid to logarithmic accuracy. $K$ has many properties in common with Shannon entropy as it should be, since both measure the information content of a string. Property (*i*) gives an upper bound on $K$, and property (*ii*) is Kraft's inequality which implies a lower bound on $K$ valid for 'most' $n$, where 'most' means that there are only $o(N)$ exceptions for $n\in\{1,...,N\}$ (Figure 2.11). Providing side information $y$ can never increase code length, requiring extra information $y$ can never decrease code length (*iii*). Coding $x$ and $y$ separately never helps (*iv*), and transforming $x$ does not increase its information content (*vi*). Property (*vi*) also shows that if $x$ codes some object $o$, switching from one coding

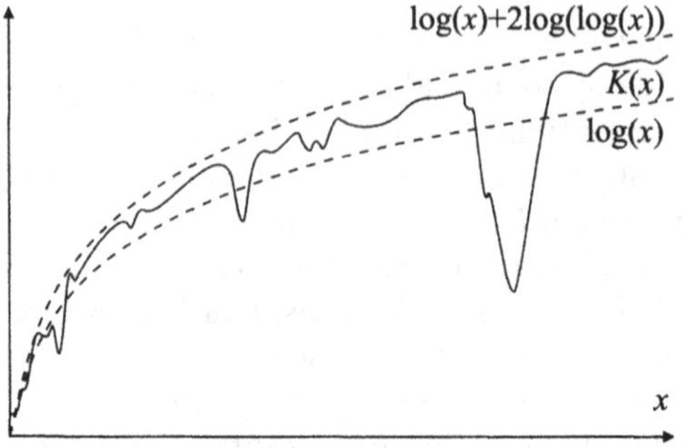


**Figure 2.11 (Kolmogorov Complexity)** Schematic graph of prefix Kolmogorov complexity $K(x)$ with string $x$ interpreted as integer. $K(x) \geq x$ for 'most' $x$ and $K(x) \leq \log_2 x + 2\log_2 \log x + c$ for all $x$ for sufficiently large constant $c$.

scheme to another by means of a recursive bijection $f$ leaves $K$ unchanged within additive $O(1)$ terms. The first nontrivial result is the symmetry of information $(v)$, which is the analogue of the chain rule (see below). Property $(vii)$ is at the heart of the MDL principle [Ris89], which approximates $K(x)$ by $-\log_2 P(x) + K(P)$.

All upper bounds on $K(z)$ are easily proven by devising *some* (effective) code for $z$ of the length of the right-hand side of the inequality and by noting that $K(z)$ is the length of the shortest code among all possible effective codes. For instance, if $T_{i_0}$ with $i_0 = O(1)$ is a Turing machine with $T_{i_0}(\epsilon`x`) = x$, then $U(\epsilon`i_0'x`) = x$; hence $K(x) \leq \ell(\epsilon`i_0'x`) \stackrel{+}{=} \ell(x`) \stackrel{+}{\leq} \ell(x) + 2\log_2 \ell(x)$, which proves $(i)$. In $(vii)$ one uses the Shannon-Fano code based on probability distribution $P$. Lower bounds are usually proven by counting arguments (easy for $(ii)$ by using (2.1) and harder for $(v)$).

### 2.2.4 Computability Concepts

We need several computability concepts weaker than can be captured by halting Turing machines.

**Definition 2.12 (Computable functions)** We consider functions $f:\mathbb{N}\to\mathbb{R}$:

$f$ is *finitely computable* or *recursive* iff there is a Turing machine $T$ with $T(x')=n'd$ and $\frac{n}{d}=f(x)$.

$f$ is *approximable* iff there is a Turing machine finitely computing $\phi(\cdot,\cdot)$ such that $\lim_{t\to\infty}\phi(x,t)=f(x)$.

$f$ is *lower semicomputable* or *enumerable* iff additionally $\phi(x,t)\leq\phi(x,t{+}1)$

$f$ is *upper semicomputable* or *co-enumerable* iff $[-f]$ is lower semicomputable.

$f$ is *semicomputable* iff $f$ is lower *or* upper semicomputable.

$f$ is *estimable* iff $f$ is lower *and* upper semicomputable.

If $f$ is estimable we can finitely compute an $\varepsilon$-approximation of $f$ by upper and lower semicomputing $f$ and terminating when differing by less than $\varepsilon$. This means that there is a Turing machine that, given $x$ and $\varepsilon$, finitely computes $\hat{y}$ such that $|\hat{y}-f(x)|<\varepsilon$. Moreover, it gives an interval estimate $f(x)\in[\hat{y}-\varepsilon,\hat{y}+\varepsilon]$. An estimable integer-valued function is finitely computable (take any $\varepsilon<1$). Note that if $f$ is only approximable or semicomputable we can still come arbitrarily close to $f(x)$, but we cannot devise a terminating algorithm that produces an $\varepsilon$-approximation. In the case of lower/upper semicomputability we can at least finitely compute lower/upper bounds to $f(x)$. In case of approximability, the weakest computability form, even this capability is lost. In analogy to lower/upper semicomputability, one may think of notions like lower/upper estimability, but they are easily shown to coincide with estimability. The following implications are valid:

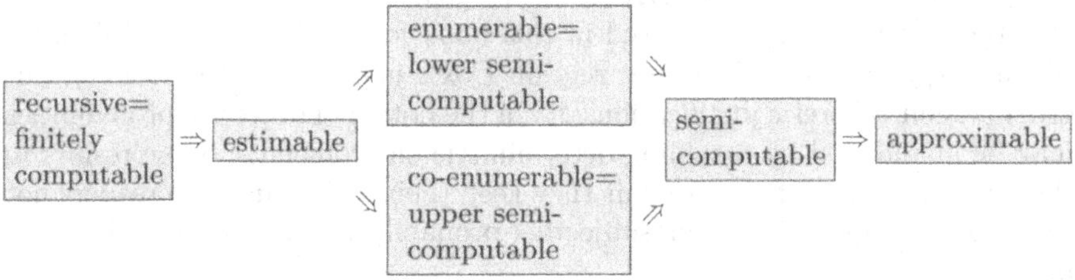


The major algorithmic property of $K$ is:

**Theorem 2.13 ((Non)computability of Kolmogorov complexity)** The Kolmogorov complexity $K:\mathbb{B}^*\to\mathbb{N}$ is co-enumerable, but not finitely computable.

Co-enumerability of $K$ is obvious from the definition of $K$. Non-computability follows from a diagonalization argument: Assume $K$ is computable. Then

$f(m) := \min\{n : K(n) \geq m\}$ exists by Theorem 2.10(*ii*) and is computable, $K(f(m)) \geq m$ by definition of $f$, and $K(f(m)) \stackrel{+}{\leq} K(m)+K(f) \stackrel{+}{\leq} 2\log_2 m$ by Theorem 2.10(*i*,*vi*). Hence, $m \leq \log_2 m + c$ for some $c$, but this is false for sufficiently large $m$.

In the following we use the term computable synonymous to finitely computable, but sometimes also generically for some of the computability forms of Definition 2.12. What we call *estimable* is often just called *computable*, but it makes sense to separate the concepts of finite computability and estimability here, since the former is conceptually easier.

## 2.3 Uncertainty & Probabilities

The aim of probability theory is to describe uncertainty. There are various sources for uncertainty and hence various interpretations of probabilities. There are at least three "schools":

- the *frequentist*: probabilities are relative frequencies. (e.g. the relative frequency of tossing head)
- the *objectivist*: probabilities are real aspects of the world. (e.g. the probability that some atom decays in the next hour)
- the *subjectivist*: probabilities describe an agent's degree of belief in something. (e.g. it is (im)plausible that ETs exist)

The following subsections describe these interpretations and discuss approaches to obtain prior probabilities.

**Remark.** In some communities the domain of applicability and the correct interpretation and form of probability theory is still controversial. For those readers we want to emphasize that probabilities could be completely abandoned from the book without trivializing its goals and results. The terminology of subjective probabilities is used in this book for motivational and illustrational purposes only. We do not rely on Cox's justification (see below), but give decision-theoretic justifications. Even the notion of objective probabilities may be abandoned by assuming deterministic environments. Some results in the book simplify in this case, but they keep their significance. So readers not believing in objective and/or subjective probabilities can still find the book interesting.

### 2.3.1 Frequency Interpretation: Counting

The *frequentist* interprets probabilities as relative frequencies. If in a sequence of $n$ independent identically distributed (i.i.d.) experiments (trials) an event occurs $k(n)$ times, the relative frequency of the event is $k(n)/n$. The limit $\lim_{n\to\infty} k(n)/n$ is *defined* as the probability of the event. This was the earliest mathematical definition of probabilities by Bernoulli, published in 1713

[Ber13]. For instance, the probability of the event *head* in a sequence of repeatedly tossing a fair coin is $\frac{1}{2}$. The frequentist position is the easiest to grasp, but it has several shortcomings:

- The frequentist obtains probabilities from physical processes as described above. To scientifically reason about probabilities one needs a mathematical theory. The problem is how to define random sequences. This is much more intricate than one might think and was only solved in the 1960s by Kolmogorov and Martin-Löf [ML66].
  The naive definition of probability is circular: The probability of an event $E$ is $p := \lim_{n\to\infty} \frac{k_n(E)}{n}$, where $k_n(E)$ is the number of occurrences of event $E$ in the first $n$ i.i.d. trials. The problem is that the limit may be anything or not even exist: e.g. a fair coin can give: head, head, head, head, ... i.e. $p=1$. Of course, this sequence is "unlikely". For a fair coin, $p=\frac{1}{2}$ with "high probability". But to make this statement rigorous we need to formally define what "high probability" means. Here is the circularity!
- Philosophically, and also often in real experiments, it is hard to justify the choice of the so-called reference class. For instance, a doctor wants to determine the chances that a patient has a particular disease by counting the frequency of the disease in "similar" patients. But if the doctor considered everything he knows about the patient (symptoms, weight, age, ancestory, ...) there would be no other comparable patients left.
- The frequency approach is limited to a (sufficiently large) sample of i.i.d. data.

### 2.3.2 Objective Interpretation: Probabilities to Describe Uncertain Events

For the *objectivist* probabilities are real aspects of the world. The outcome of an observation or an experiment is not deterministic, but involves physical random processes. The set $\Omega$ of all possible outcomes is called the *sample space*. It is said that an event $E \subset \Omega$ occurred if the outcome is in $E$. In the case of i.i.d. experiments the probabilities assigned to events should be interpretable as limiting frequencies, but the application is not limited to this case. The Kolmogorov axioms formalize the properties that probabilities should have.

**Axioms 2.14 (Kolmogorov's axioms of probability theory)** Let $\Omega$ be the sample space. Events are subsets of $\Omega$.

- If $A$ and $B$ are events, then the intersection $A \cap B$, the union $A \cup B$, and the difference $A \backslash B$ are also events.
- The sample space $\Omega$ and the empty set $\{\}$ are events.
- There is a function $p$ that assigns nonnegative reals, called probabilities, to each event.
- $p(\Omega)=1$, $p(\{\})=0$.
- $p(A \cup B)=p(A)+p(B)-p(A \cap B)$.
- For a decreasing sequence $A_1 \supset A_2 \supset A_3 ...$ of events with $\bigcap_n A_n = \{\}$, we have $\lim_{n\to\infty} p(A_n)=0$.

The function $p$ is called a *probability mass function*, or probability measure, or, more loosely, *probability distribution*. Conditional probabilities are defined in the following way:

**Definition 2.15 (Conditional probability)** If $A$ and $B$ are events with $p(A)>0$, then the probability that event $B$ will occur under the condition that event $A$ has occured is defined as

$$p(B|A) := \frac{p(A \cap B)}{p(A)}$$

It is easy to see that $p(\cdot|A)$ (as a function of the first argument) is also a probability measure, if $p(\cdot)$ satisfies the Kolmogorov axioms. One can "verify the correctness" of the Kolmogorov axioms and the definition of conditional probabilities in the case where probabilities are identified with limiting frequencies. But the idea is to take the axioms as a starting point to avoid the frequentist's problems. The relation $p(A \cap B)=p(B|A) \cdot p(A)$ is called the multiplication rule (of conditional probabilities), which is a special case of the chain rule.

**Theorem 2.16 (Bayes' rule 1)** If $A$ and $B$ are events with $p(A)>0$ and $p(B)>0$, then

$$p(B|A) = \frac{p(A|B)p(B)}{p(A)}$$

Bayes' theorem is easily proven by applying Definition 2.15 twice.

### 2.3.3 Subjective Interpretation: Probabilities to Describe Degrees of Belief

The *subjectivist* uses probabilities to characterize an agent's degree of belief in something, rather than to characterize physical random processes. This is the most relevant interpretation of probabilities in AI. We define the plausibility of an event as the degree of belief in the event, or the subjective probability of the event. The problem with the subjective view is that it is much more arguable how to define plausibilities, as compared to objective probabilities. The objectivist can motivate Kolmogorov's axioms by a frequency analysis, but there is no frequency interpretation for plausibilities. If an agent believes in extraterrestrians and assigns a plausibility of 0.9 to their existence, it does not make much sense to interpret this as "in 90 out of 100 parallel universes there are extraterrestrians" or "90 out of 100 similar agents believe in extraterrestrians". This problem has led to many different systems dealing with uncertain reasoning (see reference Section 2.5). They all have their own problems. The most consistent and successful system is, again, based on Kolmogorov's axioms, although not all would agree with this statement. It is surprising that plausibilities follow the same rules as limiting frequencies. It is possible to derive Kolmogorov's axioms from a few plausible qualitative rules they should follow. It is natural to assume that plausibilities can be represented by real numbers, that the rules qualitatively correspond to common sense, and that the rules are mathematically consistent. Cox [Cox46] starts with the following (natural) assumptions on beliefs:

**Axioms 2.17 (Cox's axioms for beliefs)**

- The degree of belief in event $B$ (plausibility of event $B$), given that event $A$ occurred, can be characterized by a real-valued function $\text{Bel}(B|A)$.
- $\text{Bel}(\Omega\setminus B|A)$ is a twice differentiable function of $\text{Bel}(B|A)$ for $A\neq\{\}$.
- $\text{Bel}(B\cap C|A)$ is a twice continuously differentiable function of $\text{Bel}(C|B\cap A)$ and $\text{Bel}(B|A)$ for $B\cap A\neq\{\}$.

Cox [Cox46] shows that every function $\text{Bel}(\cdot|\cdot)$ satisfying these axioms is isomorphic to a (conditional) probability function. One can motivate the functional relationship in Cox's axioms by analyzing all other possibilities and showing that they violate common sense [Tri69]. The somewhat strong differentiability assumptions can be weakened to more natural continuity and monotonicity assumptions [Aćz66]. Only recently, a loophole in Cox's and other's derivations has been exhibited [Par95]. Several fixes have been suggested by making additional assumptions. Most of them require the range of Bel, and hence the set of events, to be rich enough. We paraphrase these as "additional denseness conditions."

**Theorem 2.18 (Cox's theorem)** Under Axioms 2.17 and some additional denseness conditions, Bel$(\cdot|A)$ is isomorphic to a probability function in the sense that there is a continuous one-to-one function $g:\mathbb{R}\to[0,1]$ such that $p:=g\circ\text{Bel}$ satisfies Kolmogorov's Axioms 2.14 and is consistent with Definition 2.15.

Cox's result has attracted a great deal of interest, particularly in the maximum entropy and AI community. The qualitative motivation of Cox's axioms and the derivation of Cox's theorem from them is the major theoretical justification that subjective 'degrees of belief' should satisfy the same Kolmogorov axioms as limiting frequencies. Other approaches to beliefs are missing this strong theoretical foundation and consistency.

Exploiting that subjective probabilities follow the same rules as objective probabilities, we can present Bayes' rule in a particular useful form: how to update beliefs under new observations. See Problem 2.10 for a typical application.

**Theorem 2.19 (Bayes' rule 2)** Let $D$ be some possible data (i.e. $D$ is an event with $p(D)>0$) and $\{H_i\}_{i\in I}$ be a countable complete class of mutually exclusive hypotheses (i.e. $H_i$ are events with $H_i\cap H_j=\{\}\ \forall i\neq j$ and $\bigcup_{i\in I}H_i=\Omega$).
Given: $p(H_i)$ = a priori plausibility of hypotheses $H_i$ (subj. prob.)
Given: $p(D|H_i)$ = likelihood of data $D$ under hypothesis $H_i$ (obj. prob.)
Goal: $p(H_i|D)$ = a posteriori plausibility of hypothesis $H_i$ (subj. prob.)

$$\text{Solution:}\quad p(H_i|D)=\frac{p(D|H_i)p(H_i)}{\sum_{i\in I}p(D|H_i)p(H_i)}$$

**Proof.** The proof is based on all and only Axioms 2.14. $p(A\cup B)=p(A)+p(B)$ if $A\cap B=\{\}$, since $p(\{\})=0$. For finite $I$, by induction, this implies $\sum_{i\in I}p(H_i)=p(\bigcup_i H_i)=p(\Omega)=1$. For countably infinite $I=\{1,2,3,...\}$ with $S_n:=\bigcup_{i=n}^\infty H_i$ we have:

$$\sum_{i=1}^{n-1}p(H_i)+p(S_n)\;=\;p\Bigl(\bigcup_{i=1}^{n-1}H_i\cup S_n\Bigr)=p(\Omega)=1. \tag{2.20}$$

Exploiting $S_1\supset S_2\supset S_3...$, for any $\omega\in\Omega$ we have: $\exists n:\omega\in H_n \Rightarrow \omega\notin H_i\forall i>n$ $\Rightarrow\omega\notin S_i\forall i>n \Rightarrow\omega\notin\bigcap_n S_n \Rightarrow \bigcap_n S_n=\{\}$ since $\omega$ was arbitrary $\Rightarrow$ taking $n\to\infty$ in (2.20) shows $\sum_{i=1}^\infty p(H_i)=1$. Since conditional probabilities also satisfy Axioms 2.14, we also have $\sum_{i\in I}p(H_i|D)=1$ (for finite as well as infinite $I$). By Definition 2.15 of conditional probability we have

$$p(H_i|D)p(D)=p(H_i\cap D)=p(D|H_i)p(H_i).$$

Summing over all hypotheses $H_i$ gives

$$\sum_{i\in I} p(D|H_i)p(H_i) = \sum_{i\in I} p(H_i|D)\cdot p(D) = 1\cdot p(D)$$

$$\Rightarrow \quad p(H_i|D) = \frac{p(D|H_i)p(H_i)}{p(D)} = \frac{p(D|H_i)p(H_i)}{\sum_{i\in I} p(D|H_i)p(H_i)} \qquad \square$$

### 2.3.4 Determining Priors

The Kolmogorov Axioms 2.14 of probability allow us to relate probabilities and plausibilities of different events, but they do not uniquely fix a numerical value for each event, except for the sure event $\Omega$ and the empty event $\{\}$. We need new principles for determining values for at least some basis events from which others can then be computed by these Axioms. There seem to be only three general principles:

- the principle of indifference — the symmetry principle,
- the maximum entropy principle,
- Occam's razor — the simplicity principle.

Whereas the first two principles are based on the foundations of statistical physics, we will see that only Occam's razor (keep only the simplest consistent hypothesis) in combination with Epicurus' principle of multiple explanations (keep all consistent hypotheses) is general enough to assign prior probabilities in *every* situation, especially in the case of induction and other domains typical for AI. The idea is to assign high (subjective) probability to simple events, and low probability to complex events: Simple events (strings) are more plausible a priori than complex ones. This gives (approximate) justice to both Occam's razor and Epicurus' principle. In the following we also refer to this general idea as Occam's razor. Using $K$ for measuring simplicity/complexity leads to Solomonoff's universal prior $M$. In the next section we pursue this approach.

## 2.4 Algorithmic Probability & Universal Induction

In addition to the notation introduced in Section 2.2.1, we denote binary strings of length $n$ by $x = x_1 x_2 ... x_n$ with $x_t \in I\!B$, and further abbreviate $x_{1:n} := x_1 x_2 ... x_{n-1} x_n$ and $x_{<n} := x_1 ... x_{n-1}$.

### 2.4.1 The Universal Prior $M$

The prefix Kolmogorov complexity $K(x)$ has been defined as the shortest program $p$ for which the universal prefix Turing machine $U$ outputs string $x$, and similarly $K(x|y)$ in case of side information $y$ (Definition 2.9). Solomonoff [Sol64, Sol78] defined a closely related quantity, the universal prior $M(x)$.

The universal prior is defined as the probability that the output of a universal monotone Turing machine starts with $x$ when provided with fair coin flips on the input tape. Formally, $M$ can be defined as

$$M(x) \;:=\; \sum_{p\,:\,U(p)=x*} 2^{-\ell(p)}, \tag{2.21}$$

where the sum is over minimal programs $p$ for which $U$ outputs a string starting with $x$ (see Definition 2.6). Since the shortest programs $p$ dominate the sum, $M(x)$ is roughly $2^{-K(x)}$ $(M(x)=2^{-K(x)+O(K(\ell(x)))})$.

Before we can discuss the stochastic properties of $M$ we need the concept of (semi)measures for strings.

**Definition 2.22 ((Semi)measures)** $\mu(x)$ denotes the probability that a binary sequence starts with string $x$. We call $\mu\geq 0$ a semimeasure if $\mu(\epsilon)\leq 1$ and $\mu(x)\geq\mu(x0)+\mu(x1)$, and a probability measure if equalities hold.

The reason for calling $\mu$ with the above property a probability measure is that it satisfies Kolmogorov's Axioms 2.14 of probability in the following sense: The sample space is $\mathbb{B}^\infty$ with elements $\omega=\omega_1\omega_2\omega_3...\in\mathbb{B}^\infty$ being infinite binary sequences. The set of events (the $\sigma$-algebra) is defined as the set generated from the cylinder sets $\Gamma_{x_{1:n}}:=\{\omega:\omega_{1:n}=x_{1:n}\}$ by countable union and complement. A probability measure $\mu$ is uniquely defined by giving its values $\mu(\Gamma_{x_{1:n}})$ on the cylinder sets, which we abbreviate by $\mu(x_{1:n})$. We will also call $\mu$ a measure, or even more loosely a probability distribution.

The reason for extending the definition to semimeasures is that $M$ itself is unfortunately *not* a probability measure. We have $M(x0)+M(x1)<M(x)$ because there are programs $p$ that output $x$, followed neither by 0 nor 1. They just stop after printing $x$, or continue forever without any further output. In Problem 2.7 the defect is quantified. Since $M(\epsilon)=1$, $M$ is at least a semimeasure. We can now state the fundamental property of $M$.

**Theorem 2.23 (Universality of $M$)** The universal prior $M(x):=\sum_{p\,:\,U(p)=x*}2^{-\ell(p)}$ is an enumerable semimeasure which multiplicatively dominates all enumerable semimeasures in the sense that $M(x)\stackrel{\times}{\geq}2^{-K(\rho)}\cdot\rho(x)$ if $\rho$ is an enumerable semimeasure. $M$ is enumerable, but not estimable or finitely computable.

The Kolmogorov complexity of a function like $\rho$ is defined as the length of the shortest self-delimiting code of a Turing machine computing this function in the sense of Definition 2.12. Up to a multiplicative constant, $M$ assigns higher probability to all $x$ than any other computable probability distribution.

It is possible to normalize $M$ to a true probability measure $M_{norm}$ (2.30) [Sol78] with dominance still being true, but at the expense of giving up enumerability ($M_{norm}$ is still approximable). We will see that $M$ is more convenient when studying algorithmic questions, but a true probability measure like $M_{norm}$ is more convenient when studying stochastic questions.

### 2.4.2 Universal Sequence Prediction

In which sense does $M$ incorporate Occam's razor and Epicurus' principle of multiple explanations? From $M(x)\approx 2^{-K(x)}$ we see that $M$ assigns high probability to simple strings (Occam). More useful is to think of $x$ as being the observed history. We see from Definition (2.21) that every program $p$ consistent with history $x$ is allowed to contribute to $M$ (Epicurus). On the other hand, shorter programs give significantly larger contribution (Occam). How does all this affect prediction? If $M(x)$ correctly describes our (subjective) prior belief in $x$, then $M(y|x):=M(xy)/M(x)$ must be our posterior belief in $y$. From the symmetry of algorithmic information $K(x,y)\stackrel{+}{=}K(y|x,K(x))+K(x)$ (Theorem 2.10(v)), and assuming $K(x,y)\approx K(xy)$, and approximating $K(y|x,K(x))\approx K(y|x)$, $M(x)\approx 2^{-K(x)}$, and $M(xy)\approx 2^{-K(xy)}$ we get $M(y|x)\approx 2^{-K(y|x)}$. This tells us that $M$ predicts $y$ with high probability iff $y$ has an easy explanation, given $x$ (Occam & Epicurus).

The above qualitative discussion should not create the impression that $M(x)$ and $2^{-K(x)}$ always lead to predictors of comparable quality. Indeed, in the online/incremental setting studied in this book, $K(y)=O(1)$ invalidates the consideration above. The validity of Theorem 2.25 below, for instance, depends on $M$ being a semimeasure and the chain rule being exactly true, neither of them is satisfied by $2^{-K(x)}$ (see Problem 2.8).

(Binary) sequence prediction algorithms try to predict the continuation $x_n\in\mathbb{B}$ of a given sequence $x_1...x_{n-1}$. We derive the following bound:

$$\sum_{t=1}^{\infty}(1-M(x_t|x_{<t}))^2 \leq -\tfrac{1}{2}\sum_{t=1}^{\infty}\ln M(x_t|x_{<t}) = -\tfrac{1}{2}\ln M(x_{1:\infty}) \leq \tfrac{1}{2}\ln 2\cdot Km(x_{1:\infty}) \tag{2.24}$$

where the monotone complexity $Km(x_{1:\infty})$ is defined as the length of the shortest (nonhalting) program computing $x_{1:\infty}$ [ZL70]. In the first inequality we used $(1-a)^2\leq -\frac{1}{2}\ln a$ for $0\leq a\leq 1$. In the equality we exchanged the sum with the logarithm and eliminated the resulting product by the chain rule. In the last inequality we used $M(x)\geq 2^{-Km(x)}$, which follows from Definition (2.21) by dropping all terms in $\sum_p$ except for the shortest $p$ computing $x$. If $x_{1:\infty}$ is a computable sequence, then $Km(x_{1:\infty})$ is finite, which implies $M(x_t|x_{<t})\to 1$ ($\sum_{t=1}^{\infty}(1-a_t)^2<\infty\Rightarrow a_t\to 1$). This means that if the environment is a computable sequence (whichsoever, e.g. the digits of $\pi$ or $e$ in binary representation), after having seen the first few digits, $M$ correctly predicts the next digit with high probability, i.e. it recognizes the structure of the sequence.

Assume now that the true sequence is drawn from a computable probability distribution $\mu$, i.e. the true (objective) probability of $x_{1:n}$ is $\mu(x_{1:n})$.

The probability of $x_n$ given $x_{<n}$, hence, is $\mu(x_n|x_{<n}) = \mu(x_{1:n})/\mu(x_{<n})$. Solomonoff's [Sol78] central result is that $M$ converges to $\mu$:

**Theorem 2.25 (Posterior convergence of $M$ to $\mu$)**

$$\sum_{t=1}^{\infty}\sum_{x_{<t}\in\mathbb{B}^{t-1}} \mu(x_{<t})\Big(M(0|x_{<t})-\mu(0|x_{<t})\Big)^2 \stackrel{+}{\leq} \tfrac{1}{2}\ln 2\cdot K(\mu) < \infty$$

The infinite sum can only be finite if the difference $M(0|x_{<t})-\mu(0|x_{<t})$ tends to zero for $t\to\infty$ with $\mu$-probability 1 (see Definition 3.8($i$) and Problem 2.7). This holds for *any* computable probability distribution $\mu$. The reason for the astonishing property of a single (universal) function to converge to *any* computable probability distribution lies in the fact that the set of $\mu$-random sequences differ for different $\mu$. Past data $x_{<t}$ are exploited to get a (with $t\to\infty$) improving estimate $M(x_t|x_{<t})$ of $\mu(x_t|x_{<t})$.

The universality property (Theorem 2.23) is the central ingredient in the proof of Theorem 2.25. The proof of Theorem 2.23 involves the construction of a semimeasure $\xi$ whose dominance is obvious. The hard part is to show its enumerability and equivalence to $M$. Let $\mathcal{M}$ be the (countable) set of all enumerable semimeasures and define

$$\xi(x) := \sum_{\nu\in\mathcal{M}} 2^{-K(\nu)}\nu(x). \tag{2.26}$$

Then dominance

$$\xi(x) \geq 2^{-K(\nu)}\nu(x) \quad \forall\nu\in\mathcal{M} \tag{2.27}$$

is obvious (without $O(1)$ fudge). Is $\xi$ lower semicomputable? To answer this question we have to be more precise. Levin [ZL70] showed that there is a Turing machine $T$ such that for every lower semicomputable semimeasure $\nu$ there is an $i$ such that $T(i'x')$ lower semicomputes $\nu_i\equiv\nu$, i.e. $T$ enumerates *all* lower semicomputable semimeasures, possibly with repetition. For the (ordered multi) set $\mathcal{M}=\mathcal{M}_U:=\{\nu_1,\nu_2,\nu_3,...\}$ and $K(\nu_i):=K(i)$, one can easily see that $\xi$ is lower semicomputable. Finally, proving $M(x)\stackrel{\times}{=}\xi(x)$ also establishes universality of $M$.

The advantage of $\xi$ over $M$ is that it immediately generalizes to arbitrary weighted sums of (semi)measures in $\mathcal{M}$ for arbitrary countable $\mathcal{M}$. Most proofs in this book go through for generic $\mathcal{M}$ and weights. We will prove (this generalization of) Theorem 2.25 in Section 3.2.6.

### 2.4.3 Universal (Semi)Measures

What is so special about the set of all enumerable semimeasures $\mathcal{M}_U$? The larger we choose $\mathcal{M}$, the less restrictive is the assumption that $\mathcal{M}$ should contain the true distribution $\mu$, which will be essential throughout the book. Why do not restrict to the still rather general class of estimable or finitely computable (semi)measures? It is clear that for every countable set $\mathcal{M}$,

$\xi(x) := \xi_{\mathcal{M}}(x) := \sum_{\nu\in\mathcal{M}} w_\nu \nu(x)$ with $\sum_\nu w_\nu \leq 1$ and $w_\nu > 0$ dominates all $\nu \in \mathcal{M}$. This dominance is necessary for the desired convergence $\xi \to \mu$, similarly to Theorem 2.25. The question is what properties does $\xi$ possess. The distinguishing property of $\mathcal{M}_U$ is that $\xi = \xi_U \equiv \xi_{\mathcal{M}_U} \stackrel{\times}{=} M$ is itself an element of $\mathcal{M}_U$. In this book, $\xi_{\mathcal{M}} \in \mathcal{M}$ is not by itself an important property, but whether $\xi$ is computable in one of the senses of Definition 2.12. We define

$$\mathcal{M}_1 \stackrel{\times}{\geq} \mathcal{M}_2 \quad :\Leftrightarrow \text{ there is an element of } \mathcal{M}_1 \text{ that dominates all elements of } \mathcal{M}_2$$
$$:\Leftrightarrow \; \exists \rho \in \mathcal{M}_1 \; \forall \nu \in \mathcal{M}_2 \; \exists w_\nu > 0 \; \forall x : \rho(x) \geq w_\nu \nu(x).$$

Relation $\stackrel{\times}{\geq}$ is transitive (but not necessarily reflexive) in the sense that $\mathcal{M}_1 \stackrel{\times}{\geq} \mathcal{M}_2 \stackrel{\times}{\geq} \mathcal{M}_3$ implies $\mathcal{M}_1 \stackrel{\times}{\geq} \mathcal{M}_3$ and $\mathcal{M}_0 \supseteq \mathcal{M}_1 \stackrel{\times}{\geq} \mathcal{M}_2 \supseteq \mathcal{M}_3$ implies $\mathcal{M}_0 \stackrel{\times}{\geq} \mathcal{M}_3$. For the computability concepts introduced in Section 2.2.4, we have the following proper set inclusions

$$\begin{array}{ccccccc}
\mathcal{M}_{comp}^{msr} & \subset & \mathcal{M}_{est}^{msr} & \equiv & \mathcal{M}_{enum}^{msr} & \subset & \mathcal{M}_{appr}^{msr} \\
\cap & & \cap & & \cap & & \cap \\
\mathcal{M}_{comp}^{semi} & \subset & \mathcal{M}_{est}^{semi} & \subset & \mathcal{M}_{enum}^{semi} & \subset & \mathcal{M}_{appr}^{semi}
\end{array}$$

where $\mathcal{M}_c^{msr}$ stands for the set of all probability measures of appropriate computability type $c \in$ {comp=finitely computable, est=estimable, enum=enumerable, appr=approximable}, and similarly for semimeasures $\mathcal{M}_c^{semi}$. Other classes are briefly discussed in Section 3.2.9. From an enumeration of a measure $\rho$ one can construct a co-enumeration by exploiting $\rho(x_{1:n}) = 1 - \sum_{y_{1:n}\neq x_{1:n}} \rho(y_{1:n})$. This shows that every enumerable measure is also co-enumerable, and hence estimable, which proves the identity $\equiv$ above.

With this notation, [ZL70, Thm.3.3] reads $\mathcal{M}_{enum}^{semi} \stackrel{\times}{\geq} \mathcal{M}_{enum}^{semi}$. Transitivity allows to conclude, for instance, that $\mathcal{M}_{appr}^{semi} \stackrel{\times}{\geq} \mathcal{M}_{comp}^{msr}$, i.e. that there is an approximable semimeasure that dominates all computable measures.

The standard "diagonalization" way of proving $\mathcal{M}_1 \stackrel{\times}{\ngeq} \mathcal{M}_2$ is to take an arbitrary $\mu \in \mathcal{M}_1$ and "increase" it to $\rho$ such that $\mu \stackrel{\times}{\ngeq} \rho$ and show that $\rho \in \mathcal{M}_2$. There are $7\times 7$ combinations of (semi)measures $\mathcal{M}_1$ with $\mathcal{M}_2$ for which $\mathcal{M}_1 \stackrel{\times}{\geq} \mathcal{M}_2$ could be true or false. There are four basic cases, explicated in the following theorem, from which the other 49 combinations displayed in Table 2.29 follow by transitivity.

**Theorem 2.28 (Universal (semi)measures)** A semimeasure $\rho$ is said to be universal for $\mathcal{M}$ if it multiplicatively dominates all elements of $\mathcal{M}$ in the sense $\forall\nu\exists w_\nu>0:\rho(x)\geq w_\nu\nu(x)\forall x$. The following holds true:

(o) $\exists\rho:\{\rho\}\stackrel{\times}{\geq}\mathcal{M}$: For every countable set of (semi)measures $\mathcal{M}$, there is a (semi)measure that dominates all elements of $\mathcal{M}$.

(i) $\mathcal{M}_{enum}^{semi}\stackrel{\times}{\geq}\mathcal{M}_{enum}^{semi}$: The class of enumerable semimeasures *contains* a universal element.

(ii) $\mathcal{M}_{appr}^{msr}\stackrel{\times}{\geq}\mathcal{M}_{enum}^{semi}$: There *is* an approximable measure that dominates all enumerable semimeasures.

(iii) $\mathcal{M}_{est}^{semi}\stackrel{\times}{\not\geq}\mathcal{M}_{comp}^{msr}$: There is *no* estimable semimeasure that dominates all computable measures.

(iv) $\mathcal{M}_{appr}^{semi}\stackrel{\times}{\not\geq}\mathcal{M}_{appr}^{msr}$: There is *no* approximable semimeasure that dominates all approximable measures.

**Table 2.29 (Existence of universal (semi)measures)** The entry in row $r$ and column $c$ indicates whether there is a $r$-able (semi)measure $\rho$ for the set $\mathcal{M}$ that contains all $c$-able (semi)measures, where $r,c\in\{$comput, estimat, enumer, approxim$\}$. Enumerable measures are estimable. This is the reason why the enum. row and column in case of measures are missing. The superscript indicates from which part of Theorem 2.28 the answer follows: for the boldface entries directly, for the others using transitivity of $\stackrel{\times}{\geq}$.

| $\nwarrow$ | $\mathcal{M}$ | semimeasure | | | | measure | | |
|---|---|---|---|---|---|---|---|---|
| $\rho$ | $\searrow$ | comp. | est. | enum. | appr. | comp. | est. | appr. |
| s | comp. | no$^{iii}$ | no$^{iii}$ | no$^{iii}$ | no$^{iv}$ | no$^{iii}$ | no$^{iii}$ | no$^{iv}$ |
| e | est. | no$^{iii}$ | no$^{iii}$ | no$^{iii}$ | no$^{iv}$ | **no$^{\mathbf{iii}}$** | no$^{iii}$ | no$^{iv}$ |
| m | enum. | yes$^{i}$ | yes$^{i}$ | **yes$^{\mathbf{i}}$** | no$^{iv}$ | yes$^{i}$ | yes$^{i}$ | no$^{iv}$ |
| i | appr. | yes$^{i}$ | yes$^{i}$ | yes$^{i}$ | no$^{iv}$ | yes$^{i}$ | yes$^{i}$ | **no$^{\mathbf{iv}}$** |
| m | comp. | no$^{iii}$ | no$^{iii}$ | no$^{iii}$ | no$^{iv}$ | no$^{iii}$ | no$^{iii}$ | no$^{iv}$ |
| s | est. | no$^{iii}$ | no$^{iii}$ | no$^{iii}$ | no$^{iv}$ | no$^{iii}$ | no$^{iii}$ | no$^{iv}$ |
| r | appr. | yes$^{ii}$ | yes$^{ii}$ | **yes$^{\mathbf{ii}}$** | no$^{iv}$ | yes$^{ii}$ | yes$^{ii}$ | no$^{iv}$ |

If we ask for a universal (semi)measure that at least satisfies the weakest form of computability, namely being approximable, we see that the largest dominated set among the seven sets defined above is the set of enumerable semimeasures. This is the reason why $\mathcal{M}_{enum}^{semi}$ plays a special role in this (and other) works. On the other hand, $\mathcal{M}_{enum}^{semi}$ is not the largest set dominated by an approximable semimeasure, and indeed no such largest set exists. One may, hence, ask for "natural" larger sets $\mathcal{M}$. One such set, namely the set of cumulatively enumerable semimeasures $\mathcal{M}_{\text{CEM}}$, was recently discovered by Schmidhuber [Sch00, Sch02a], for which even $\xi_{\text{CEM}}\in\mathcal{M}_{\text{CEM}}$ holds.

Theorem 2.28 also holds for *discrete (semi)measures* $P$ defined as

$$P : \mathbb{N} \to [0,1] \quad \text{with} \quad \sum_{x\in\mathbb{N}} P(x) \overset{(\leq)}{=} 1.$$

We first prove the theorem for this discrete case, since it contains the essential ideas in a cleaner form. We then present the proof for "continuous" (semi)measures $\mu$ (Definition 2.22). The proofs naturally generalize from binary to arbitrary finite alphabet. The value of $x$ that minimizes $f(x)$ is denoted by $\text{argmin}_x f(x)$. Ties are broken in an arbitrary but computable way (e.g. by taking the smallest $x$).

**Proof (discrete case).** $(o)$ $Q(x) := \sum_{P\in\mathcal{M}} w_P P(x)$ with $w_P > 0$ obviously dominates all $P \in \mathcal{M}$ (with constant $w_P$). With $\sum_P w_P = 1$ and all $P$ being discrete (semi)measures also $Q$ is a discrete (semi)measure.

$(i)$ See [LV97, Thm.4.3.1].

$(ii)$ Let $P$ be the universal element in $\mathcal{M}_{enum}^{semi}$ and $\alpha := \sum_x P(x)$. We normalize $P$ by $Q(x) := \frac{1}{\alpha}P(x)$. Since $\alpha \leq 1$ we have $Q(x) \geq P(x)$, hence $Q \geq P \overset{\times}{\geq} \mathcal{M}_{enum}^{semi}$. As a ratio between two enumerable functions, $Q$ is still approximable, hence $\mathcal{M}_{appr}^{msr} \overset{\times}{\geq} \mathcal{M}_{enum}^{semi}$.

$(iii)$ [3]Let $P \in \mathcal{M}_{comp}^{semi}$. We partition $\mathbb{N}$ into chunks $I_n := \{2^{n-1},...,2^n-1\}$ $(n \geq 1)$ of increasing size. With $x_n := \text{argmin}_{x\in I_n} P(x)$ we define $Q(x_n) := \frac{1}{n(n+1)} \forall n$, and $Q(x) := 0$ for all other $x$. Exploiting the fact that a minimum is smaller than an average, we get

$$P(x_n) = \min_{x\in I_n} P(x) \leq \frac{1}{|I_n|}\sum_{x\in I_n} P(x) \leq \frac{1}{|I_n|} = \frac{1}{2^{n-1}} = \frac{n(n+1)}{2^{n-1}} Q(x_n).$$

Since $\frac{n(n+1)}{2^{n-1}} \to 0$ for $n \to \infty$, $P$ cannot dominate $Q$ $(P \overset{\times}{\not\geq} Q)$. With $P$ also $Q$ is computable. Since $P$ was an arbitrary computable semimeasure and $Q$ is a computable measure ($\sum Q(x) = \sum[\frac{1}{n(n+1)}] = \sum[\frac{1}{n} - \frac{1}{n+1}] = 1$) this implies $\mathcal{M}_{comp}^{semi} \overset{\times}{\not\geq} \mathcal{M}_{comp}^{msr}$.

Assume now that there is an estimable semimeasure $S \overset{\times}{\geq} \mathcal{M}_{comp}^{msr}$. We construct a finitely computable semimeasure $P \overset{\times}{\geq} S$ as follows. Choose an initial $\varepsilon > 0$ and finitely compute an $\varepsilon$-approximation $\hat{S}$ of $S(x)$. If $\hat{S} > 2\varepsilon$ define $P(x) := \frac{1}{2}\hat{S}$, else halve $\varepsilon$ and repeat the process. Since $S(x) > 0$ (otherwise it could not dominate, e.g. $T(x) := \frac{1}{x(x+1)} \in \mathcal{M}_{comp}^{msr}$) the loop terminates after finite time. So $P$ is finitely computable. Inserting $\hat{S} = 2P(x)$ and $\varepsilon < \frac{1}{2}\hat{S} = P(x)$ into $|S(x) - \hat{S}| < \varepsilon$, we get $|S(x) - 2P(x)| < P(x)$, which implies $S(x) \geq P(x)$ and $S(x) \leq 3P(x)$. The former implies $\sum_x P(x) \leq \sum_x S(x) \leq 1$, i.e. $P$ is a

[3] The proof of $\mathcal{M}_{comp}^{semi} \overset{\times}{\not\geq} \mathcal{M}_{comp}^{semi}$ in [LV97, p249] contains minor errors and is not extensible to $\mathcal{M}_{est}^{semi} \overset{\times}{\not\geq} \mathcal{M}_{comp}^{msr}$.

semimeasure. The latter implies $P\geq\frac{1}{3}S\stackrel{\times}{\geq}\mathcal{M}_{comp}^{msr}$. Hence $P$ is a computable semimeasure dominating all computable measures, which contradicts what we proved in the first half of $(iii)$. Hence the assumption on $S$ was wrong, which establishes $\mathcal{M}_{est}^{semi}\stackrel{\times}{\not\geq}\mathcal{M}_{comp}^{msr}$.

$(iv)$ Assume $P\in\mathcal{M}_{appr}^{semi}\stackrel{\times}{\geq}\mathcal{M}_{appr}^{msr}$. We construct an approximable measure $Q$ that is not dominated by $P$, thus contradicting the assumption. Let $P_1,P_2,...$ be a sequence of recursive functions converging to $P$. We construct $x_1,x_2,...$ such that $\forall c>0\ \exists n\in I\!N : P(x_n)\not\geq c\cdot Q(x_n)$. For this we recursively define sequences $x_n^1,x_n^2,...$ converging to $x_n$, and from them $Q_1,Q_2,...$ converging to $Q$. Let $I_n:=\{2^{n-1},...,2^n-1\}$ and $x_n^1=2^{n-1}\forall n$. If $P_t(x_n^{t-1})>n^{-3}$ then $x_n^t:=\text{argmin}_{x\in I_n}P_t(x)$ else $x_n^t:=x_n^{t-1}$. We show that $x_n^t$ converges for $t\to\infty$ by assuming the contrary and showing a contradiction. Since $x_n^t\in I_n$ some value, say $x_n^*$, is assumed infinitely often. Nonconvergence implies that the sequence leaves and returns to $x_n^*$ infinitely often. $x_n^*$ is only left ($x_n^{t-1}=x_n^*\neq x_n^t$) if $P_t(x_n^*)>n^{-3}$. On the other hand, at the time where $x_n^t$ returns to $x_n^*$ ($x_n^{t-1}\neq x_n^*=x_n^t$) we have $P_t(x_n^*)=P_t(x_n^t)=\min_{x\in I_n}P_t(x)\leq|I_n|^{-1}=2^{-n+1}$. Hence $P_t(x_n^*)$ oscillates (for $n\geq 12$) infinitely often between $\leq 2^{-n+1}$ and $\geq n^{-3}$, which contradicts the assumption that $P_t$ converges. Hence the assumption of a nonconvergent $x_n^t$ was wrong. $x_n^t$ converges to $x_n^*$ and $P_t(x_n^*)$ to a value $\leq n^{-3}$. With $x_n^t$ also the measure $Q_t(x_n^t):=\frac{1}{n(n+1)}$ (and $Q_t(x)=0$ for all other $x$) converges. Since $P(x_n^*)\leq n^{-3}$ does not dominate $Q(x_n^*)$, we have $P\stackrel{\times}{\not\geq}Q$. Since $P\in\mathcal{M}_{appr}^{semi}$ was arbitrary and $Q$ is an approximable measure we get $\mathcal{M}_{appr}^{semi}\stackrel{\times}{\not\geq}\mathcal{M}_{appr}^{msr}$. □

**Proof (continuous case).** The major difference from the discrete case is that one also has to take care that $\rho(x)\geq\rho(x0)+\rho(x1)$, $x\in I\!B^*$, is respected. On the other hand, the chunking $I_n:=I\!B^n$ is more natural here.

$(o)$ $\rho(x):=\sum_{\nu\in\mathcal{M}}w_\nu\nu(x)$ with $w_\nu>0$ obviously dominates all $\nu\in\mathcal{M}$ (with domination constant $w_\nu$). With $\sum_\nu w_\nu=1$ and all $\nu$ being (semi)measures also $\rho$ is a (semi)measure.

$(i)$ See [LV97, Th4.5.1].

$(ii)$ Let $\xi$ be a universal element in $\mathcal{M}_{enum}^{semi}$. We define [Sol78]

$$\xi_{norm}(x_{1:n}) \;:=\; \prod_{t=1}^{n}\frac{\xi(x_{1:t})}{\xi(x_{<t}0)+\xi(x_{<t}1)}. \qquad (2.30)$$

By induction one can show that $\xi_{norm}$ is a measure and that $\xi_{norm}(x)\geq\xi(x)\forall x$, hence $\xi_{norm}\geq\xi\stackrel{\times}{\geq}\mathcal{M}_{enum}^{semi}$. As a ratio of enumerable functions, $\xi_{norm}$ is still approximable, hence $\mathcal{M}_{appr}^{msr}\stackrel{\times}{\geq}\mathcal{M}_{enum}^{semi}$.

$(iii)$ [4]Let $\mu\in\mathcal{M}_{comp}^{semi}$. We recursively define the sequence $x_{1:\infty}^*$ by $x_k^*:=\text{argmin}_{x_k}\mu(x_{<k}^*x_k)$ and the measure $\rho$ by $\rho(x_{1:k}^*)=1\forall k$ and $\rho(x)=0$ for all $x$

[4] The proof in [LV97, p276] only applies to infinite alphabet and not to the binary/finite case considered here.

that are not prefixes of $x^*_{1:\infty}$. Exploiting the fact that a minimum is smaller than an average and that $\mu$ is a semimeasure, we get

$$\mu(x^*_{1:k}) = \min_{x_k} \mu(x^*_{<k}x_k) \;\leq\; \tfrac{1}{2}[\mu(x^*_{<k}0)+\mu(x^*_{<k}1)] \;\leq\; \tfrac{1}{2}\mu(x^*_{<k}).$$

Hence $\mu(x^*_{1:n}) \leq (\frac{1}{2})^n = (\frac{1}{2})^n \rho(x^*_{1:n})$, which demonstrates that $\mu$ does not dominate $\rho$. Since $\mu \in \mathcal{M}^{semi}_{comp}$ was arbitrary and $\rho$ is a computable measure this implies $\mathcal{M}^{semi}_{comp} \not\stackrel{\times}{\geq} \mathcal{M}^{msr}_{comp}$.

Assume now that there is an estimable semimeasure $\sigma \stackrel{\times}{\geq} \mathcal{M}^{msr}_{comp}$. We construct a finitely computable function $\mu \stackrel{\times}{\geq} \sigma$ as follows. Choose an initial $\varepsilon > 0$ and finitely compute an $\varepsilon$-approximation $\hat{\sigma}$ of $\sigma(x)$. If $\hat{\sigma} > 4\varepsilon$ define $\mu(x) := \hat{\sigma}$, else halve $\varepsilon$ and repeat the process. Since $\sigma(x) > 0$ (otherwise it could not dominate, e.g. $2^{-\ell(x)}$) the loop terminates after finite time. So $\mu$ is finitely computable. Inserting $\hat{\sigma} = \mu(x)$ and $\varepsilon < \frac{1}{4}\hat{\sigma} = \frac{1}{4}\mu(x)$ into $|\sigma(x) - \hat{\sigma}| < \varepsilon$ we get $|\sigma(x) - \mu(x)| < \frac{1}{4}\mu(x)$, which implies $\frac{3}{4}\mu(x) \leq \sigma(x) \leq \frac{5}{4}\mu(x)$. Unfortunately $\mu$ is not a semimeasure, but it still satisfies the weaker inequality $\mu(x0) + \mu(x1) \leq \frac{4}{3}[\sigma(x0) + \sigma(x1)] \leq \frac{4}{3}\sigma(x) \leq \frac{4}{3}\cdot\frac{5}{4}\mu(x) = \frac{5}{3}\mu(x)$. This is sufficient for the first half of the proof of $(iii)$ to go through with $\frac{1}{2}$ replaced by $\frac{1}{2}\cdot\frac{5}{3} = \frac{5}{6} < 1$, which shows that $\mu \not\stackrel{\times}{\geq} \mathcal{M}^{msr}_{comp}$. However, this contradicts $\mu \geq \frac{4}{5}\sigma \stackrel{\times}{\geq} \mathcal{M}^{msr}_{comp}$ showing that our assumed estimable semimeasure $\sigma$ does not exist, i.e. $\mathcal{M}^{semi}_{est} \not\stackrel{\times}{\geq} \mathcal{M}^{msr}_{comp}$.

$(iv)$ Assume $\mu \in \mathcal{M}^{semi}_{appr} \stackrel{\times}{\geq} \mathcal{M}^{msr}_{appr}$. We construct an approximable measure $\rho$ which is not dominated by $\mu$, thus contradicting the assumption. Let $\mu_1,\mu_2,...$ be a sequence of recursive functions converging to $\mu$. We recursively (in $t$ *and* $n$) define sequences $y^1_n, y^2_n, ..., y^t_n$ converging to $y_n$ and from them $\rho_1,\rho_2,...$ converging to $\rho$. Let $y^1_n = 0\,\forall n$. If $\mu_t(y^t_{<n}y^{t-1}_n) > \frac{2}{3}\mu_t(y^t_{<n})$, then $y^t_n := \mathrm{argmin}_{x_n}\mu_t(y^t_{<n}x_n)$, else $y^t_n := y^{t-1}_n$. We show that $y^t_n$ converges for $t \to \infty$ by assuming the contrary and showing a contradiction. Assume that $k$ is the smallest $n$ for which $y^t_n \not\to y_n$. Since $y^t_n \to y_n$ for all $n < k$ and $y^t_n \in I\!B$ is discrete there is a $t_0$ such that $y^t_{<k} = y_{<k}\,\forall t > t_0$. Assume $t > t_0$ in the following. Since $y^t_k \in I\!B$, some value, say $\tilde{y}_k$, is assumed infinitely often. Nonconvergence implies that the sequence leaves and enters to $\tilde{y}_k$ infinitely often. If $\tilde{y}_k$ is left ($y^{t-1}_k = \tilde{y}_k \neq y^t_k$), we have

$$\mu_t(y_{<k}\tilde{y}_k) \;=\; \mu_t(y^t_{<k}y^{t-1}_k) \;>\; \tfrac{2}{3}\mu_t(y^t_{<k}) \;=\; \tfrac{2}{3}\mu_t(y_{<k}) \;\stackrel{t\to\infty}{\longrightarrow}\; \tfrac{2}{3}\mu(y_{<k}).$$

If $\tilde{y}_k$ is entered ($y^{t-1}_k \neq \tilde{y}_k = y^t_k$) we have

$$\mu_t(y_{<k}\tilde{y}_k) \;=\; \mu_t(y^t_{<k}y^t_k) \;=\; \min_{x_k}\mu_t(y^t_{<k}x_k) \;\leq\; \tfrac{1}{2}[\mu_t(y^t_{<k}0) + \mu_t(y^t_{<k}1)] \;\leq$$

$$\leq\; \tfrac{1}{2}\mu_t(y^t_{<k}) \;=\; \tfrac{1}{2}\mu_t(y_{<k}) \;\stackrel{t\to\infty}{\longrightarrow}\; \tfrac{1}{2}\mu(y_{<k}).$$

Hence $\mu_t(y_{<k}\tilde{y}_k)$ oscillates infinitely often between $> \frac{2}{3}\mu(y_{<k})$ and $\leq \frac{1}{2}\mu(y_{<k})$, which contradicts the assumption that $\mu_t$ converges. Hence the assumption

of a nonconvergent $y_k^t$ was wrong. With $y_k^t$ also the measure $\rho_t(y^t_{1:n}):=1$ (and $\rho_t(x)=0$ for all other $x$ that are not prefixes of $y^t_{1:\infty}$) converges. For all sufficiently large $t$ we have $y_{1:n}=y^t_{1:n}$, hence $\mu_t(y_{1:n})=\mu_t(y^t_{1:n})\leq\frac{2}{3}\mu_t(y^t_{<n})\leq ...\leq(\frac{2}{3})^n$. Since $\mu(y_{1:n})\leq(\frac{2}{3})^n$ does not dominate $\rho(y_{1:n})=1$ ($\forall t>t_0$), we have $\mu \not\overset{\times}{\geq} \rho$. Since $\mu\in\mathcal{M}^{semi}_{appr}$ was arbitrary and $\rho$ is an approximable measure, we get $\mathcal{M}^{semi}_{appr}\not\overset{\times}{\geq}\mathcal{M}^{msr}_{appr}$. □

### 2.4.4 Martin-Löf Randomness

Martin-Löf randomness is a very important concept of randomness of individual sequences that is closely related to Kolmogorov complexity and Solomonoff's universal prior. Since we refer to this concept only in Section 3.2.7 we will be very brief here. We give a characterization equivalent to Martin-Löf's original definition, in order to bypass the necessity of giving a formal definition of 'effective randomness tests' [Lev73a]:

**Theorem 2.31 (Martin-Löf random sequences)** A sequence $x_{1:\infty}$ is called $\mu$-Martin-Löf random ($\mu$.M.L.) iff there is a constant $c$ such that $M(x_{1:n})\leq c\cdot\mu(x_{1:n})$ for all $n$.

An equivalent formulation for computable $\mu$ is:

$$x_{1:\infty} \text{ is } \mu\text{.M.L.-random} \quad\Leftrightarrow\quad Km(x_{1:n}) \overset{+}{=} -\log_2\mu(x_{1:n}) \;\forall n, \qquad (2.32)$$

where $Km(x_{1:n})$ is the length of the shortest (possibly nonhalting) program computing a string starting with $x_{1:n}$. Theorem 2.31 follows from (2.32) by exponentiation, "using $2^{-Km}\approx M$" and noting that $M\overset{\times}{\geq}\mu$ follows from universality of $M$. Consider the special case of $\mu$ being a fair coin, i.e. $\mu(x_{1:n})=2^{-n}$, then $x_{1:\infty}$ is M.L. random *iff* $Km(x_{1:n})\overset{+}{=}n$, i.e. if $x_{1:n}$ is incompressible. For general $\mu$, $-\log_2\mu(x_{1:n})$ is the length of the Shannon-Fano code of $x_{1:n}$, hence $x_{1:\infty}$ is $\mu$.M.L.-random *iff* the Shannon-Fano code is optimal.

One can show that a $\mu$.M.L.-random sequence $x_{1:\infty}$ passes *all* thinkable effective randomness tests, e.g. the law of large numbers, the law of the iterated logarithm, etc. In particular, the set of all $\mu$.M.L.-random sequences has $\mu$-measure 1. The following generalization is natural when considering general Bayes mixtures $\xi$:

**Definition 2.33 ($\mu/\xi$-random sequences)** A sequence $x_{1:\infty}$ is called $\mu/\xi$-random ($\mu.\xi$.r.) iff there is a constant $c$ such that $\xi(x_{1:n})\leq c\cdot\mu(x_{1:n})$ for all $n$.

Typically, $\xi$ is a mixture over some $\mathcal{M}$ as defined in (2.26), in which case the reverse inequality $\xi(x)\overset{\times}{\geq}\mu(x)$ is also true (for all $x$). For finite $\mathcal{M}$ or if $\xi\in\mathcal{M}$, the definition of $\mu/\xi$-randomness depends only on $\mathcal{M}$, and not on the specific weights used in $\xi$. For $\mathcal{M}=\mathcal{M}_U$, $\mu/\xi$-randomness is just $\mu$.M.L.-randomness.

The larger $\mathcal{M}$, the more patterns are recognized as nonrandom. Roughly speaking, those regularities characterized by some $\nu \in \mathcal{M}$ are recognized by $\mu/\xi$-randomness, i.e. for $\mathcal{M} \subset \mathcal{M}_U$ some $\mu/\xi$-random strings may not be M.L. random. Other randomness concepts, e.g. those by Schnorr, Ko, van Lambalgen, Lutz, Kurtz, von Mises, Wald, and Church (see [Wan96, Lam87, Sch71]), could possibly also be characterized in terms of $\mu/\xi$-randomness for particular choices of $\mathcal{M}$.

## 2.5 History & References

Most notation is taken over from [LV97]. The general theory of coding and prefix codes can be found in [Gal68], the important Kraft inequality is due to Kraft [Kra49].

**Algorithmic information theory.** Turing introduced the concept of a Turing machine and demonstrated that the halting problem is undecidable in [Tur36]. Turing machines are formally equivalent to partial recursive functions (see [Rog67, Odi89, Odi99] for an introduction). The halting problem corresponds to Gödel's incompleteness theorem [Göd31, Sho67] whose proof is based on a diagonal argument invented by Cantor [Can74, Dau90]. The short compiler Assumption 2.5 is an effective version of Kolmogorov's assumption that complexities based on different "reasonable" universal "Turing" machines coincide reasonably well [Kol65]. The works [Göd31, Kle36, Tur36, Pos44, ZL70, Sch02a] show the importance of the various computability concepts defined in (2.12). The consideration of (and naming for) estimable functions in the context of universal priors is from [Hut03b].

A coarse picture of the early history of algorithmic information theory could be drawn as follows: Kolmogorov [Kol65] and Chaitin [Cha66, Cha69], suggested defining the information content of an object as the length of the shortest program computing a representation of it. Solomonoff [Sol64] independently invented the closely related universal prior probability distribution and used it for binary sequence prediction [Sol64, Sol78]. Levin worked out most of the mathematical details [ZL70, Lev74] and invented the fastest algorithm for function inversion and optimization, save for a (huge) constant factor [Lev73b]. These papers may be regarded as the invention of what is now called algorithmic information theory. The invariance (2.8) is due to [Sol64, Kol65, Cha69], Theorem 2.10(*vii*) is due to [Lev74], the symmetry of information (*v*) due to [ZL70, Gác74, Kol83], (*ii*) is due to [Lev74], the other parts are elementary.

The short introduction we gave in this chapter necessarily described only the key ideas, ignoring many related and especially newer developments. Some references are given in the following.

There are many variants of "Kolmogorov" complexity. The prefix Kolmogorov complexity $K$ we defined here [Lev74, Gác74, Cha75], the earliest form, "plain" Kolmogorov complexity $C$ [Kol65], process complexity [Sch73], monotone complexity $Km$ [Lev73a], and uniform complexity [Lov69b, Lov69a], Solomonoff's universal prior $M = 2^{-KM}$ [Sol64, Sol78], Chaitin's complexity $Kc$ [Cha75], extension semimeasure $Mc$ [Cov74], and some others. They often differ from $K$ only by $O(\log K)$, but otherwise have similar properties. For an introduction to Shannon's information theory [Sha48] and its relation to Kolmogorov complexity, see [Kol65, Kol83, ZL70, CT91].

The main drawback of all these variants of Kolmogorov complexity is that they are not finitely computable [Kol65, Sol64]. They may be approximated from above [Kol65, Sol64], but no accuracy guarantee can be given, and what is worse, the best upper bound for the runtime until one has reasonable accuracy for $K(n)$ grows faster than any computable function in $n$. This led to the development of time-bounded complexity/probability that is finitely computable, or more general resource-bounded complexity/probability (e.g. space) [Dal73, Dal77, FMG92, Ko86, PF97, Sch02b].

For an excellent introduction to algorithmic information theory, and a more accurate treatment of its history (more than 500 references), and many applications one should consult the authoritative book of Li and Vitányi [LV97].

**Foundations of probability theory.** Although games of chance date back at least to around 300 B.C., the first mathematical analysis of probabilities appears to be much later. Important breakthroughs have been achieved (in chronological order and with significant simplification) by Cardano [Car63], a systematic way of calculating probabilities by Pascal (in correspondence with Fermat) and conditional probability [Pas54], Bayes' rule [Bay63], the distinction between subjective and objective interpretation of probabilities and the weak law of large numbers by Bernoulli [Ber13], equi-probability due to symmetry and other things by Laplace [Lap12], the principle of indifference by Keynes [Key21], Kolmogorov's axioms of probability theory [Kol33], early attempts to define the notion of randomness of individual objects/sequences by von Mises, Wald and Church [Mis19, Wal37, Chu40], finally successful by Martin-Löf [ML66], the notion of a universal a priori probability by Solomonoff [Sol64] and its mathematical investigation by Levin [ZL70, Lev74].

There is an ongoing debate between objective and subjective probability, which became sharper in the 20th century (not only in AI). Prominent advocates of the relative frequency or objective interpretation were Kolmogorov [Kol63], Fisher [Fis22], and von Mises [Mis28]. There are many advocates of probabilities as degrees of belief [Pop34, Ram31, Fin37, Cox46, Sav54, Jef83]. Carnap [Car48, Car50] tried to supplement logic with probability theory to so-called inductive logic. This works fine for propositional logic [Jay03], but not for predicate logic [Put63]. The closely related reference class problem is addressed in [Rei49, Kyb77, Kyb83, BGHK92].

There are many books on probability theory with different focus. For a thorough treatment of the early history of the concept of probability the reader is referred to the books by Hacking [Hac75] and Hald [Hal90], and for the foundations developed in the 20th century to the book by Schnorr [Sch71] and the PhD theses by van Lambalgen [Lam87] and Wang [Wan96]. A good standard textbook is by Feller [Fel68]. A pleasant to read book with a philosophical touch is by Jaynes [Jay03]. It treats probability theory as a natural extension to (Boolean) logical reasoning, emphasizes the "full" Bayesian approach with priors determined by the maximum entropy principle, and discusses various historical paradoxes and how these pitfalls could have been avoided by not becoming addicted to measure theory, but by sticking to elementary discrete math. The historic battle between different schools is treated at (over)length in a rather polemic way. Gelman [GCSR95] is a modern and more practical book on Bayesian data analysis.

**Alternatives to probability theory.** Given the success story of Bayesian probability theory, it is somewhat surprising that so many alternatives have been considered in AI. Many reasons why probability theory is unsuitable for AI have been stated: strict numerical values are not appropriate for a qualitative reasoning system, probability theory cannot deal with impreciseness, or vagueness, or subjective beliefs, or is just impractical. Setbacks caused by naive and/or inconsistent application are also responsible for Bayesian probabilistic reasoning falling out of favor in the 1970s. Default reasoning [Rei80], nonmonotonic logic [MD80], and circumscription [McC80] treat conclusions or events not as "believed to a certain degree" but as "believed by default until a better reason is found to believe something else" (see the anthology [Gin87]). Certainty factors ("fudge factors") have been introduced into classical rule-based expert systems to accommodate uncertainty [Sho76, BS84]. Dempster-Shafer theory uses probability intervals for probability values if they themselves are not perfectly known, usually because they have been estimated from a finite amount of data [Dem68, Sha76]. More generally, this approach goes under the name imprecise probabilities [Wal91]. In the full Bayesian treatment one defines a (second-order) probability distribution over probability values to deal with this kind of ignorance or beliefs. Fuzzy logic deals with vaguely defined events (Fuzzy sets), which are only "sort of" true, like the "Eiffel tower is high" [Zad65, Zim91]. Possibility theory has been introduced to handle uncertainty in fuzzy systems [Zad78]. See [Fin73, Wal91] or [RN95, Chp.15.6] for a more detailed account of various uncertain reasoning systems. Finally, quantum systems must be described with complex-valued probability amplitudes resulting in strange interference effects. A time may come (e.g. for nanobots) when quantum logic [Hug89] will be needed in AI.

All these alternate approaches have their problems: Either they have unclear semantics, or they are not self-consistent, or they don't scale up, or have other problems. It is not that Bayesian probability theory leaves no wishes open, but it is the most consistent system developed so far. Imprecise probability theory, a probabilistic robust or worst-case reasoning approach, is quite consistent and a possibly useful extension of probability theory in game theory and safety critical areas. The other approaches may survive as useful (efficient) approximations to a full Bayesian treatment. Although probability theory slowly (re)covers AI, the debate still goes on [Che85, Che88].

**Cox's axioms and theorem.** In [Cox46] Cox shows that every function $\mathrm{Bel}(\cdot|\cdot)$ satisfying his Axioms 2.17 is isomorphic to a (conditional) probability function. This (with considerable delay) gave a significant boost in using standard probability theory for dealing with subjective beliefs and uncertainty. Cheeseman [Che88] has called Cox's derivation the "strongest argument for use of standard (Bayesian) probability theory". Similar sentiments are expressed by Jaynes [Jay78, p24]; indeed, Cox's theorem is one of the cornerstones of Jaynes' recent book [Jay03]. Horvitz, Heckerman, and Langlotz [HHL86] used it as a basis for comparison of probability and other nonprobabilistic approaches to reasoning about uncertainty. Heckerman [Hec88] used it as a basis for providing an axiomatization for belief updates. Various variants of Cox's axioms have been considered in the literature [Rei49, Acz66, Hec88, Jay03, Fin73, Tri69], which simplify the derivation, or weaken, replace or better motivate the assumption. A loophole in all these derivations was only recently discovered [Par95]. They are all related to the following unwariness. The function $F$ of Cox's axioms mapping $\mathrm{Bel}(C|B\cap A)$ and $\mathrm{Bel}(B|A)$ to $\mathrm{Bel}(B\cap C|A)$ is proven to be associative, i.e. $F(x,F(y,z))=F(F(x,y),z)$, but actually associativity is only proven for $(x,y,z)$ of the form $x=\mathrm{Bel}(D|C\cap B\cap A)$, $y=\mathrm{Bel}(C|B\cap A)$, and $z=\mathrm{Bel}(B|A)$ for some events $A$, $B$, $C$, and $D$. If the set of such triples $(x,y,z)$ is dense in $[0,1]^3$, then by continuity, $F$ is associative. Paris provided a rigorous proof of Cox's result, assuming that the range of Bel is contained in [0,1] and using assumptions similar to [HHL86]. However, he and all others who tried to fix the proof needed to make additional assumptions that are not very appealing. Usually they demand or imply that the belief values are dense in a certain subset of $I\!R$, which excludes systems with a finite number of events. It remains an open question whether there is an appropriate strengthening of the assumptions that lead to Cox's result in finite settings. See [Hal99] and references therein for details.

**Algorithmic probability & universal induction.** The notion of universal (enumerable semi)measures was introduced in [Sol64, ZL70, LV77]. Levin [ZL70] defined universal a priori probability as one dominating all enumerable semimeasures. The dominance (2.23) and the equivalence $M\stackrel{\times}{=}\xi_U$ is due to Levin [ZL70]. Convergence of $M$ to $\mu$ in the conditional mean squared sense (Theorem 2.25) is due to Solomonoff [Sol78] (who normalizes $M$ as in (2.30) by giving up enumerability). The elementary proof of $M(x_t|x_{<t})\stackrel{t\to\infty}{\longrightarrow}1$ for computable $x_{1:\infty}$ is not based on any source. The direct study of predictions based on past observations without discussing models was called 'the prequential approach' by Dawid [Daw84]. Good reviews on universal induction with

[3] This is an exponent, not a footnote!

a philosophical touch are found in [LV92b, Sol97]. For an older, but general review of inductive inference see Angluin [AS83].

Schmidhuber [Sch00, Sch02a] constructed a natural hierarchy of generalizations of algorithmic probability and complexity and introduced more general, approximable and universal cumulatively enumerable semimeasures. The restriction to time-bounded universal probability is treated in [LV91, LV97, Sch00, Sch02b, Sch03a, Sch04] and is closely related to resource-bounded complexity and universal Levin search.

Other topics related to universal induction are the weighted majority algorithm by Littlestone and Warmuth [LW94], universal forecasting by Vovk [Vov92], Levin search [Lev73b], PAC-learning introduced by Valiant [Val84], the minimum message and description length principles [WB68, Ris78, Ris89, LV92a, Grü98, VL00], and Occam's razor, learnability and VC dimension [BEHW87, BEHW89].

Randomness of individual objects in terms of randomness tests was defined by Martin-Löf [ML66] and is closely related to Kolmogorov complexity and algorithmic probability [LV97, Cal02]. Another interesting randomness criterion for individual sequences by Vovk in terms of the Hellinger distance can be found in [Vov87]. Randomness of individual sequences in a wider context is exhaustively analyzed in the survey papers by Uspenskii et al. [KU87, USS90].

**Applications of Kolmogorov complexity and Levin search.** Schmidhuber [Sch00, Sch02b] defined the *speed prior*, closely related to Levin search, and derives a *computable* strategy for optimal inductive reasoning. He analyzed consequences for computable universes sampled from such priors. Good numerical approximations to Kolmogorov complexity are computationally expensive. But the ongoing decrease of processing costs has permitted the first successful implementations and applications [Sch97, SZW97]. A derivate of Levin's universal search algorithm was used in [Sch97] to discover neural nets with low Levin complexity, low Kolmogorov complexity, and high generalization capability. Adaptive Levin search (ALS) and the optimal ordered problem solver (OOPS) extend Levin search by making its underlying probability distribution on program space adaptive and by improving it according to experience [SZW97, Sch03a, Sch04]. This can significantly speed up the discovery of algorithmic solutions.

There are numerous applications of MDL, which can be viewed as an applied form of Kolmogorov complexity [LV97]. Apart from that, there are various other "direct" approximations, implementations, or practical applications. The universal similarity metric by Vitányi et al. based on string compression has been successfully applied to a plethora of clustering problems, including automatic music classification, and phylogeny and language tree reconstruction [Ben98, Li 03, CV03]. Conte et al. [Con97] evolved short Lisp programs to estimate Kolmogorov complexity. Chaitin [Cha91] speculated on the computational power of the evolutionary information gathering process and its relation to algorithmic information. Schmidt [Sch99] argued that (time-bounded)

Kolmogorov complexity helps and not prevents the search for extra terrestrial intelligence (SETI). Vovk [VW98] described universal portfolio selection schemes.

## 2.6 Problems

**2.1 ((Un)natural Turing machines)** [C10] Show that for every string $x$ there exists a universal Turing machine $U'$ such that $K_{U'}(x)=1$. Arguments of this sort are often used to demonstrate the arbitrariness or non-absolute character of algorithmic information. Argue that $U'$ is not a natural Turing machine if $x$ is complex. Elaborate on the difficulties in rigorously proving such a statement.

**2.2 (Exact $\xi$ correspondence)** [C20] We showed that $M(x) := \sum_{p:U(p)=x*}2^{-\ell(p)}$ equals $\xi_U(x) := \sum_{\nu\in\mathcal{M}_U}2^{-K(\nu)}\nu(x)$ within a multiplicative constant, i.e. $M \stackrel{\times}{=} \xi_U$. Improve this result to an exact equality in the sense that $M(x) = \xi_w(x) := \sum_{\nu\in\mathcal{M}_U}w_\nu\nu(x)$ for *some* weights with $w_\nu \geq 2^{-K(\nu)-O(1)}$ (solution due to Paul Vitányi, private communication). $M \stackrel{\times}{=} \xi_w$ is true for *any* choice of the weights $w_\nu > 0\,\forall\nu\in\mathcal{M}_U$. Show that equality (within a constant) no longer holds for a similarly generalized $M$ with $2^{-\ell(p)}$ replaced by arbitrary $w_p>0$.

**2.3 (Martin-Löf random sequences and convergence)** [C45soi] Show that a theorem true for all $\mu$-random sequences (see Theorem 2.31) is also true with $\mu$-probability 1. Under what conditions is the reverse direction true? In particular, is $\sum_{t=1}^\infty(\mu(x_t|x_{<t})-M(x_t|x_{<t}))^2 < \infty$ true for every individual $\mu$-random sequence? (cf. Theorem 3.19($i$)). It has been shown that $M(x_t|x_{<t})/\mu(x_t|x_{<t}) \stackrel{t\to\infty}{\longrightarrow} 1$ w.$\mu$.p.1 (see [LV97, Thm.5.2.2] and Theorem 3.19($v$) and Problem 3.10). Does the stronger statement of convergence individually for all Martin-Löf $\mu$-random sequences hold? The argument given in [LV97, Thm.5.2.2] and [VL00, Thm.10] is incomplete.[5] The implication "$M(x_{1:n})\leq c\cdot\mu(x_{1:n})\forall n\Rightarrow\lim_{n\to\infty}M(x_{1:n})/\mu(x_{1:n})$ exists" has been used, but not proven, and is indeed wrong. Construct a universal Turing machine $U$ and a uniformly M.L.-random sequence $\alpha_{1:\infty}$ on which $M(\alpha_t|\alpha_{<t})\not\to\frac{1}{2}$. What about for generic $U$ or for $\xi_U$? Construct an enumerable semimeasure $W$ (not necessarily universal) which converges to $\mu$ on all $\mu$.M.L.-random sequences for all computable measures $\mu$ [HM04].

[5] The formulation of their theorem is quite misleading in general: "*Let $\mu$ be a positive recursive measure. If the length of $y$ is fixed and the length of $x$ grows to infinity, then $M(y|x)/\mu(y|x)\to 1$ with $\mu$-probability one. The infinite sequences $\omega$ with prefixes $x$ satisfying the displayed asymptotics are precisely ['$\Rightarrow$' and '$\Leftarrow$'] the $\mu$-random sequences.*" First, for off-sequence $y$ convergence w.p.1 does not hold ($xy$ must be demanded to be a prefix of $\omega$). Second, the proof of '$\Leftarrow$' has loopholes (see main text). Last, '$\Rightarrow$' is given without proof and is wrong.

**2.4 (Oracle properties of Kolmogorov complexity)** [C20s] A function or problem $A$ is said to be Turing-reducible to $B$ if there exists a Turing machine (finitely) computing or solving $A$ provided $B$ is given as an oracle [HMU01]. Let $K:\mathbb{B}^*\to\mathbb{N}$ be the Kolmogorov complexity, $H:\mathbb{B}^*\to\mathbb{B}$ be the halting sequence ($H(p)=1\Leftrightarrow U(p)$ halts), and $\Omega:=\sum_{p:U(p)\ halts}2^{-\ell(p)}\in\mathbb{R}\cong[\mathbb{N}\to\mathbb{B}]$ be the halting probability, sometimes call 'the number of wisdom' [Cha75, BG79]. Show that $K$, $H$, and $\Omega$ are Turing-reducible to each other (cf. [LV97, p175]).

**2.5 (Weakly forgetful environments)** [C15u] Consider two sequences $x^1_{1:\infty}$ and $x^2_{1:\infty}$, "typical" in the sense that both are $\mu$.M.L.-random. Assume a different early history ($x^1_{<k}\neq x^2_{<k}$, $k$ fixed), continued by the same observations ($x^1_{k:n-1}=x^2_{k:n-1}=x_{k:n-1}$) for a long time $n$. Show that for computable $\mu$ the future is not affected by the far back history $x^i_{<k}$ in the sense that $\mu(x_n|x^1_{<k}x_{k:n-1})-\mu(x_n|x^2_{<k}x_{k:n-1})\to 0$ for $n\to\infty$. Hint: show $M(x_n|x_{k:n-1})\to\mu(x_n|x^i_{<k}x_{k:n-1})$ for $i=1$ *and* $i=2$. This property of $\mu$ for "typical" sequences may be considered as a weak form of forgetfulness. Argue that it is more appropriate to define forgetfulness as asymptotic independence of $x^1_{<k}$ for *all* environments (cf. definition in Section 5.3.6). Suggestion: compare ergodic Markov processes (see Definition 5.37) with $\mu$ defined by $\mu(1|x_{<n}):=i/3$ for $x_1=i-1\in\mathbb{B}$.

**2.6 (Complexity increase)** [C25u/C45oi] We are interested in good upper bounds on the increase in complexity when elongating a string $y:=x_{<t}$ to $yx:=x_{<t}x_{t:n}=x_{1:n}$. From Theorem 2.10(*iv*) we know that $K(yx)-K(y)\leq K(x|y)+O(1)$. Later (cf. Problem 3.13) we need similar bounds with $K$ on the l.h.s. replaced by $KM(x):=-\log_2 M(x)$. Furthermore, let $C(x)$ be the plain Kolmogorov complexity, defined as the length of the shortest plain (as opposed to prefix) program computing $x$ (see [LV97, Chp.2]). We have no particular demands on the r.h.s. of the inequality. So let us consider $\tilde{K}(x|y)$ defined as the length of the shortest plain *or* prefix, halting *or* nonhalting program, computing $x$ *or* a string starting with $x$, given $y$ *or* a string starting with $y$. The only important property of $\tilde{K}(x|y)$ is that it corresponds to the length of a shortest program computing $x$ from $y$. We do not want $\tilde{K}(x|y)$ to be *defined* as a difference $\tilde{K}(yx)-\tilde{K}(y)$. Prove the following inequalities:

$$(i)\qquad C(yx)-C(y)\ \leq\ \tilde{K}(x|y)+O(?)$$
$$(ii)\quad KM(yx)-KM(y)\ \leq\ \tilde{K}(x|y)+O(?)$$
$$(iii)\quad KM(yx)-KM(y)\ \leq\ \tilde{K}(\mu|y)-\log_2\mu(x|y)+O(?)$$

Since $C$, $K$, $\tilde{K}$, and $KM$ coincide within additive logarithmic terms, all inequalities follow from Theorem 2.10(*iv*) to logarithmic accuracy $O(\log\ell(xy))$. Improve the bounds to $O(?)\stackrel{\pm}{=}\{K(C(y)),K(\ell(y)),K(\ell(y))\}$, respectively, independent of $x$ for suitable $\tilde{K}$. It is an open question whether the bounds hold within an additive constant independent of $x$ *and* $y$ for any of the $\tilde{K}$.

**2.7 (Lower convergence bounds and defect of $M$)** [C15u/C35o] Prove

$$\left.\begin{array}{r}|1-M(x_t|x_{<t})| \\ \sum_{x_t}|M(x_t|x_{<t})-\mu(x_t|x_{<t})|\end{array}\right\} \geq 1-\sum_{x_t} M(x_t|x_{<t}) = \frac{\sum_{p:U(p)\rightsquigarrow x_{<t}} 2^{-\ell(p)}}{\sum_{p:U(p)=x_{<t}*} 2^{-\ell(p)}} \stackrel{\times}{\geq} 2^{-K(t)}$$

for all $t$ and $x_{1:t}$, where $\sum_{p:U(p)\rightsquigarrow x}$ sums over all halting *and* nonhalting prefix programs printing $x$ (and no more). Bounds (2.24) and (2.25) show rapid convergence of $M$ in a cumulative sense (see Section 3.2.4). The lower bound above shows that the instantaneous prediction quality is lower-bounded by the defect $1-\sum_{x_t} M(x_t|x_{<t})$ of $M$ from a proper measure, which itself is lower-bounded by $2^{-K(t)}$. This shows that occasionally, namely for simply describable $t$, the prediction quality is poor. Show that $2^{-K(t)}$ converges to zero slower than any computable to zero converging function, i.e. $2^{-K(t)} \neq O(f(t))$ for any computable $f$ with $f(t)\to 0$, for instance, $2^{-K(t)} \neq O([\ln t]^{-1})$. Show that similar lower bounds hold when performance is measured by (3.14)–(3.17) instead of (3.13) by exploiting their relations stated in Problem 3.16. This shows that the assertion in [LV97, Thm.5.2.1] that $S_t := \mathbf{E}[\sum_{x'_t}(\mu(x'_t|x_{<t})-M(x'_t|x_{<t}))^2]$ converges to zero faster than $1/t$ is not correct. Do similar lower performance bounds hold for measure $M_{norm}(x_t|x_{<t}) := M(x_t|x_{<t})/\sum_{x_t} M(x_t|x_{<t})$, which has no defect? For which sequences does equality $\stackrel{\times}{=} 2^{-K(t)}$ hold?

**2.8 (Predictions based on monotone complexity)** [C40s] It is easy to see that a predictor based on complexity $K$ fails, since for any sequence $x$ the predictor will essentially be indifferent about predicting 0 versus 1 due to $K(x1) \stackrel{+}{=} K(x0)$. Clearly, for e.g. $x=0^n$, any reasonable predictor should favor 0. The monotone complexity $Km(x) := \min_p\{\ell(p) : U(p) = x*\}$ does not suffer from this problem. Further, the leading contribution $m(x) := 2^{-Km(x)}$ in $M(x)$ is extremely close to $M(x)$ [Gác83]. Hence $m = 2^{-Km}$ is more promising for prediction than $K$. Show that for computable environments, $m(x_t|x_{<t}) := \frac{m(x_{1:t})}{m(x_{<t})}$ converges rapidly on-sequence $(a)$,$(b)$, but may converge $(c)$ slowly $(e)$ off-sequence $(\bar{x}_t \neq x_t)$, and (in disagreement with [LV97, Cor.5.2.2]) not at all for probabilistic environments $(f)$:

$(a)$ $\sum_{t=1}^n |1-m(x_t|x_{<t})| \leq \frac{1}{2} Km(x_{1:n})$, $\quad m(x_t|x_{<t}) \stackrel{fast}{\longrightarrow} 1$ for computable $x_{1:\infty}$

$(b)$ Indeed, $m(x_t|x_{<t}) \neq 1$ at most $Km(x_{1:\infty})$ times

$(c)$ $\sum_{t=1}^n m(\bar{x}_t|x_{<t}) \leq 2^{Km(x_{1:n})}$, $\quad m(\bar{x}_t|x_{<t}) \stackrel{slow?}{\longrightarrow} 0$ for computable $x_{1:\infty}$

$(d)$ $\sum_{t=1}^n m(\bar{x}_t|x_{<t}) \stackrel{\times}{\leq} [Km(x_{1:n})]^3$, $\quad m(\bar{x}_t|x_{<t}) \stackrel{fast?}{\longrightarrow} 0$ for computable $x_{1:\infty}$

$(e)$ $\forall s\, \exists U, x_{1:\infty} : Km(x_{1:\infty}) = s$ and $\sum_{t=1}^\infty m(\bar{x}_t|x_{<t}) \geq 2^s - 2$

$(f)$ $\exists \mu \in \mathcal{M}_{comp}^{msr} \setminus \mathcal{M}_{det} : m(x_t|x_{<t}) \stackrel{t\to\infty}{\not\to} \mu(x_t|x_{<t})\ \forall x_{1:\infty}$

Explain why cubic upper bound $(d)$ does not contradict exponential lower bound $(e)$. Show that $m$ is not a semimeasure, but normalization $m_{norm}(x_t|x_{<t}) := m(x_t|x_{<t})/\sum_{x_t} m(x_t|x_{<t})$ does not improve $(c)$–$(e)$ and even deteriorates $(a)$ and $(b)$. Show that the loss of the Bayes-predictor based on $m$ displays the same behavior as $m_{norm}$ (see Section 3.4). See [Hut03d, Hut04a] for solutions.

**2.9 (Prediction of selected bits)** [C35oi] Assume we observe sequence $x_{1:\infty}$, but we only need to predict the even bits. The odd bits shall be kept, since they may provide useful information for predicting the even bits (see below). Sure, (2.24) implies $\sum_{even\ t}(1-M(x_t|x_{<t}))^2 \leq \frac{1}{2}Km(x_{1:\infty})$. Assume that the even bits coincide with their preceding odd bits, i.e. $x_t = x_{t-1}$ for even $t$, and that $Km(x_{1:\infty}) \stackrel{+}{=} Km(x_1 x_3 ...)$ is large (or even infinite). The task of recognizing that 'even bit=odd bit' seems quite easy (of complexity $O(1)$), so a reasonable prediction scheme should make only $O(1)$ errors on the even bits, e.g. $\sum_{even\ t}(1-M(x_t|x_{<t}))^2 \leq O(1)$? Prove or disprove a bound of this form. Generalize the positive/negative result to probabilistic $\mu$ like (2.25), and to other more complex selection rules (cf. Problem 3.18).

**2.10 (Application of Bayes' rule)** [C05] Assume the prevalence of a certain disease in the general population is 1%. Assume there exists a quite reliable test for the disease, say, the test on a diseased/healthy person is positive/negative with 99% probability. If the test is positive, what is the chance of having the disease? Many medical doctors say about 99%. Show with Bayes' rule (2.19) that the true answer is scanty 50%.

**2.11 (Bayes' and Laplace' rule for Bernoulli sequences)** [C15] Let $x = x_1 x_2 ... x_n \in \mathbb{B}^n$ be generated by a biased coin with head probability $\theta \in [0,1]$, i.e. the likelihood of $x$ under hypothesis $\theta$ is $p(x|\theta) = \theta^{n_1}(1-\theta)^{n-n_1}$, where $n_1 = x_1 + ... + x_n$. Bayes [Bay63] assumed a uniform prior density $p(\theta) = 1$. (Since $\theta$ is continuous we must normalize $\int_0^1 p(\theta)\, d\theta = 1$ instead of $\sum_\theta p(\theta) = 1$.) Show that $p(x) = \int_0^1 p(x|\theta)p(\theta)\, d\theta = \frac{n_1!(n-n_1)!}{(n+1)!}$ and that the posterior probability density $p(\theta|x) = \frac{p(x|\theta)p(\theta)}{p(x)}$ of $\theta$ after seeing $x$ is strongly peaked around the frequency estimate $\theta = \frac{n_1}{n}$ for large $n$. Laplace [Lap12] asked about the predictive probability $p(1|x)$ of observing $x_{n+1} = 1$ after having seen $x = x_1 ... x_n$. Show that $p(1|x) = \frac{p(x1)}{p(x)} = \frac{n_1+1}{n+2}$. Laplace believed that the sun had risen for 1826213 days since creation, so he concluded that the probability that the sun won't rise tomorrow is $\frac{1}{1826215}$.

Pierre Laplace
(1749–1827)

*Probability theory is nothing but common sense reduced to calculation* — Pierre Laplace

# 3 Universal Sequence Prediction

In this chapter we investigate Solomonoff's universal induction scheme in detail. More generally, we consider a universal (or mixture) distribution $\xi$, defined as a weighted sum or integral of distributions $\nu\in\mathcal{M}$, where $\mathcal{M}$ is any countable or continuous set of distributions including the true distribution $\mu$. This is a generalization of Solomonoff induction, in which $\mathcal{M}$ is the set of all enumerable semimeasures. We show for several performance measures that using the universal $\xi$ as a prior is nearly as good as using the unknown true distribution $\mu$. In a sense, this solves the problem of the unknown prior in a universal way. All results are obtained for general finite alphabet. Convergence of $\xi$ to $\mu$ in a conditional mean squared sense and of $\xi/\mu\to 1$ with $\mu$-probability 1 is proven. The number of additional errors $E_\xi$ made by the optimal universal prediction scheme based on $\xi$ minus the number of errors $E_\mu$ of the optimal informed prediction scheme based on $\mu$ is proven to be bounded by $O(\sqrt{E_\mu})$. The prediction framework is generalized to arbitrary loss functions. A system is allowed to take an action $y_t$, given $x_1...x_{t-1}$ and receives loss $\ell_{x_t y_t}$ if $x_t$ is the next symbol of the sequence. No assumptions on $\ell$ are necessary, besides boundedness. Optimal universal $\Lambda_\xi$ and optimal informed $\Lambda_\mu$ prediction schemes are defined, and the total loss of $\Lambda_\xi$ is bounded in terms of the total loss of $\Lambda_\mu$, similar to the error bounds. We show that the bounds are tight and that no other predictor can lead to significantly smaller bounds. Furthermore, for various performance measures we show Pareto optimality of $\xi$ in the sense that there is no other predictor that performs better or equal in all environments $\nu \in \mathcal{M}$ and strictly better in at least one. So, optimal predictors can (w.r.t. to most performance measures in expectation) be based on the mixture $\xi$. Finally, we give an Occam's razor argument that the choice $w_\nu \sim 2^{-K(\nu)}$ for the weights is optimal, where $K(\nu)$ is the length of the shortest program describing $\nu$. Furthermore, games of chance, defined as a sequence of bets, observations, and rewards, are studied. The average profit achieved by the $\Lambda_\xi$ scheme rapidly converges to the best possible profit. The time needed to reach the winning zone is proportional to the relative entropy of $\mu$ and $\xi$. The prediction schemes presented here are compared to predictors based on expert advice. Although the algorithms, the settings, and the proofs are quite different, the bounds of both schemes have a very similar structure. Extensions to infinite alphabets, partial, multistep, delayed, and probabilistic prediction, classification, and more active systems are discussed.

## 3.1 Introduction

**Induction.** Many problems are of the induction type in which statements about the future have to be made, based on past observations. What is the probability of rain tomorrow, given the weather observations of the last few days? Is the Dow Jones stock index likely to rise tomorrow, given the chart of the last years and possibly additional newspaper information? Can we reasonably doubt that the sun will rise tomorrow? Indeed, one definition of science is to predict the future, where, as an intermediate step, one tries to understand the past by developing theories and, as a consequence of prediction, one tries to manipulate the future. All induction problems may be studied in the Bayesian framework. The probability of observing $x_t$ at time $t$, given the observations $x_1...x_{t-1}$ can be computed with the chain rule, if we know the true probability distribution, which generates the observed sequence $x_1x_2x_3...$. The problem is that in many cases we do not even have a reasonable guess of the true distribution $\mu$. What is the true probability of weather sequences, stock charts, or sunrises?

**Universal sequence prediction.** In order to overcome the problem of the unknown true distribution, one can define a mixture distribution $\xi$ as a $w_\nu$-weighted sum or integral over distributions $\nu \in \mathcal{M}$, where $\mathcal{M}$ is any discrete or continuous (hypothesis) set including $\mu$. $\mathcal{M}$ is assumed to be known and to contain the true distribution, i.e. $\mu \in \mathcal{M}$. Since the probability $\xi$ can be shown to converge rapidly to the true probability $\mu$ in a conditional sense, making decisions based on $\xi$ is often nearly as good as the infeasible optimal decision based on the unknown $\mu$ [MF98]. Solomonoff [Sol64] had the idea to define a universal mixture $M$ (Section 2.4) as a weighted average over deterministic programs. Lower weights were assigned to longer programs. He unified Epicurus' principle of multiple explanations and Occam's razor (simplicity) principle into one formal theory (see [LV97] for this interpretation of [Sol64]). Inspired by Solomonoff's work, Levin [ZL70] had the idea to define the closely related universal prior $\xi_U$ as a weighted average over *all* semicomputable semiprobability distributions. If the environment possesses some effective structure at all, Solomonoff's posterior [Sol78] "finds" this structure and allows for a good prediction. In a sense, this solves the induction problem in a universal way, i.e. without making problem-specific assumptions.

## 3.2 Setup and Convergence

In this section we show that the mixture $\xi$ converges rapidly to the true distribution $\mu$. After defining basic notation, we introduce the *universal or mixture distribution* $\xi$ as the $w_\nu$-weighted sum of probability distributions $\nu$ of a set $\mathcal{M}$ that includes the true distribution $\mu$. No structural assumptions are made on the $\nu$. A posterior representation of $\xi$ with incremental weight update is also presented. We introduce various convergence concepts for random sequences together with their interrelations, and various distance measures between $\xi$ and $\mu$ and their relations. We show that the relative entropy between $\mu$ and $\xi$ is bounded by $\ln w_\mu^{-1}$, which implies that $\xi$ rapidly converges to $\mu$ in difference and in ratio. The difficulty in establishing convergence results on individual (Martin-Löf) random sequences, and the case $\mu \notin \mathcal{M}$ are briefly discussed. The section concludes with a discussion of various standard sets $\mathcal{M}$ of probability measures, including computable, enumerable, cumulatively enumerable, approximable, finite-state, and Markov (semi)measures.

### 3.2.1 Random Sequences

We denote strings over a finite alphabet $\mathcal{X}$ by $x_1x_2...x_n$ with $x_t\in\mathcal{X}$ and $t,n,N\in\mathbb{N}$ and $N=|\mathcal{X}|$. We further use the abbreviations $\epsilon$ for the empty string, $x_{t:n}:=x_tx_{t+1}...x_{n-1}x_n$ for $t\leq n$ and $\epsilon$ for $t>n$, and $x_{<t}:=x_1...x_{t-1}$, and $\omega=x_{1:\infty}$ for infinite sequences. We use Greek letters for probability distributions (or measures). Let $\rho(x_1...x_n)$ be the probability that an (infinite) sequence starts with $x_1...x_n$:

$$\sum_{x_{1:n}\in\mathcal{X}^n}\rho(x_{1:n})=1,\quad \sum_{x_t\in\mathcal{X}}\rho(x_{1:t})=\rho(x_{<t}),\quad \rho(\epsilon)=1. \tag{3.1}$$

We also need conditional probabilities. Presuming they exist, we have

$$\rho(x_t|x_{<t}) = \rho(x_{1:t})/\rho(x_{<t}), \tag{3.2}$$

$$\rho(x_1...x_n) = \rho(x_1)\cdot\rho(x_2|x_1)\cdot...\cdot\rho(x_n|x_1...x_{n-1}), \tag{3.3}$$

called multiplication rule (of conditional probabilities), or chain rule. The first equation states that the probability that a string $x_1...x_{t-1}$ is followed by $x_t$

is equal to the probability that a string starts with $x_1...x_t$ divided by the probability that a string starts with $x_1...x_{t-1}$.

The second equation is the first, applied $n$ times. Whereas $\rho$ might be any probability distribution, $\mu$ denotes the true (unknown) generating distribution of the sequences. We denote expectations by $\mathbf{E}$. The (conditional) expected value of a function $f:\mathcal{X}^t\to I\!R$, dependent on $x_{1:t}$, independent of $x_{t+1:\infty}$, (given $x_{<t}$) is

$$\mathbf{E}[f]=\sum_{x_{1:n}\in\mathcal{X}^n}{}^{'}\mu(x_{1:n})f(x_{1:t}),\qquad \mathbf{E}_t[f]:=E[f(x_{1:t})|x_{<t}]=\sum_{x_t\in\mathcal{X}}{}^{'}\mu(x_t|x_{<t})f(x_{1:t}) \tag{3.4}$$

for any choice of $n\geq t$. Expectations $\mathbf{E}$ are *always* w.r.t. the true distribution $\mu$. The prime in $\sum'$ denotes that the sum is restricted to $x_{1:n}$ with $\mu(x_{1:t})\neq 0$ ($\mu(x_t|x_{<t})\neq 0$). If $\mu(x_{<t})=0$, then $\mu(x_t|x_{<t})$ and hence $\mathbf{E}_t$ is undefined. Since the sum in $\mathbf{E}$ is restricted to $\mu(x_{1:t})\neq 0$, $\mathbf{E}[\mathbf{E}_t[..]]=\mathbf{E}[..]$ is valid in any case (by the chain rule).

In a more probabilistic terminology we have a sample space $\Omega=\mathcal{X}^\infty$ with elements $\omega=\omega_1\omega_2\omega_3...\in\Omega$ being infinite sequences over the finite alphabet $\mathcal{X}$. The cylinder sets $\Gamma_{x_{1:n}}:=\{\omega:\omega_{1:n}=x_{1:n}\}$ are events. We define the $\sigma$-algebra $\mathcal{F}$ as the smallest set containing all cylinder sets and which is closed under complement and countable union. A probability measure $\mu$ is uniquely defined by giving its values $\mu(\Gamma_{x_{1:n}})$ on the cylinder sets, which we abbreviate by $\mu(x_{1:n})$. See Section 2.3 for a brief introduction to probability theory or [LV97, Doo53] or any statistics book for a more thorough treatment. Now, $f$ may be interpreted as a random variable or measurable function. Two functions differing on a set of measure zero have the same expectation. So if we "undefine" $f$ for some $x_{1:t}$ with $\mu(x_{1:t})=0$, the expectation should not be affected. Hence, $\sum'$ is the correct definition for partial functions. The prime is ineffective and can be ignored for total functions. Many expressions in this book are undefined on a set of measure zero. Henceforth we will not mention this anymore. See Section 3.9.1 for alternative ways of treating $\mu=0$.

Finally, the probability of an event $A\subseteq\Omega$ is $\mathbf{P}[A]=\mathbf{E}[\chi_A]$, where $\chi$ is the characteristic function of $A$, i.e. $\chi_A(\omega)=1$ if $\omega\in A$, and $\chi(\omega)=0$ otherwise.

### 3.2.2 Universal Prior Probability Distribution

Every inductive inference problem can be brought into the following form: Given a string $x_{<t}$, take a guess at its continuation $x_t$. We will assume that the strings that have to be continued are drawn from a probability distribution $\mu$. This includes deterministic environments as a special case, in which the probability distribution $\mu$ is 1 for some sequence $x_{1:\infty}$ and 0 for all others. We call probability distributions of this kind *deterministic*. The maximal prior information a prediction algorithm can possess is the exact knowledge of $\mu$, but in many cases (like for the probability of sun tomorrow) the true generating distribution is not known. Instead, the prediction is based on a guess $\rho$ of

$\mu$. We expect that a predictor based on $\rho$ performs well, if $\rho$ is close to $\mu$ or converges, in a sense, to $\mu$. Let $\mathcal{M}:=\{\nu_1,\nu_2,...\}$ be a countable set of candidate probability distributions on strings. Results are generalized to continuous sets $\mathcal{M}$ in Section 3.7.2. We define a weighted average on $\mathcal{M}$

$$\xi(x_{1:n}) \equiv \xi_{\mathcal{M}}(x_{1:n}) := \sum_{\nu\in\mathcal{M}} w_\nu\cdot\nu(x_{1:n}), \quad \sum_{\nu\in\mathcal{M}} w_\nu = 1, \quad w_\nu > 0. \tag{3.5}$$

It is easy to see that $\xi$ is a probability distribution as the weights $w_\nu$ are positive and normalized to 1 and the $\nu\in\mathcal{M}$ are probabilities.[1] For finite $\mathcal{M}$ a possible choice for the $w$ is to give all $\nu$ equal weight ($w_\nu=\frac{1}{|\mathcal{M}|}$). We call $\xi$ universal relative to $\mathcal{M}$, as it multiplicatively dominates all distributions in $\mathcal{M}$

$$\xi(x_{1:n}) \geq w_\nu\cdot\nu(x_{1:n}) \quad \text{for all} \quad \nu\in\mathcal{M}. \tag{3.6}$$

In the following, we assume that $\mathcal{M}$ is known and contains the true distribution, i.e. $\mu\in\mathcal{M}$. If $\mathcal{M}$ is chosen sufficiently large, then $\mu\in\mathcal{M}$ is not a serious constraint.

### 3.2.3 Universal Posterior Probability Distribution

All prediction schemes in this book are based on the conditional probabilities $\rho(x_t|x_{<t})$. It is possible to also express the conditional probability $\xi(x_t|x_{<t})$ as a weighted average over the conditional $\nu(x_t|x_{<t})$, but now with time-dependent weights:

$$\xi(x_t|x_{<t}) = \sum_{\nu\in\mathcal{M}} w_\nu(x_{<t})\nu(x_t|x_{<t}), \quad w_\nu(x_{1:t}) := w_\nu(x_{<t})\frac{\nu(x_t|x_{<t})}{\xi(x_t|x_{<t})}, \tag{3.7}$$

and $w_\nu(\epsilon) := w_\nu$. The denominator just ensures correct normalization $\sum_\nu w_\nu(x_{1:t})=1$. By induction and the chain rule we see that $w_\nu(x_{<t}) = w_\nu\nu(x_{<t})/\xi(x_{<t})$. Inserting this into $\sum_\nu w_\nu(x_{<t})\nu(x_t|x_{<t})$ using (3.5) gives $\xi(x_t|x_{<t})$, which proves the equivalence of (3.5) and (3.7). If $w_\nu$ is interpreted as the prior (subjective) belief in $\nu$, then $w_\nu(x_{<t})$ is the posterior belief in $\nu$ after having seen $x_{<t}$. The expressions (3.7) can be used to give an intuitive, but non-rigorous, argument why $\xi(x_t|x_{<t})$ converges to $\mu(x_t|x_{<t})$: The weight $w_\nu(x_{<t})$ of $\nu$ in $\xi$ increases/decreases if $\nu$ assigns a high/low probability to the new symbol $x_t$, given $x_{<t}$. For a $\mu$-random sequence $x_{1:t}$, $\mu(x_{1:t})\gg\nu(x_{1:t})$ if $\nu$ (significantly) differs from $\mu$. We expect the total weight for all $\nu$ consistent with $\mu$ to converge to 1, and all other weights to converge to 0 for $t\to\infty$. Therefore we expect $\xi(x_t|x_{<t})$ to converge to $\mu(x_t|x_{<t})$ for $\mu$-random strings $x_{1:\infty}$.

[1] The weight $w_\nu$ may be interpreted as the initial degree of belief in $\nu$ and $\xi(x_1...x_n)$ as the degree of belief in $x_1...x_n$. If the existence of true randomness is rejected on philosophical grounds, one may consider $\mathcal{M}$ containing only deterministic environments. $\xi$ still represents belief probabilities. See Section 2.3 for details..

Expression (3.7) seems to be more suitable than (3.5) for studying convergence and loss bounds of the universal predictor $\xi$, but it will turn out that (3.6) is all we need, with the sole exception in the proof of Theorem 3.66 and in Section 5.5. Expression (3.7) is useful when one tries to understand the learning aspect in $\xi$.

### 3.2.4 Convergence of Random Sequences

A classical (nonrandom) real-valued sequence $a_t$ is defined to converge to $a_*$, short $a_t \to a_*$, if $\forall\varepsilon\exists t_0\forall t \geq t_0 : |a_t - a_*| < \varepsilon$. We are interested in convergence properties of random sequences $z_t(\omega)$ for $t \to \infty$ (e.g. $z_t(\omega) = \xi(\omega_t|\omega_{<t}) - \mu(\omega_t|\omega_{<t})$). We define six convergence concepts for random sequences and relate them.

**Definition 3.8 (Convergence of random sequences)** Let $z_1(\omega)$, $z_2(\omega)$, ... be a sequence of real-valued random variables (and probability measure be $\mu$). $z_t$ is said to converge for $t \to \infty$ to (random variable) $z_*$

(*i*) with probability 1 (w.p.1) $:\Leftrightarrow \mathbf{P}[\{\omega : z_t(\omega) \to z_*(\omega)\}] = 1$
$\Leftrightarrow \forall\varepsilon > 0 : \mathbf{P}[\sup_{s \geq t}|z_s - z_*| \geq \varepsilon] \to 0$ for $t \to \infty$,

(*ii*) in the mean (i.m.) $:\Leftrightarrow \mathbf{E}[(z_t - z_*)^2] \to 0$ for $t \to \infty$,

(*iii*) in mean sum (i.m.s.) $:\Leftrightarrow \sum_{t=1}^{\infty} \mathbf{E}[(z_t - z_*)^2] < \infty$,

(*iv*) in probability (i.p.) $:\Leftrightarrow \forall\varepsilon > 0 : \mathbf{P}[|z_t - z_*| \geq \varepsilon] \to 0$ for $t \to \infty$,

(*v*) for every $\mu$-Martin-Löf random sequence ($\mu$.M.L.) $:\Leftrightarrow$
$\forall\omega : [\exists c\forall n : \xi_U(\omega_{1:n}) \leq c\mu(\omega_{1:n})]$ implies $z_t(\omega) \to z_*(\omega)$ for $t \to \infty$,

(*vi*) for every $\mu/\xi$-random sequence ($\mu.\xi$.r.) $:\Leftrightarrow$
$\forall\omega : [\exists c\forall n : \xi(\omega_{1:n}) \leq c\mu(\omega_{1:n})]$ implies $z_t(\omega) \to z_*(\omega)$ for $t \to \infty$.

See Section 2.4 for a definition of $\xi_U \equiv \xi_{\mathcal{M}_{enum}^{semi}} \stackrel{\times}{=} M$ and Martin-Löf randomness. In statistics, (*i*) is the "default" characterization of convergence of random sequences. Definitions (*ii*), (*iii*), and (*iv*) are also well known and are often more convenient to deal with than (*i*). Further, convergence i.m.s. is very strong: it provides a "rate" of convergence in the sense that the expected number of times $t$ in which $z_t$ deviates more than $\varepsilon$ from $z_*$ is finite and bounded by $c/\varepsilon^2$ and the probability that the number of $\varepsilon$-deviations exceeds $\frac{c}{\varepsilon^2\delta}$ is smaller than $\delta$, where $c := \sum_{t=1}^{\infty} \mathbf{E}[(z_t - z_*)^2]$. Definition (*v*) uses Martin-Löf's notion of randomness of *individual* sequences to define convergence M.L. Since this chapter mainly deals with general Bayes mixtures $\xi$, we generalized in (*vi*) the definition of convergence M.L. based on $\xi_U$ to convergence $\mu.\xi$.r. based on $\xi$ in a natural way. For finite $\mathcal{M}$ or if $\xi \in \mathcal{M}$, the definition of $\mu/\xi$-randomness depends only on $\mathcal{M}$, and not on the specific weights used in $\xi$. Convergence in one sense often implies convergence in another. The following implications for convergence of random sequences are true. Unconnected criteria (in the transitive hull) are incomparable (see also Problem 3.9).

**Lemma 3.9 (Relations between random convergence criteria)**

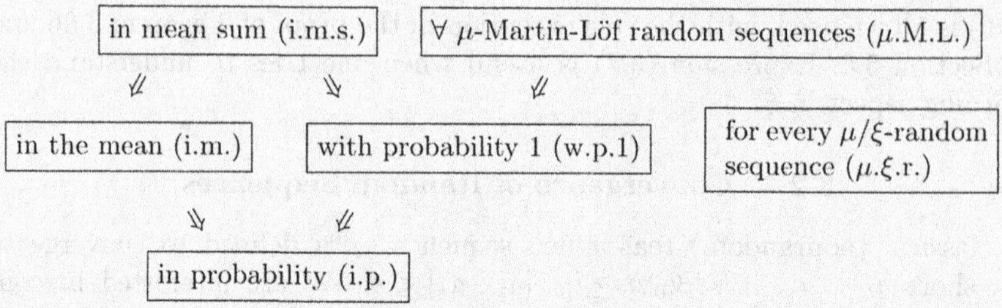


### 3.2.5 Distance Measures between Probability Distributions

We need several distance measures between vectors $\boldsymbol{y}=(y_i)$ and $\boldsymbol{z}=(z_i)$ in general, and probability distributions/vectors for which $y_i\geq 0$, $z_i\geq 0$, and $\sum_i y_i=\sum_i z_i=1$ in particular, $i=1,...,N$, namely the[2]

$$\begin{aligned}
\text{absolute (or Manhattan) distance:}\quad & a(\boldsymbol{y},\boldsymbol{z}) := \textstyle\sum_i |y_i - z_i| \\
\text{quadratic (or squared Euclidian) distance:}\quad & s(\boldsymbol{y},\boldsymbol{z}) := \textstyle\sum_i (y_i - z_i)^2 \\
\text{(squared) Hellinger distance:}\quad & h(\boldsymbol{y},\boldsymbol{z}) := \textstyle\sum_i (\sqrt{y_i}-\sqrt{z_i})^2 \qquad (3.10)\\
\text{relative entropy or KL divergence:}\quad & d(\boldsymbol{y},\boldsymbol{z}) := \textstyle\sum_i y_i \ln \frac{y_i}{z_i} \\
\text{absolute divergence:}\quad & b(\boldsymbol{y},\boldsymbol{z}) := \textstyle\sum_i y_i |\ln \frac{y_i}{z_i}|
\end{aligned}$$

The relative entropy is not a metric, but for probability distributions, for which it is defined, it is at least nonnegative and zero if and only if $\boldsymbol{y}=\boldsymbol{z}$. All bounds we prove in this chapter rely heavily on the following inequalities:

**Lemma 3.11 (Entropy inequalities)** Let $\{y_i\}$ and $\{z_i\}$ be two probability distributions, i.e. $y_i\geq 0$, $z_i\geq 0$, and $\sum_i y_i=\sum_i z_i=1$, and $f$ be a convex and even ($f(x)=f(-x)$) function with $f(0)\leq 0$. Then the following inequalities hold:[3]

$$\tfrac{1}{2}\Sigma f\leq f\sqrt{\tfrac{1}{2}d}:\quad \frac{1}{2}\sum_i f(y_i-z_i) \stackrel{(f)}{\leq} f\Bigg(\sqrt{\frac{1}{2}\sum_i y_i \ln\frac{y_i}{z_i}}\Bigg)$$

$$s\leq d:\quad \sum_i (y_i-z_i)^2 \stackrel{(s)}{\leq} \sum_i y_i \ln\frac{y_i}{z_i}$$

$$b-d\leq a\leq\sqrt{2d}:\quad \sum_i y_i\left|\ln\frac{y_i}{z_i}\right| - \sum_i y_i\ln\frac{y_i}{z_i} \stackrel{(b)}{\leq} \sum_i |y_i-z_i| \stackrel{(a)}{\leq} \sqrt{2\sum_i y_i\ln\frac{y_i}{z_i}}$$

$$h\leq d:\quad \sum_i (\sqrt{y_i}-\sqrt{z_i})^2 \stackrel{(h)}{\leq} \sum_i y_i\ln\frac{y_i}{z_i}$$

[2] $0\ln\frac{0}{z}:=0\,\forall z\geq 0$ and $y\ln\frac{y}{0}:=\infty\,\forall y>0$.

A proof of the lemma is deferred to Section 3.9. Inequality (3.11$s$) is a generalization of the binary $N=2$ case used in [Sol78, Hut01c, LV97]. If we insert

$$\mathcal{X}=\{1,...,N\}, \quad N=|\mathcal{X}|, \quad i=x_t, \quad y_i=\mu(x_t|x_{<t}), \quad z_i=\xi(x_t|x_{<t}) \tag{3.12}$$

into (3.10) we get various *instantaneous distances* (at time $t$) between $\mu$ and $\xi$. If we take the expectation $\mathbf{E}$ and sum over $\sum_{t=1}^n$ we get various *total distances* between $\mu$ and $\xi$:

$$a_t(x_{<t}) := \sum_{x_t}\Big|\mu(x_t|x_{<t})-\xi(x_t|x_{<t})\Big|, \qquad A_n := \sum_{t=1}^n \mathbf{E}[a_t(x_{<t})] \tag{3.13}$$

$$s_t(x_{<t}) := \sum_{x_t}\Big(\mu(x_t|x_{<t})-\xi(x_t|x_{<t})\Big)^2, \qquad S_n := \sum_{t=1}^n \mathbf{E}[s_t(x_{<t})] \tag{3.14}$$

$$h_t(x_{<t}) := \sum_{x_t}\Big(\sqrt{\mu(x_t|x_{<t})}-\sqrt{\xi(x_t|x_{<t})}\Big)^2, \qquad H_n := \sum_{t=1}^n \mathbf{E}[h_t(x_{<t})] \tag{3.15}$$

$$d_t(x_{<t}) := \sum_{x_t}\mu(x_t|x_{<t})\ln\frac{\mu(x_t|x_{<t})}{\xi(x_t|x_{<t})}, \qquad D_n := \sum_{t=1}^n \mathbf{E}[d_t(x_{<t})] \tag{3.16}$$

$$b_t(x_{<t}) := \sum_{x_t}\mu(x_t|x_{<t})\Big|\ln\frac{\mu(x_t|x_{<t})}{\xi(x_t|x_{<t})}\Big|, \qquad B_n := \sum_{t=1}^n \mathbf{E}[b_t(x_{<t})] \tag{3.17}$$

For $D_n$ the following can be shown [Sol78, Hut01c, LV97]:

$$\begin{aligned} D_n &= \sum_{t=1}^n \mathbf{E}[d_t(x_{<t})] = \sum_{t=1}^n \mathbf{E}\Big[\mathbf{E}_t\Big[\ln\frac{\mu(x_t|x_{<t})}{\xi(x_t|x_{<t})}\Big]\Big] \\ &= \mathbf{E}\Big[\ln\prod_{t=1}^n\frac{\mu(x_t|x_{<t})}{\xi(x_t|x_{<t})}\Big] = \mathbf{E}\Big[\ln\frac{\mu(x_{1:n})}{\xi(x_{1:n})}\Big] \leq \ln w_\mu^{-1} =: k_\mu \end{aligned} \tag{3.18}$$

In the first line we inserted (3.16) and used the definition of $\mathbf{E}_t$. Using $\mathbf{E}[\mathbf{E}_t[..]]=\mathbf{E}[..]$, the $t$ sum can thereafter be exchanged with the expectation $\mathbf{E}$ and transforms to a product inside the logarithm. In the last equality we used the chain rule (3.3) for $\mu$ and $\xi$. Using universality (3.6) of $\xi$, i.e. $\ln[\mu(x_{1:n})/\xi(x_{1:n})]\leq\ln w_\mu^{-1}$ for $\mu\in\mathcal{M}$ yields the final inequality in (3.18).

[3] In ($b$) if some $z=0$ we define $y|\ln\frac{y}{0}|-y\ln\frac{y}{0}:=0$.

### 3.2.6 Convergence of $\xi$ to $\mu$

**Theorem 3.19 (Convergence of $\xi$ to $\mu$)** Let there be sequences $x_1x_2...$ over a finite alphabet $\mathcal{X}$ drawn with probability $\mu(x_{1:n})$ for the first $n$ symbols. The universal posterior probability $\xi(x'_t|x_{<t})$ of the next symbol $x'_t$ given $x_{<t}$ is related to the true posterior probability $\mu(x'_t|x_{<t})$ in the following way:

$$(i)\ \sum_{t=1}^{n}\mathbf{E}\Big[\sum_{x'_t}\Big(\mu(x'_t|x_{<t})-\xi(x'_t|x_{<t})\Big)^2\Big] \equiv S_n \leq D_n \leq \ln w_\mu^{-1} < \infty$$

$$(ii)\ \sum_{x'_t}\Big(\mu(x'_t|x_{<t})-\xi(x'_t|x_{<t})\Big)^2 \equiv s_t(x_{<t}) \leq d_t(x_{<t}) \stackrel{t\to\infty}{\longrightarrow} 0 \quad \text{w.}\mu\text{.p.1}$$

$$(iii)\ \xi(x'_t|x_{<t})-\mu(x'_t|x_{<t}) \to 0 \quad \text{for } t\to\infty \text{ w.}\mu\text{.p.1 (and i.m.s) for any } x'_t$$

$$(iv)\ \sum_{t=1}^{n}\mathbf{E}\left[\left(\sqrt{\frac{\xi(x_t|x_{<t})}{\mu(x_t|x_{<t})}}-1\right)^2\right] \leq H_n \leq D_n \leq \ln w_\mu^{-1} < \infty$$

$$(v)\ \sqrt{\frac{\xi(x_t|x_{<t})}{\mu(x_t|x_{<t})}} \to 1 \text{ i.m.s and } \frac{\xi(x_t|x_{<t})}{\mu(x_t|x_{<t})} \to 1 \text{ w.}\mu\text{.p.1 for } t\to\infty$$

$$(vi)\ b_t(x_{<t})-d_t(x_{<t}) \leq a_t(x_{<t}) \leq \sqrt{2d_t(x_{<t})}, \quad B_n-D_n \leq A_n \leq \sqrt{2nD_n}$$

where $d_t$ and $D_n$ are the relative entropies (3.16), $w_\mu$ is the weight (3.5) of $\mu$ in $\xi$, and $x'_{1:\infty}$ an arbitrary (nonrandom) sequence.

**Proof.** Inequality $(ii)$ follows from the Definitions (3.14) and (3.16) and from the entropy inequality (3.11$s$). From the definition and finiteness of $D_\infty$ (3.18), and from $d_t(x_{<t}) \geq 0$, one sees that $\sqrt{d_t(x_{<t})} \to 0$ for $t\to\infty$ i.m.s., which implies $d_t(x_{<t}) \to 0$ w.$\mu$.p.1. The inequality $(i)$ follows from $(ii)$ by taking the $\mathbf{E}$ expectation and the $\sum_{t=1}^{n}$ sum. Convergence $(iii)$ follows from $(i)$ by dropping $\sum_{x'_t}$. The reason for the astonishing property of a single (universal) function $\xi$ to converge to *any* $\mu\in\mathcal{M}$ lies in the fact that the sets of $\mu$-random sequences differ for different $\mu$. Properties $(iv)$ and $(v)$ are related to $(i)$ and $(iii)$, but are incomparable convergence results. To prove $(iv)$ we use the abbreviations $\mu_t = \mu(x_t|x_{<t})$ and $\xi_t = \xi(x_t|x_{<t})$. For $\mu(x_{<t}) \neq 0$ we have

$$\mathbf{E}_t\Big[\Big(\sqrt{\tfrac{\xi_t}{\mu_t}}-1\Big)^2\Big] = \sum_{x_t}{}' \mu(x_t|x_{<t})\Big(\sqrt{\tfrac{\xi_t}{\mu_t}}-1\Big)^2 \tag{3.20}$$

$$= \sum_{x_t}{}' \big(\sqrt{\xi_t}-\sqrt{\mu_t}\big)^2 \leq h_t(x_{<t}) \leq d_t(x_{<t}).$$

The two inequalities follow from (3.15) and (3.11$h$). Bound $(iv)$ now follows by taking the $\mathbf{E}$ expectation and the $\sum_{t=1}^{n}$ sum. Convergence $(v)$ follows from

$(iv)$ from Definition 3.8$(iii)$ of convergence i.m.s., which implies convergence w.$\mu$.p.1. The first two inequalities in $(vi)$ immediately follow from inequalities (3.11$a$,$b$) and Definitions (3.13), (3.16) and (3.17). The third inequality follows from the first by linearity of $\mathbf{E}$ and $\sum$. The last inequality $A_n \leq \sqrt{2nD_n}$ follows from

$$\tfrac{1}{n}A_n \;\equiv\; \frac{1}{n}\sum_{t=1}^{n}\mathbf{E}[a_t] \;\leq\; \frac{1}{n}\sum_{t=1}^{n}\mathbf{E}[\sqrt{2d_t}] \;\leq\; \sqrt{\frac{1}{n}\sum_{t=1}^{n}\mathbf{E}[2d_t]} \;\equiv\; \sqrt{\tfrac{2}{n}D_n}, \quad (3.21)$$

where we have used Jensen's inequality for exchanging the averages ($\frac{1}{n}\sum_{t=1}^{n}$ and $\mathbf{E}$) with the concave function $\sqrt{\cdot}$. □

Since the conditional probabilities are the basis of all prediction algorithms considered in this book and $\xi$ converges rapidly to $\mu$ (see Problem 3.11), we expect a good prediction performance if we use $\xi$ as a guess of $\mu$. Performance measures are defined in the following sections.

Relations $(i)-(iii)$ generalize Solomonoff's result [Sol78] to an arbitrary finite alphabet. Without the use of the Hellinger distance $(iv)$, a somewhat weaker statement than $(v)$ can be derived from $(vi)$:

$$\mathbf{E}\left|\ln\frac{\mu(x_t|x_{<t})}{\xi(x_t|x_{<t})}\right| \;=\; \mathbf{E}[b_t] \;\leq\; \mathbf{E}[d_t]+\mathbf{E}[\sqrt{2d_t}] \;\leq\; \mathbf{E}[d_t]+\sqrt{2\mathbf{E}[d_t]} \;\stackrel{t\to\infty}{\longrightarrow}\; 0,$$

since $\mathbf{E}[d_t] \to 0$. I.e. $|\ln\frac{\mu(x_t|x_{<t})}{\xi(x_t|x_{<t})}|^{1/2} \to 0$ i.m., which implies $\xi(x_t|x_{<t})/\mu(x_t|x_{<t}) \to 1$ i.p. The explicit appearance of $n$ in the last expression of $(vi)$ prevents proving stronger convergence w.$\mu$.p.1 from $(vi)$.

The elementary proof for $(v)$ w.$\mu$.p.1 given here does not rely on the semi-martingale convergence theorem [Doo53, p324–325] as the proof of Gács [LV97, Thm.5.2.2]. Furthermore, $(iv)$ gives the "speed" of convergence. Note the subtle difference between $(iii)$ and $(v)$. For *any* sequence $x'_{1:\infty}$ (possibly constant and not necessarily $\mu$-random), $\mu(x'_t|x_{<t})-\xi(x'_t|x_{<t})$ converges to zero w.p.1 (referring to $x_{1:\infty}$), but no statement is possible for $\xi(x'_t|x_{<t})/\mu(x'_t|x_{<t})$, since $\liminf\mu(x'_t|x_{<t})$ could be zero. On the other hand, if we stay *on* the $\mu$-random sequence ($x'_{1:\infty}=x_{1:\infty}$), $(v)$ shows that $\xi(x_t|x_{<t})/\mu(x_t|x_{<t})\to 1$ (whether $\inf\mu(x_t|x_{<t})$ tends to zero or not does not matter). Indeed, it is easy to give an example where $\xi(x'_t|x_{<t})/\mu(x'_t|x_{<t})$ diverges. If we choose

$$\mathcal{M}=\{\mu_1,\mu_2\},\quad \mu\equiv\mu_1,\quad \mu_1(1|x_{<t})=\tfrac{1}{2}t^{-3} \quad\text{and}\quad \mu_2(1|x_{<t})=\tfrac{1}{2}t^{-2},$$

the contribution of $\mu_2$ to $\xi$ causes $\xi$ to fall off like $\mu_2\sim t^{-2}$, much slower than $\mu\sim t^{-3}$, causing the quotient to diverge:

$$\mu_1(0_{1:n}) \;=\; \prod_{t=1}^{n}(1-\tfrac{1}{2}t^{-3}) \;\stackrel{n\to\infty}{\longrightarrow}\; c_1=0.450...>0$$

$$\Rightarrow\; 0_{1:\infty} \text{ is a } \mu\text{-random sequence,}$$

$$\mu_2(0_{1:n}) = \prod_{t=1}^{n}(1-\tfrac{1}{2}t^{-2}) \stackrel{n\to\infty}{\longrightarrow} c_2 = 0.358... > 0$$

$$\Rightarrow \xi(0_{1:n}) \to w_1 c_1 + w_2 c_2 =: c_\xi > 0,$$

$$\xi(0_{<t}1) = w_1\mu_1(1|0_{<t})\mu_1(0_{<t}) + w_2\mu_2(1|0_{<t})\mu_2(0_{<t}) \to \tfrac{1}{2}w_2 c_2 t^{-2}$$

$$\Rightarrow \xi(1|0_{<t}) = \frac{\xi(0_{<t}1)}{\xi(0_{<t})} \to \frac{w_2 c_2}{2c_\xi}t^{-2} \Rightarrow \frac{\xi(1|0_{<t})}{\mu(1|0_{<t})} \to \frac{w_2 c_2}{c_\xi}t \to \infty \quad \text{diverges.}$$

### 3.2.7 Convergence in Martin-Löf Sense

An interesting open question is whether $\xi$ converges to $\mu$ (in the sense of *(iii)* or *(v)*) individually for all Martin-Löf random sequences, short $\xi \stackrel{\text{M.L.}}{\longrightarrow} \mu$ (see Problem 3.10). Clearly, convergence $\mu$.M.L. implies *(iii)* and *(v)* w.$\mu$.p.1 by Lemma 3.9, but the converse may fail on a set of sequences with $\mu$-measure zero. A convergence M.L. result would be particularly interesting and natural for the universal prior $\xi_U$, since M.L. randomness can be defined in terms of $\xi_U \stackrel{\times}{=} M$. Attempts to convert *(ii)* or *(iv)* to effective $\mu$.M.L.-randomness tests fail, since $\xi_U(x_t|x_{<t})$, and hence *(ii)* and *(iv)* are not enumerable. The argument given in [LV97, Thm.5.2.2] and [VL00, Thm.10] is incomplete. The implication "$\xi_U(x_{1:n}) \leq c\cdot\mu(x_{1:n})\forall n \Rightarrow \lim_{n\to\infty}\xi_U(x_{1:n})/\mu(x_{1:n})$ exists" has been used, but not proven, and may indeed be wrong (cf. Problem 2.3). Vovk [Vov87] shows that for two finitely computable semimeasures $\mu$ and $\rho$ and $x_{1:\infty}$ being $\mu$ *and* $\rho$ M.L. random that

$$\sum_{t=1}^{\infty}\sum_{x'_t}\Big(\sqrt{\mu(x'_t|x_{<t})} - \sqrt{\rho(x'_t|x_{<t})}\Big)^2 < \infty \quad \text{and} \quad \sum_{t=1}^{\infty}\left(\frac{\rho(x_t|x_{<t})}{\mu(x_t|x_{<t})} - 1\right)^2 < \infty.$$

If $\xi_U$ were recursive, then $\xi_U \to \mu$ and $\xi_U/\mu \to 1$ for every $\mu$.M.L.-random sequence $x_{1:\infty}$, since *every* sequence is $\xi_U$.M.L. random (see Definition 2.31). Since $\xi_U$ is *not* recursive, Vovk's theorem cannot be applied and it is not obvious how to generalize it. Indeed, one can show that M.L.-convergence fails for some choices of weights in $\xi_U$, but there are *non*-universal semimeasures which M.L.-converge to $\mu$ [HM04]. It is unknown whether there exists a M.L.-converging universal semimeasure. More generally, one may ask whether $\xi_\mathcal{M} \to \mu$ for every $\mu/\xi_\mathcal{M}$-random sequence. It turns out that this is true for some $\mathcal{M}$, but false for others.

**Theorem 3.22 ($\mu/\xi$-convergence of $\xi$ to $\mu$)** Let $\mathcal{X} = \mathbb{B}$ be binary and $\mathcal{M}_\Theta := \{\mu_\theta : \mu_\theta(1|x_{<t}) = \theta \forall t,\ \theta \in \Theta\}$ be the set of Bernoulli($\theta$) distributions with parameters $\theta \in \Theta$. Let $\Theta_D$ be a countable dense subset of $[0,1]$, e.g. $[0,1] \cap \mathbb{Q}$, and let $\Theta_G$ be a closed countable subset of $(0,1)$ with a gap, e.g. $\{\frac{1}{4},\frac{1}{2}\}$ or $\mathbb{Q} \cap ([\frac{1}{5},\frac{2}{5}] \cup [\frac{3}{5},\frac{4}{5}])$. Then

(*i*) If $x_{1:\infty}$ is $\mu/\xi_{\mathcal{M}_{\Theta_D}}$ random with $\mu \in \mathcal{M}_{\Theta_D}$, then $\xi_{\mathcal{M}_{\Theta_D}}(x_t|x_{<t}) \to \mu(x_t|x_{<t})$.

(*ii*) There are $\mu \in \mathcal{M}_{\Theta_G}$ and $\mu/\xi_{\mathcal{M}_{\Theta_G}}$ random $x_{1:\infty}$ for which $\xi_{\mathcal{M}_{\Theta_G}}(x_t|x_{<t}) \not\to \mu(x_t|x_{<t})$.

Our original or main motivation of studying $\mu/\xi$-randomness is the implication of Theorem 3.22 that $\xi_U \xrightarrow{\text{M.L.}} \mu$ cannot be decided from $\xi$ being a mixture distribution (3.5) or from the dominance property (3.6) alone. Further structural properties of $\mathcal{M}_U$ have to be employed. For Bernoulli sequences, convergence $\mu.\xi_{\mathcal{M}_\Theta}$.r. is related to denseness of $\mathcal{M}_\Theta$. Maybe a denseness characterization of $\mathcal{M}_{enum}^{semi}$ can solve the question of convergence M.L. of $\xi_U$. The property $\xi_U \in \mathcal{M}_U$ is also not sufficient to resolve the question, since there are $\mathcal{M} \ni \xi$ for which $\xi \xrightarrow{\mu.\xi.r} \mu$ and $\mathcal{M} \ni \xi$ for which $\xi \overset{\mu.\xi.r}{\not\to} \mu$ (see also Problem 3.10). Theorem 3.22 can be generalized to i.i.d. sequences over general finite alphabet $\mathcal{X}$.

The idea to prove (*ii*) is to construct a sequence $x_{1:\infty}$ that is $\mu_{\theta_0}/\xi$-random *and* $\mu_{\theta_1}/\xi$-random for $\theta_0 \neq \theta_1$. This is possible if and only if $\Theta$ contains a gap and $\theta_0$ and $\theta_1$ are the boundaries of the gap. Obviously, $\xi$ cannot converge to $\theta_0$ *and* $\theta_1$, thus proving nonconvergence. For no $\theta \in [0,1]$ will this $x_{1:\infty}$ be $\mu_\theta$ M.L.-random. Finally, the proof of Theorem 3.22 makes essential use of the mixture representation of $\xi$, as opposed to the proof of Theorem 3.19, which only needs dominance $\xi \overset{\times}{\geq} \mathcal{M}$.

An example for (*ii*) is $\mathcal{M} = \{\mu_0, \mu_1\}$, $\mu_0(1|x_{<t}) = \mu_1(0|x_{<t}) = \frac{1}{4}$, $x_{1:\infty} = (01)^\infty = 01010101... \Rightarrow \mu_0(x_{1:2n}) = \mu_1(x_{1:2n}) = \xi(x_{1:2n}) = (\frac{1}{4})^n(\frac{3}{4})^n \Rightarrow x_{1:\infty}$ is $\mu_0/\xi$-random *and* $\mu_1/\xi$-random, but $\mu_0(x_{2n}|x_{<2n}) = \frac{1}{4}$, $\mu_0(x_{2n+1}|x_{1:2n}) = \frac{3}{4}$, $\mu_1(x_{2n}|x_{<2n}) = \frac{3}{4}$, $\mu_1(x_{2n+1}|x_{1:2n}) = \frac{1}{4}$ and $\xi(x_{2n}|x_{<2n}) = \frac{3}{8}$, $\xi(x_{2n+1}|x_{1:2n}) = \frac{1}{2}$ for $w_0 = w_1 = \frac{1}{2} \Rightarrow \xi(x_n|x_{<n}) \not\to \mu_{0/1}(x_n|x_{<n})$.

**Proof.** Let $\mathcal{X} = \mathbb{B}$ and $\mathcal{M} = \{\mu_\theta : \theta \in \Theta\}$ with countable $\Theta \subset [0,1]$ and $\mu_\theta(1|x_{1:n}) = \theta = 1 - \mu_\theta(0|x_{1:n})$, which implies

$$\mu_\theta(x_{1:n}) = \theta^{n_1}(1-\theta)^{n-n_1}, \qquad n_1 := x_1 + ... + x_n, \qquad \hat\theta \equiv \hat\theta_n := \frac{n_1}{n}.$$

$\hat\theta$ depends on $n$; all other used/defined $\theta$ will be independent of $n$. We assume $\theta_{..} \in \Theta$, where .. stands for some (possible empty) index, and $\ddot\theta \in [0,1]$ (possibly $\notin \Theta$), where ¨ stands for some superscript, i.e. $\mu_{\theta_{..}}$ and $w_{\theta_{..}}$ make sense, whereas $\mu_{\ddot\theta}$ and $w_{\ddot\theta}$ do not. $\xi$ is defined in the standard way as

$$\xi(x_{1:n}) = \sum_{\theta \in \Theta} w_\theta \mu_\theta(x_{1:n}) \quad \Rightarrow \quad \xi(x_{1:n}) \geq w_\theta \mu_\theta(x_{1:n}), \tag{3.23}$$

where $\sum_\theta w_\theta = 1$ and $w_\theta > 0\,\forall\theta$. In the following let $\mu = \mu_{\theta_0} \in \mathcal{M}$ be the true environment.

$$\omega = x_{1:\infty} \text{ is } \mu/\xi\text{-random} \quad \Leftrightarrow \quad \exists c_\omega : \xi(x_{1:n}) \leq c_\omega \cdot \mu_{\theta_0}(x_{1:n})\ \forall n \tag{3.24}$$

For binary alphabet it is sufficient to establish whether $\xi(1|x_{1:n}) \stackrel{n\to\infty}{\longrightarrow} \theta_0 \equiv \mu(1|x_{1:n})$ for $\mu/\xi$-random $x_{1:\infty}$ in order to decide $\xi(x_n|x_{<n}) \to \mu(x_n|x_{<n})$. We need the following posterior representation of $\xi$:

$$\xi(1|x_{1:n}) = \sum_{\theta\in\Theta} w_n^\theta \mu_\theta(1|x_{1:n}), \quad w_n^\theta := w_\theta \frac{\mu_\theta(x_{1:n})}{\xi(x_{1:n})} \leq \frac{w_\theta}{w_{\theta_0}} \frac{\mu_\theta(x_{1:n})}{\mu_{\theta_0}(x_{1:n})}, \quad \sum_{\theta\in\Theta} w_n^\theta = 1 \tag{3.25}$$

The ratio $\mu_\theta/\mu_{\theta_0}$ can be represented as follows:

$$\frac{\mu_\theta(x_{1:n})}{\mu_{\theta_0}(x_{1:n})} = \frac{\theta^{n_1}(1-\theta)^{n-n_1}}{\theta_0^{n_1}(1-\theta_0)^{n-n_1}} = \left[\left(\frac{\theta}{\theta_0}\right)^{\hat\theta_n} \left(\frac{1-\theta}{1-\theta_0}\right)^{1-\hat\theta_n}\right]^n \tag{3.26}$$

$$= e^{n[D(\hat\theta_n||\theta_0) - D(\hat\theta_n||\theta)]}, \quad \text{where} \quad D(\hat\theta||\theta) = \hat\theta \ln\tfrac{\hat\theta}{\theta} + (1-\hat\theta)\ln\tfrac{1-\hat\theta}{1-\theta}$$

is the relative entropy between $\hat\theta$ and $\theta$, which is continuous in $\hat\theta$ and $\theta$, and is 0 if and only if $\hat\theta = \theta$. We also need the following implication for sets $\Omega \subseteq \Theta$:

$$\text{If} \quad w_n^\theta \leq w_\theta g_\theta(n) \stackrel{n\to\infty}{\longrightarrow} 0 \quad \text{and} \quad g_\theta(n) \leq c\,\forall\theta\in\Omega, \tag{3.27}$$

$$\text{then} \quad \sum_{\theta\in\Omega} w_n^\theta \mu_\theta(1|x_{1:n}) \leq \sum_{\theta\in\Omega} w_n^\theta \stackrel{n\to\infty}{\longrightarrow} 0,$$

which easily follows from boundedness $\sum_\theta w_n^\theta \leq 1$ and $\mu_\theta \leq 1$ (Lemma 5.28*ii*). We now prove Theorem 3.22. We leave the special considerations necessary when $0,1 \in \Theta$ to the reader and assume henceforth $0,1 \notin \Theta$.

(*i*) Let $\Theta$ be a countable dense subset of (0,1) and $x_{1:\infty}$ be $\mu/\xi$-random. Using (3.23) and (3.24) in (3.26) for $\theta \in \Theta$ to be determined later we can bound

$$e^{n[D(\hat\theta_n||\theta_0) - D(\hat\theta_n||\theta)]} = \frac{\mu_\theta(x_{1:n})}{\mu_{\theta_0}(x_{1:n})} \leq \frac{c_\omega}{w_\theta} =: c < \infty. \tag{3.28}$$

Let us assume that $\hat\theta \equiv \hat\theta_n \not\to \theta_0$. This implies that there exists a cluster point $\tilde\theta \neq \theta_0$ of sequence $\hat\theta_n$. That is, $\hat\theta_n$ is infinitely often in an $\varepsilon$-neighborhood of $\tilde\theta$, e.g. $D(\hat\theta_n||\tilde\theta) \leq \varepsilon$ for infinitely many $n$. $\tilde\theta \in [0,1]$ may be outside $\Theta$. Since $\tilde\theta \neq \theta_0$ this implies that $\hat\theta_n$ must be "far" away from $\theta_0$ infinitely often. For example, for $\varepsilon = \frac{1}{4}(\tilde\theta - \theta_0)^2$, using $D(\hat\theta||\tilde\theta) + D(\hat\theta||\theta_0) \geq (\tilde\theta - \theta_0)^2$, we get $D(\hat\theta||\theta_0) \geq 3\varepsilon$. We now choose $\theta \in \Theta$ so near to $\tilde\theta$ such that $|D(\hat\theta||\theta) - D(\hat\theta||\tilde\theta)| \leq \varepsilon$ (here we use denseness of $\Theta$). Chaining all inequalities we get $D(\hat\theta||\theta_0) - D(\hat\theta||\theta) \geq 3\varepsilon - \varepsilon - \varepsilon = \varepsilon > 0$. This, together with (3.28) implies $e^{n\varepsilon} \leq c$ for infinitely many $n$, which is impossible. Hence, the assumption $\hat\theta_n \not\to \theta_0$ was wrong.

Now, $\hat\theta_n \to \theta_0$ implies that for arbitrary $\theta \neq \theta_0$, $\theta \in \Theta$, and for sufficiently large $n$ there exists $\delta_\theta > 0$ such that $D(\hat\theta_n||\theta) \geq 2\delta_\theta$ (since $D(\theta_0||\theta) \neq 0$) and $D(\hat\theta_n||\theta_0) \leq \delta_\theta$. This implies

$$w_n^\theta \;\leq\; \frac{w_\theta}{w_{\theta_0}} e^{n[D(\hat\theta_n||\theta_0) - D(\hat\theta_n||\theta)]} \;\leq\; \frac{w_\theta}{w_{\theta_0}} e^{-n\delta_\theta} \stackrel{n\to\infty}{\longrightarrow} 0,$$

where we used (3.25) and (3.26) in the first inequality, and the second inequality holds for sufficiently large $n$. Hence $\sum_{\theta \neq \theta_0} w_n^\theta \to 0$ by (3.27) and $w_n^{\theta_0} \to 1$ by normalization (3.25), which finally gives

$$\xi(1|x_{1:n}) = w_n^{\theta_0} \mu_{\theta_0}(1|x_{1:n}) + \sum_{\theta \neq \theta_0} w_n^\theta \mu_\theta(1|x_{1:n}) \stackrel{n\to\infty}{\longrightarrow} \mu_{\theta_0}(1|x_{1:n}).$$

***(ii)*** We first consider the case $\Theta = \{\theta_0, \theta_1\}$: Let us choose $\bar\theta$ ($= \ln(\frac{1-\theta_0}{1-\theta_1}) / \ln(\frac{\theta_1}{\theta_0}\frac{1-\theta_0}{1-\theta_1}) \notin \Theta$) in the (KL) middle of $\theta_0$ and $\theta_1$ such that

$$D(\bar\theta||\theta_0) = D(\bar\theta||\theta_1), \qquad 0 < \theta_0 < \bar\theta < \theta_1 < 1, \tag{3.29}$$

and choose $x_{1:\infty}$ such that $\hat\theta_n := \frac{n_1}{n}$ satisfies $|\hat\theta_n - \bar\theta| \leq \frac{1}{n}$ $\quad (\Rightarrow \hat\theta_n \stackrel{n\to\infty}{\longrightarrow} \bar\theta)$.
We will show that $x_{1:\infty}$ is $\mu_{\theta_0}/\xi$-random *and* $\mu_{\theta_1}/\xi$-random. Obviously no $\xi$ can converge to $\theta_0$ *and* $\theta_1$, thus proving $\mu/\xi$-nonconvergence. ($x_{1:\infty}$ is obviously not $\mu_{\theta_{0/1}}$ M.L.-random, since the relative frequency $\hat\theta_n \not\to \theta_{0/1}$. $x_{1:\infty}$ is not even $\mu_{\bar\theta}$ M.L.-random, since $\hat\theta_n$ converges too fast ($\sim \frac{1}{n}$). $x_{1:\infty}$ is indeed very regular, whereas $\frac{n_1}{n}$ of a truly $\mu_{\bar\theta}$ M.L.-random sequence has fluctuations of the order $1/\sqrt{n}$. The fast convergence is necessary for doubly $\mu/\xi$-randomness. The reason that $x_{1:\infty}$ is $\mu/\xi$-random, but not M.L.-random is that $\mu/\xi$-randomness is a weaker concept than M.L.-randomness for $\mathcal{M} \subset \mathcal{M}_{enum}^{semi}$. Only regularities characterized by $\nu \in \mathcal{M}$ are recognized by $\mu/\xi$-randomness.)

In the following we assume that $n$ is sufficiently large such that $\theta_0 \leq \hat\theta_n \leq \theta_1$. We need

$$|D(\hat\theta||\theta) - D(\bar\theta||\theta)| \leq c|\hat\theta - \bar\theta| \quad \forall \theta, \hat\theta, \bar\theta \in [\theta_0, \theta_1] \quad \text{with} \quad c := \ln\frac{\theta_1(1-\theta_0)}{\theta_0(1-\theta_1)} < \infty, \tag{3.30}$$

which follows for $\hat\theta \geq \bar\theta$ (similarly $\hat\theta \leq \bar\theta$) from

$$D(\hat\theta||\theta) - D(\bar\theta||\theta) = \int_{\bar\theta}^{\hat\theta} [\ln\tfrac{\theta'}{\theta} - \ln\tfrac{1-\theta'}{1-\theta}]d\theta' \leq \int_{\bar\theta}^{\hat\theta} [\ln\tfrac{\theta_1}{\theta_0} - \ln\tfrac{1-\theta_1}{1-\theta_0}]d\theta' = c\cdot(\hat\theta - \bar\theta),$$

where we have increased $\theta'$ to $\theta_1$ and decreased $\theta$ to $\theta_0$ in the inequality. Using (3.30) in (3.26) twice we get

$$\frac{\mu_{\theta_1}(x_{1:n})}{\mu_{\theta_0}(x_{1:n})} = e^{n[D(\hat\theta_n||\theta_0) - D(\hat\theta_n||\theta_1)]} \leq e^{n[D(\bar\theta||\theta_0) + c|\hat\theta_n - \bar\theta| - D(\bar\theta||\theta_1) + c|\hat\theta_n - \bar\theta|]} \leq e^{2c}, \tag{3.31}$$

where we used (3.29) in the last inequality. Now, (3.31) and (3.25) lead to

$$w_n^{\theta_0} = w_{\theta_0}\frac{\mu_{\theta_0}(x_{1:n})}{\xi(x_{1:n})} = [1+\frac{w_{\theta_1}}{w_{\theta_0}}\frac{\mu_{\theta_1}(x_{1:n})}{\mu_{\theta_0}(x_{1:n})}]^{-1} \geq [1+\frac{w_{\theta_1}}{w_{\theta_0}}\mathrm{e}^{2c}]^{-1} =: c_0 > 0, \tag{3.32}$$

which shows that $x_{1:\infty}$ is $\mu_{\theta_0}/\xi$-random by (3.24). Exchanging $\theta_0 \leftrightarrow \theta_1$ in (3.31) and (3.32) we similarly get $w_n^{\theta_1} \geq c_1 > 0$, which implies (using $w_n^{\theta_0} + w_n^{\theta_1} = 1$)

$$\xi(1|x_{1:n}) = \sum_{\theta\in\{\theta_0,\theta_1\}} w_n^\theta \mu_\theta(1|x_{1:n}) = w_n^{\theta_0}\cdot\theta_0 + w_n^{\theta_1}\cdot\theta_1 \neq \theta_0 = \mu_{\theta_0}(1|x_{1:n}). \tag{3.33}$$

This shows $\xi(1|x_{1:n}) \overset{n\to\infty}{\not\to} \mu(1|x_{1:n})$. One can show that $\xi(1|x_{1:n})$ does not only not converge to $\theta_0$ (and $\theta_1$), but that it does not converge at all. The fast convergence demand $|\hat\theta_n - \bar\theta| \leq \frac{1}{n}$ on $x_{1:\infty}$ can be weakened to $\hat\theta_n \leq \bar\theta + O(\frac{1}{n})\,\forall n$ and $\hat\theta_n \geq \bar\theta - O(\frac{1}{n})$ for infinitely many $n$, then $x_{1:\infty}$ is still $\mu_{\theta_0}/\xi$-random, and $w_n^{\theta_1} \geq c_1' > 0$ for infinitely many $n$, which is sufficient to prove $\xi \not\to \mu$.

We now consider general $\Theta$ with gap in the sense that there exist $0<\theta_0<\theta_1<1$ with $[\theta_0,\theta_1]\cap\Theta = \{\theta_0,\theta_1\}$: We show that all $\theta \neq \theta_0,\theta_1$ give asymptotically no contribution to $\xi(1|x_{1:n})$, i.e. (3.33) still applies. Let $\theta \in \Theta\setminus\{\theta_0,\theta_1\}$; all other definitions as before. Then $\delta_\theta := D(\bar\theta||\theta) - D(\bar\theta||\theta_{0/1}) > 0$, since $\theta$ is farther than $\theta_{0/1}$ away from $\bar\theta$ ($|\theta-\bar\theta| > |\theta_{0/1}-\bar\theta|$). Similarly to (3.31) with $\theta$ instead $\theta_1$, we get

$$\frac{\mu_\theta(x_{1:n})}{\mu_{\theta_0}(x_{1:n})} = \mathrm{e}^{n[D(\hat\theta_n||\theta_0)-D(\hat\theta_n||\theta)]} \leq \mathrm{e}^{2c}\cdot\mathrm{e}^{n[D(\bar\theta||\theta_0)-D(\bar\theta||\theta)]} = \mathrm{e}^{2c}\mathrm{e}^{-n\delta_\theta} \overset{n\to\infty}{\longrightarrow} 0.$$

Hence $w_n^\theta \leq \frac{w_\theta}{w_{\theta_0}}\mathrm{e}^{2c}\mathrm{e}^{-n\delta_\theta} \to 0$ from (3.25) and $\varepsilon_n := \sum_{\theta\in\Theta\setminus\{\theta_0,\theta_1\}} w_n^\theta \mu_\theta(1|x_{1:n}) \overset{n\to\infty}{\longrightarrow} 0$ from (3.27). Hence $\xi(1|x_{1:n}) = w_n^{\theta_0}\cdot\theta_0 + w_n^{\theta_1}\cdot\theta_1 + \varepsilon_n \neq \theta_0 = \mu_{\theta_0}(1|x_{1:n})$ for sufficiently large $n$, since $\varepsilon_n \to 0$, $w_n^{\theta_1} \geq c_1' > 0$ and $\theta_0 \neq \theta_1$. □

### 3.2.8 The Case where $\mu \notin \mathcal{M}$

In the following we discuss two cases in which $\mu \notin \mathcal{M}$, but most parts of this book still apply. Actually all theorems remain valid for $\mu$ being a finite linear combination $\mu(x_{1:n}) = \sum_{\nu\in\mathcal{L}} v_\nu \nu(x_{1:n})$ of $\nu$'s in $\mathcal{L} \subseteq \mathcal{M}$. Dominance $\xi(x_{1:n}) \geq w_\mu \cdot \mu(x_{1:n})$ is still ensured with $w_\mu := \min_{\nu\in\mathcal{L}} \frac{w_\nu}{v_\nu} \geq \min_{\nu\in\mathcal{L}} w_\nu$. More generally, if $\mu$ is an infinite linear combination, dominance is still ensured if $w_\nu$ itself dominates $v_\nu$ in the sense that $w_\nu \geq \alpha v_\nu$ for some $\alpha > 0$ (then $w_\mu \geq \alpha$).

Another possibly interesting situation is when the true generating distribution $\mu \notin \mathcal{M}$, but a "nearby" distribution $\hat\mu$ with weight $w_{\hat\mu}$ is in $\mathcal{M}$. If we measure the distance of $\hat\mu$ to $\mu$ with the Kullback-Leibler divergence $D_n(\mu||\hat\mu) := \sum_{x_{1:n}} \mu(x_{1:n}) \ln\frac{\mu(x_{1:n})}{\hat\mu(x_{1:n})}$ and assume that it is bounded by a constant $c$, then

$$D_n = \mathbf{E}\left[\ln\frac{\mu(x_{1:n})}{\xi(x_{1:n})}\right] = \mathbf{E}\left[\ln\frac{\hat\mu(x_{1:n})}{\xi(x_{1:n})}\right] + \mathbf{E}\left[\ln\frac{\mu(x_{1:n})}{\hat\mu(x_{1:n})}\right] \leq \ln w_{\hat\mu}^{-1} + c.$$

So $D_n \leq \ln w_\mu^{-1}$ remains valid if we define $w_\mu := w_{\hat\mu}\cdot e^{-c}$.

### 3.2.9 Probability Classes $\mathcal{M}$

In the following we describe some well-known and some less-known probability classes $\mathcal{M}$. This relates our setting to other works in this area, embeds it into the historical context, illustrates the type of classes we have in mind, and discusses computational issues.

We get a rather wide class $\mathcal{M}$ if we include *all* computable probability distributions in $\mathcal{M}$. In this case, the assumption $\mu \in \mathcal{M}$ is very weak, as it only assumes that the strings are drawn from *any computable* distribution; and all valid physical theories (and hence all environments) *are* computable to arbitrary precision (estimable in a probabilistic sense).

It is favorable to assign high weights $w_\nu$ to the $\nu$. Simplicity should be favored over complexity, according to Occam's razor. In our context this means that a high weight should be assigned to simple $\nu$. The prefix Kolmogorov complexity $K(\nu)$ is a universal complexity measure [Kol65, Lev74, Gác74, Cha75, LV97]. It is defined as the length of the shortest self-delimiting program on a universal Turing machine computing $\nu(x_{1:n})$ given $x_{1:n}$ (cf. Section 2.2). If we define

$$w_\nu := 2^{-K(\nu)}$$

then distributions that can be calculated by short programs have high weights. The relative entropy is bounded by the Kolmogorov complexity of $\mu$ in this case ($D_n \leq K(\mu)\cdot\ln 2$). Solomonoff [Sol64, Eq.(13)] considered the class $\mathcal{M} = \mathcal{M}_{comp}^{msr}$ of all computable measures, which unfortunately leads to an inapproximable mixture $\xi$. Levin obtained the universal semimeasure $\xi_U$ by taking $\mathcal{M} = \mathcal{M}_U = \mathcal{M}_{enum}^{semi}$ to be the (multi)set enumerated by a Turing machine that enumerates all enumerable semimeasures [ZL70, LV97] (see Section 2.4 for details). Recently, $\mathcal{M}$ was further enlarged to include all cumulatively enumerable semimeasures [Sch02a]. In the enumerable and cumulatively enumerable cases, $\xi$ is not estimable, but can be approximated to arbitrary but not pre-specifiable precision. If we consider *all* approximable (i.e. asymptotically computable) distributions, then the universal distribution $\xi$, although still well defined, is not even approximable (like $\xi_{\mathcal{M}_{comp}^{msr}}$). An interesting and quickly approximable distribution is the Speed prior $S$ defined in [Sch02b]. It is related to Levin complexity and Levin search [Lev73b, Lev84] (Section 7.1.2), but it is unclear for now which distributions are dominated by $S$ (Problem 3.2). If one considers only finite-state automata instead of general Turing machines, $\xi$ is related to the quickly computable, universal finite-state prediction scheme of Feder et al. [FMG92], which itself is related to the famous Lempel-Ziv data compression algorithm. If one has extra knowledge on the source generating the sequence, one might further reduce $\mathcal{M}$ and increase $w$. Note that $\xi \in \mathcal{M}$

in the enumerable and cumulatively enumerable case, but $\xi\notin\mathcal{M}$ in the computable, approximable and finite-state case. If $\xi$ is itself in $\mathcal{M}$, it is called a universal element of $\mathcal{M}$ [LV97]. As we do not need this property here, $\mathcal{M}$ may be *any* countable set of distributions. In the following we consider generic $\mathcal{M}$ and $w$.

We have discussed various discrete classes $\mathcal{M}$, which are sufficient from a constructive or computational point of view. On the other hand, it is convenient to also allow for continuous classes $\mathcal{M}$. For instance, the class of *all* Bernoulli processes with parameter $\theta\in[0,1]$ and uniform prior $w_\theta\equiv 1$ (Problems 2.11 and 5.14) is much easier to deal with than computable $\theta$ only, with prior $w_\theta=2^{-K(\theta)}$. Other important continuous classes are the class of i.i.d. and Markov processes. Continuous classes $\mathcal{M}$ are briefly considered in Section 3.7.2.

## 3.3 Error Bounds

In this section we derive error bounds for predictors based on the mixture $\xi$. We introduce the concept of Bayes optimal predictors $\Theta_\rho$ that minimize $\rho$-expected error. We bound $E^{\Theta_\xi}-E^{\Theta_\mu}$ by $O(\sqrt{E^{\Theta_\mu}})$, where $E^{\Theta_\xi}$ is the expected number of errors made by the optimal universal predictor $\Theta_\xi$, and $E^{\Theta_\mu}$ is the expected number of errors made by the optimal informed prediction scheme $\Theta_\mu$.

### 3.3.1 Bayes Optimal Predictors

We start with a very simple error measure: making a wrong prediction counts as one error, making a correct prediction counts as no error. This serves as an introduction to the more complicated model with arbitrary loss function. Let $\Theta_\mu$ be the optimal prediction scheme when the strings are drawn from the probability distribution $\mu$, i.e. the probability of $x_t$ given $x_{<t}$ is $\mu(x_t|x_{<t})$, and $\mu$ is known. $\Theta_\mu$ predicts (by definition) $x_t^{\Theta_\mu}$ when observing $x_{<t}$. The prediction is erroneous if the true $t^{th}$ symbol is not $x_t^{\Theta_\mu}$. The probability of this event is $1-\mu(x_t^{\Theta_\mu}|x_{<t})$. It is minimized if $x_t^{\Theta_\mu}$ maximizes $\mu(x_t^{\Theta_\mu}|x_{<t})$. More generally, let $\Theta_\rho$ be a prediction scheme predicting $x_t^{\Theta_\rho}:=\text{argmax}_{x_t}\rho(x_t|x_{<t})$ for some distribution $\rho$. Every deterministic predictor can be interpreted as maximizing some distribution.

### 3.3.2 Total Expected Numbers of Errors

The $\mu$-probability of making a wrong prediction for the $t^{th}$ symbol and the total $\mu$-expected number of errors in the first $n$ predictions of predictor $\Theta_\rho$ are

$$e_t^{\Theta_\rho}(x_{<t}) \;:=\; 1-\mu(x_t^{\Theta_\rho}|x_{<t}), \qquad E_n^{\Theta_\rho} \;:=\; \sum_{t=1}^n \mathbf{E}[e_t^{\Theta_\rho}(x_{<t})]. \tag{3.34}$$

If $\mu$ is known, $\Theta_\mu$ is obviously the best prediction scheme in the sense of making the least number of expected errors

$$E_n^{\Theta_\mu} \;\leq\; E_n^{\Theta_\rho} \quad \text{for any} \quad \Theta_\rho, \tag{3.35}$$

since

$$e_t^{\Theta_\mu}(x_{<t}) = 1-\mu(x_t^{\Theta_\mu}|x_{<t}) = \min_{x_t}\{1-\mu(x_t|x_{<t})\} \leq 1-\mu(x_t^{\Theta_\rho}|x_{<t}) = e_t^{\Theta_\rho}(x_{<t})$$

for any $\rho$. Of special interest is the universal predictor $\Theta_\xi$. As $\xi$ converges to $\mu$ the prediction of $\Theta_\xi$ might converge to the prediction of the optimal $\Theta_\mu$. Hence, $\Theta_\xi$ may not make many more errors than $\Theta_\mu$ and, hence, than any other predictor $\Theta_\rho$. Note that $x_t^{\Theta_\rho}$ is a discontinuous function of $\rho$, and $x_t^{\Theta_\xi}\to x_t^{\Theta_\mu}$ does not follow from $\xi\to\mu$. Indeed, this problem occurs in related prediction schemes, where the predictor has to be regularized so that it is continuous [FMG92]. Fortunately this is not necessary here. We prove the following error bound.

**Theorem 3.36 (Error bound)** Let there be sequences $x_1x_2...$ over a finite alphabet $\mathcal{X}$ drawn with probability $\mu(x_{1:n})$ for the first $n$ symbols. The $\Theta_\rho$-system predicts by definition $x_t^{\Theta_\rho}\in\mathcal{X}$ from $x_{<t}$, where $x_t^{\Theta_\rho}$ maximizes $\rho(x_t|x_{<t})$. $\Theta_\xi$ is the universal prediction scheme based on the universal prior $\xi$, and $\Theta_\mu$ the optimal informed prediction scheme. The total $\mu$-expected number of prediction errors $E_n^{\Theta_\xi}$ and $E_n^{\Theta_\mu}$ of $\Theta_\xi$ and $\Theta_\mu$ as defined in (3.34) is bounded in the following way[4]

$$\begin{aligned} 0 \;\leq\; E_n^{\Theta_\xi} - E_n^{\Theta_\mu} \;&\leq\; \sqrt{2(E_n^{\Theta_\xi}+E_n^{\Theta_\mu})S_n} \\ &\leq\; S_n + \sqrt{4E_n^{\Theta_\mu}S_n + S_n^2} \;\leq\; 2S_n + 2\sqrt{E_n^{\Theta_\mu}S_n}, \end{aligned}$$

where $S_n\leq D_n\leq \ln w_\mu^{-1}$. $S_n$ is the squared distance (3.14), $D_n$ is the relative entropy (3.18), and $w_\mu$ is the weight (3.5) of $\mu$ in $\xi$.

The first bound actually contains $E_n^{\Theta_\xi}$ on the r.h.s., so it is not particularly useful, but this is the major bound we will prove; the others follow easily. Furthermore, it has a somewhat nicer structure than the second bound. In Section 3.6 we show that the second bound is optimal. The last bound, which we discuss in the following, has the same asymptotics as the second bound.

First, we observe that the number of errors $E_\infty^{\Theta_\xi}$ of the universal $\Theta_\xi$ predictor is finite if the number of errors $E_\infty^{\Theta_\mu}$ of the informed $\Theta_\mu$ predictor is finite.

[4] Note that the error bound in terms of $S_n$ actually holds for any two distributions $\xi$ and $\mu$. The mixture property of $\xi$ is only used to bound $S_n$ by $\ln w_\mu^{-1}$.

This is especially the case for deterministic $\mu$, as $E_n^{\Theta_\mu}\equiv 0$ in this case[5], i.e. $\Theta_\xi$ makes only a finite number of errors on deterministic environments. This can also be proven by elementary means. Assume $x_1x_2...$ is the sequence generated by $\mu$ and $\Theta_\xi$ makes a wrong prediction $x_t^{\Theta_\xi}\neq x_t$. Since $\xi(x_t^{\Theta_\xi}|x_{<t})\geq\xi(x_t|x_{<t})$, this implies $\xi(x_t|x_{<t})\leq\frac{1}{2}$. Hence $e_t^{\Theta_\xi}=1\leq -\ln\xi(x_t|x_{<t})/\ln 2=d_t/\ln 2$. If $\Theta_\xi$ makes a correct prediction, $e_t^{\Theta_\xi}=0\leq d_t/\ln 2$ is obvious. Using (3.18) this proves $E_\infty^{\Theta_\xi}\leq D_\infty/\ln 2\leq\log_2 w_\mu^{-1}$. A combinatoric argument given in Section 3.6 shows that there are $\mathcal{M}$ and $\mu\in\mathcal{M}$ with $E_\infty^{\Theta_\xi}\geq\log_2|\mathcal{M}|$. This shows that the upper bound $E_\infty^{\Theta_\xi}\leq\log_2|\mathcal{M}|$ for uniform $w$ is sharp. From Theorem 3.36 we get the slightly weaker bound $E_\infty^{\Theta_\xi}\leq 2S_\infty\leq 2D_\infty\leq 2\ln w_\mu^{-1}$. For more complicated probabilistic environments, where even the ideal informed system makes an infinite number of errors, the theorem ensures that the error regret $E_n^{\Theta_\xi}-E_n^{\Theta_\mu}$ is only of order $\sqrt{E_n^{\Theta_\mu}}$. The regret is quantified in terms of the information content $D_n$ of $\mu$ (relative to $\xi$), or the weight $w_\mu$ of $\mu$ in $\xi$. This ensures that the error densities $E_n/n$ of both systems converge to each other. Actually, the theorem ensures more, namely that the quotient converges to 1, and also gives the speed of convergence $E_n^{\Theta_\xi}/E_n^{\Theta_\mu}=1+O((E_n^{\Theta_\mu})^{-1/2})\longrightarrow 1$ for $E_n^{\Theta_\mu}\to\infty$. If we increase the first occurrence of $E_n^{\Theta_\mu}$ in the theorem to $E_n^{\Theta}$ and the second $E_n^{\Theta_\mu}$ to $E_n^{\Theta_\xi}$ we get the bound $E_n^{\Theta}\geq E_n^{\Theta_\xi}-2\sqrt{E_n^{\Theta_\xi}S_n}$, which shows that *no* (causal) predictor $\Theta$ whatsoever makes significantly fewer errors than $\Theta_\xi$. In Section 3.6 we show that the second bound for $E_n^{\Theta_\xi}-E_n^{\Theta_\mu}$ given in Theorem 3.36 can, in general, not be improved, i.e. for every predictor $\Theta$ (and especially $\Theta_\xi$) there exist $\mathcal{M}$ and $\mu\in\mathcal{M}$ such that the upper bound is essentially achieved. See [Hut01c] for some further discussion and bounds for binary alphabet.

### 3.3.3 Proof of Theorem 3.36

The first inequality in Theorem 3.36 has already been proven (3.35). For the second inequality, let us start more modestly and consider constants $A>0$ and $B>0$ that satisfy the linear inequality

$$E_n^{\Theta_\xi}-E_n^{\Theta_\mu}\;\leq\; A(E_n^{\Theta_\xi}+E_n^{\Theta_\mu})+BS_n. \tag{3.37}$$

If we could show

$$e_t^{\Theta_\xi}(x_{<t})-e_t^{\Theta_\mu}(x_{<t})\;\leq\; A[e_t^{\Theta_\xi}(x_{<t})+e_t^{\Theta_\mu}(x_{<t})]+Bs_t(x_{<t}) \tag{3.38}$$

for all $t\leq n$ and all $x_{<t}$, (3.37) would follow immediately by summation and the definition of $E_n$ and $S_n$. With the abbreviations (3.12) and the abbreviations $m=x_t^{\Theta_\mu}$ and $s=x_t^{\Theta_\xi}$ the, various error functions can then be expressed by $e_t^{\Theta_\xi}=1-y_s$, $e_t^{\Theta_\mu}=1-y_m$ and $s_t=\sum_i(y_i-z_i)^2$. Inserting this into (3.38) we get

[5] Remember that we named a probability distribution *deterministic* if it is 1 for exactly one sequence and 0 for all others.

$$y_m - y_s \leq A[2-(y_m+y_s)] + B\sum_{i=1}^{N}(y_i - z_i)^2. \tag{3.39}$$

By definition of $x_t^{\Theta_\mu}$ and $x_t^{\Theta_\xi}$ we have $y_m \geq y_i$ and $z_s \geq z_i$ for all $i$. We prove a sequence of inequalities which show that

$$B\sum_{i=1}^{N}(y_i - z_i)^2 + A[2-(y_m+y_s)] - (y_m - y_s) \geq \ldots \tag{3.40}$$

is positive for suitable $A \geq 0$ and $B \geq 0$, which proves (3.39). For $m=s$ (3.40) is obviously positive. So we will assume $m \neq s$ in the following. From the square we keep only contributions from $i=m$ and $i=s$.

$$\ldots \geq B[(y_m - z_m)^2 + (y_s - z_s)^2] + A[2-(y_m+y_s)] - (y_m - y_s) \geq \ldots$$

By definition of $y$, $z$, $\mathcal{M}$ and $s$ we have the constraints $y_m + y_s \leq 1$, $z_m + z_s \leq 1$, $y_m \geq y_s \geq 0$ and $z_s \geq z_m \geq 0$. From the latter two it is easy to see that the square terms (as a function of $z_m$ and $z_s$) are minimized by $z_m = z_s = \frac{1}{2}(y_m + y_s)$. Furthermore, we define $x := y_m - y_s$ and increase $(y_m + y_s)$ to 1.

$$\ldots \geq \tfrac{1}{2}Bx^2 + A - x \geq \ldots \tag{3.41}$$

Expression (3.41) is quadratic in $x$ and minimized by $x^* = \frac{1}{B}$. Inserting $x^*$ gives

$$\ldots \geq A - \frac{1}{2B} \geq 0 \quad \text{for} \quad 2AB \geq 1. \tag{3.42}$$

Inequality (3.37) therefore holds for any $A > 0$, provided we insert $B = \frac{1}{2A}$. Thus we might minimize the r.h.s. of (3.37) w.r.t. $A$, leading to the upper bound

$$E_n^{\Theta_\xi} - E_n^{\Theta_\mu} \leq \sqrt{2(E_n^{\Theta_\xi} + E_n^{\Theta_\mu})S_n} \qquad \text{for} \qquad A^2 = \frac{S_n}{2(E_n^{\Theta_\xi} + E_n^{\Theta_\mu})}, \tag{3.43}$$

which is the first bound in Theorem 3.36. For the second bound we have to prove

$$\sqrt{2(E_n^{\Theta_\xi} + E_n^{\Theta_\mu})S_n} - S_n \leq \sqrt{4E_n^{\Theta_\mu}S_n + S_n^2}. \tag{3.44}$$

If we square both sides of this expression and simplify we just get (3.43). Hence, (3.43) implies (3.44). The last inequality in Theorem 3.36 is a simple triangle inequality. This completes the proof of Theorem 3.36. □

Note that also the third bound implies the second one:

$$\begin{aligned}
& E_n^{\Theta_\xi} - E_n^{\Theta_\mu} \leq \sqrt{2(E_n^{\Theta_\xi} + E_n^{\Theta_\mu})S_n} \\
\Leftrightarrow\quad & (E_n^{\Theta_\xi} - E_n^{\Theta_\mu})^2 \leq 2(E_n^{\Theta_\xi} + E_n^{\Theta_\mu})S_n \\
\Leftrightarrow\quad & (E_n^{\Theta_\xi} - E_n^{\Theta_\mu} - S_n)^2 \leq 4E_n^{\Theta_\mu}S_n + S_n^2 \\
\Leftrightarrow\quad & E_n^{\Theta_\xi} - E_n^{\Theta_\mu} - S_n \leq \sqrt{4E_n^{\Theta_\mu}S_n + S_n^2}
\end{aligned}$$

where we only have used $E_n^{\Theta_\xi} \geq E_n^{\Theta_\mu}$. Nevertheless, the bounds are not equal. In Section 3.9 we give an alternative direct proof of the second bound.

## 3.4 Loss Bounds

We now generalize the prediction framework to an arbitrary loss functions. A system is allowed to take an action $y_t$, given $x_1...x_{t-1}$ and receives loss $\ell_{x_t y_t}$ if $x_t$ is the next symbol of the sequence. No assumptions on $\ell$ are necessary, besides boundedness. Bayes optimal universal $\Lambda_\xi$ and Bayes optimal informed $\Lambda_\mu$ prediction schemes are defined and the total loss of $\Lambda_\xi$ is bounded in terms of the total loss of $\Lambda_\mu$, similar to the error bounds. Convergence of instantaneous losses is also established. Various example loss functions are discussed.

### 3.4.1 Unit Loss Function

A prediction is very often the basis for some decision. The decision results in an action, which itself leads to some reward or loss. If the action itself can influence the environment we enter the domain of acting agents, which will be analyzed in the context of universal probability in later chapters. To stay in the framework of (passive) prediction we have to assume that the action itself does not influence the environment. Let $\ell_{x_t y_t} \in I\!R$ be the received loss when taking action $y_t \in \mathcal{Y}$, and $x_t \in \mathcal{X}$ is the $t^{th}$ symbol of the sequence. We assume that $\ell$ is bounded, which is trivially satisfied for finite $\mathcal{X}$ and $\mathcal{Y}$. Without loss of generality we normalize $\ell$ by linear scaling such that $0 \leq \ell_{x_t y_t} \leq 1$. For instance, if we make a sequence of weather forecasts $\mathcal{X} = \{$*sunny*, *rainy*$\}$ and base our decision, whether to take an umbrella or wear sunglasses $\mathcal{Y} = \{$*umbrella*, *sunglasses*$\}$ on it, the action of taking the umbrella or wearing sunglasses does not influence the future weather (ignoring the butterfly effect). The losses might be

| Loss | *sunny* | *rainy* |
|---|---|---|
| *umbrella* | 0.1 | 0.3 |
| *sunglasses* | 0.0 | 1.0 |

Note the loss assignment even when making the right decision to take an umbrella when it rains because sun is still preferable to rain.

In many cases the prediction of $x_t$ can be identified or is already the action $y_t$. The forecast *sunny* can be identified with the action *wear sunglasses*, and *rainy* with *take umbrella*. $\mathcal{X} \equiv \mathcal{Y}$ in these cases. The error assignment of the previous section falls into this class together with a special loss function. It assigns unit loss to an erroneous prediction ($\ell_{x_t y_t} = 1$ for $x_t \neq y_t$) and no loss to a correct prediction ($\ell_{x_t x_t} = 0$).

For convenience we name an action a prediction in the following, even if $\mathcal{X} \neq \mathcal{Y}$. The true probability of the next symbol being $x_t$, given $x_{<t}$, is $\mu(x_t|x_{<t})$. The expected loss when predicting $y_t$ is $\mathbf{E}_t[\ell_{x_t y_t}]$. The goal is to minimize the expected loss. More generally, we define the $\Lambda_\rho$ prediction scheme

$$y_t^{\Lambda_\rho} := \arg\min_{y_t\in\mathcal{Y}} \sum_{x_t} \rho(x_t|x_{<t})\ell_{x_t y_t}, \tag{3.45}$$

which minimizes the $\rho$-expected loss.[6] As the true distribution is $\mu$, the actual $\mu$-expected loss when $\Lambda_\rho$ predicts the $t^{th}$ symbol and the total $\mu$-expected loss in the first $n$ predictions are

$$l_t^{\Lambda_\rho}(x_{<t}) := \mathbf{E}_t[\ell_{x_t y_t^{\Lambda_\rho}}], \qquad L_n^{\Lambda_\rho} := \sum_{t=1}^n \mathbf{E}[l_t^{\Lambda_\rho}(x_{<t})]. \tag{3.46}$$

Let $\Lambda$ be *any* (causal) prediction scheme (deterministic or probabilistic) with no constraint at all, predicting *any* $y_t^\Lambda \in \mathcal{Y}$ with losses $l_t^\Lambda$ and $L_n^\Lambda$ similarly defined as (3.46). If $\mu$ is known, $\Lambda_\mu$ is obviously the best prediction scheme in the sense of achieving minimal expected loss

$$L_n^{\Lambda_\mu} \leq L_n^\Lambda \quad \text{for any} \quad \Lambda, \tag{3.47}$$

since

$$l_t^{\Lambda_\mu}(x_{<t}) = \mathbf{E}_t \ell_{x_t y_t^{\Lambda_\mu}} = \min_{y_t} \mathbf{E}_t \ell_{x_t y_t} \leq \mathbf{E}_t \ell_{x_t y_t^\Lambda} = l_t^\Lambda(x_{<t})$$

for any $\Lambda$. The predictor $\Lambda_\xi$, based on the universal distribution $\xi$, is, again, of special interest. Theorem 3.36 generalizes to arbitrary loss functions.

**Theorem 3.48 (Unit loss bound)** Let there be sequences $x_1x_2...$ over a finite alphabet $\mathcal{X}$ drawn with probability $\mu(x_{1:n})$ for the first $n$ symbols. A system taking action (or predicting) $y_t \in \mathcal{Y}$ given $x_{<t}$ receives loss $\ell_{x_t y_t} \in [0,1]$ if $x_t$ is the true $t^{th}$ symbol of the sequence. The $\Lambda_\rho$-system (3.45) acts (or predicts) as to minimize the $\rho$-expected loss. $\Lambda_\xi$ is the universal prediction scheme based on the universal prior $\xi$, and $\Lambda_\mu$ the optimal informed prediction scheme. The total $\mu$-expected losses $L_n^{\Lambda_\xi}$ of $\Lambda_\xi$ and $L_n^{\Lambda_\mu}$ of $\Lambda_\mu$ as defined in (3.46) are bounded in the following way:

$$0 \leq L_n^{\Lambda_\xi} - L_n^{\Lambda_\mu} \leq D_n + \sqrt{4L_n^{\Lambda_\mu} D_n + D_n^2} \leq 2D_n + 2\sqrt{L_n^{\Lambda_\mu} D_n},$$

where $D_n \leq \ln w_\mu^{-1}$ is the relative entropy (3.18), and $w_\mu$ is the weight (3.5) of $\mu$ in $\xi$.

The loss bounds have the same form as the error bounds when substituting $S_n \leq D_n$ in Theorem 3.36, so most of the discussion of Theorem 3.36 also applies here. Replacing $D_n$ by $S_n$ in Theorem 3.48 gives an invalid bound, so the general bound is slightly weaker. For instance, for $\mathcal{X} = \mathcal{Y} = \{0,1\}$, $\ell_{00} =$

[6] $\text{argmin}_y(\cdot)$ is defined as the $y$ that minimizes the argument. A tie is broken arbitrarily. If $\mathcal{Y}$ is finite, then $y_t^{\Lambda_\rho}$ always exists. For an infinite action space $\mathcal{Y}$ we assume that a minimizing $y_t^{\Lambda_\rho} \in \mathcal{Y}$ exists. This is, for instance, the case if $\mathcal{Y}$ is compact and $\ell_{xy}$ is continuous in $y$, or for $\mathcal{Y} = I\!N$, if $\lim_{y\to\infty} \ell_{xy}$ exists for all $x$ and is larger or equal to $\ell_{xy}$ for most $y$.

$\ell_{11}=0$, $\ell_{10}=1$, $\ell_{01}=c<\frac{1}{4}$, $\mu(1)=0$, $\nu(1)=2c$, and $w_\mu=w_\nu=\frac{1}{2}$, we get $\xi(1)=c$, $s_1=2c^2$, $y_1^{\Lambda_\mu}=0$, $l_1^{\Lambda_\mu}=\ell_{00}=0$, $y_1^{\Lambda_\xi}=1$, $l_1^{\Lambda_\xi}=\ell_{01}=c$; hence $L_1^{\Lambda_\xi}-L_1^{\Lambda_\mu}=c\not\leq 4c^2=2S_1+2\sqrt{L_1^{\Lambda_\mu}S_1}$. For convenience we collect the most important consequences of Theorem 3.48 in the following corollary.

**Corollary 3.49 (Unit loss bound)** Under the same conditions as in Theorem 3.48 the following relations hold:

$$\begin{aligned}
&(i)\ L_\infty^{\Lambda_\xi} \text{ is finite} \iff L_\infty^{\Lambda_\mu} \text{ is finite,}\\
&(ii)\ L_\infty^{\Lambda_\xi} \leq 2D_\infty \leq 2\ln w_\mu^{-1} \text{ for deterministic } \mu \text{ if } \forall x\exists y \ell_{xy}=0,\\
&(iii)\ L_n^{\Lambda_\xi}/L_n^{\Lambda_\mu} = 1+O((L_n^{\Lambda_\mu})^{-1/2}) \longrightarrow 1 \text{ for } L_n^{\Lambda_\mu}\to\infty.\\
&(iv)\ L_n^{\Lambda_\xi}-L_n^{\Lambda_\mu} = O(\sqrt{L_n^{\Lambda_\mu}}),
\end{aligned}$$

Let $\Lambda$ be *any* prediction scheme.

$$\begin{aligned}
&(v)\ L_n^{\Lambda_\mu} \leq L_n^{\Lambda}, \qquad l_t^{\Lambda_\mu}(x_{<t}) \leq l_t^{\Lambda}(x_{<t}),\\
&(vi)\ L_n^{\Lambda} \geq L_n^{\Lambda_\xi} - 2\sqrt{L_n^{\Lambda_\xi}D_n},\\
&(vii)\ L_n^{\Lambda_\xi}/L_n^{\Lambda} \leq 1+O((L_n^{\Lambda})^{-1/2}).
\end{aligned}$$

### 3.4.2 Loss Bound of Merhav & Feder

The first general loss bound with no structural assumptions on $\mu$ and $\ell$ (except boundedness) was derived in a survey paper by Merhav and Feder in [MF98, Sec.III.A.2]. They showed that the regret $L_n^{\Lambda_\xi}-L_n^{\Lambda_\mu}$ is bounded by $\ell_{max}\sqrt{2nD_n}$ for $\ell\in[0,\ell_{max}]$. Assuming $\ell_{max}=1$ (general $\ell_{max}$ can be recovered by scaling) their bound reads (in our notation)

$$L_n^{\Lambda_\xi}-L_n^{\Lambda_\mu} \leq A_n \leq \sqrt{2nD_n}. \tag{3.50}$$

Later in Theorem 3.59 we prove

$$l_t^{\Lambda_\xi}(x_{<t})-l_t^{\Lambda_\mu}(x_{<t}) \leq a_t(x_{<t}) \leq \sqrt{2d_t(x_{<t})}.$$

Taking the expectation $\mathbf{E}$ and the average $\frac{1}{n}\sum_{t=1}^n$ and using Jensen's inequality for the concave square root similarly to (3.21) or directly Theorem 3.19$(vi)$ shows (3.50).

Bound (3.50) and our bound (Theorem 3.48) are in general incomparable. Since $2D_\infty$ is finite and $L_n^{\Lambda_\mu}\leq n$, bound (3.50) can be at best a factor $\sqrt{2}$ and an additive constant better than our bound. On the other hand, for large $n$ and for $L_n^{\Lambda_\mu}<\frac{n}{2}$ our bound is tighter. The latter condition is satisfied if the best predictor $\Lambda_\mu$ suffers small instantaneous loss $<\frac{1}{2}$ on average. Significant improvement occurs if $L_n^{\Lambda_\mu}$ does not grow linearly with $n$, but is, for instance, finite (see Corollary 3.49, especially $(i)$ and $(ii)$).

### 3.4.3 Example Loss Functions

The case $\mathcal{X}\equiv\mathcal{Y}$ with unit error assignment $\ell_{xy}=1-\delta_{xy}$ ($\delta_{xy}=1$ for $x=y$ and $\delta_{xy}=0$ for $x\neq y$) has already been discussed and proven in Section 3.3.

$$y_t^{\Lambda_\rho} \;=\; \arg\min_{y_t}\sum_{x_t}\rho(x_t|x_{<t})(1-\delta_{x_t y_t}) \;=\; \arg\max_{x_t}\rho(x_t|x_{<t}) \;=\; x_t^{\Theta_\rho}$$

In this case $L_n^{\Lambda_\rho}\equiv E_n^{\Theta_\rho}$ is the total expected number of prediction errors. For $\mathcal{X}=\mathcal{Y}=\{0,1\}$, like in the weather example above, $\Lambda_\rho$ is a threshold strategy with $y_t^{\Lambda_\rho}=\mathrm{argmin}_{y\in\{0,1\}}\{\rho_1\ell_{1y}+\rho_0\ell_{0y}\}=0/1$ for $\rho_1\gtrless\gamma$, where $\gamma:=\frac{\ell_{01}-\ell_{00}}{\ell_{01}-\ell_{00}+\ell_{10}-\ell_{11}}$ and $\rho_i=\rho(i|x_{<t})$. In the special error case $\ell_{xy}=1-\delta_{xy}$, the bit with the highest $\rho$-probability is predicted ($\gamma=\frac{1}{2}$). In the following we consider some standard loss functions for binary outcome $\mathcal{X}=\{0,1\}$ and continuous action $y$ in the unit interval $\mathcal{Y}=[0,1]$. The *absolute loss* is defined as $\ell_{xy}=|x-y|\in[0,1]$. The $\Lambda_\rho$ scheme predicts $y_t^{\Lambda_\rho}=\mathrm{argmin}_{y\in[0,1]}\{\rho_1(1-y)+\rho_0 y\}=0/1$ for $\rho_0\gtrless\rho_1$. Since all predictions $y$ lie in the subset $\{0,1\}\subset[0,1]$, and $|x-y|=1-\delta_{xy}$ for $y\in\{0,1\}$, this case coincides with the binary error case above. The same holds for the $\alpha$-loss $|x-y|^\alpha$ with $0<\alpha\leq 1$. The $\mu$-expected loss is $l_t^{\Lambda_\rho}=\mu(i|x_{<t})$ for the $i$ with $\rho_i>\frac{1}{2}$. For the *quadratic loss* $\ell_{xy}=(x-y)^2\in[0,1]$ the action/prediction $y_t^{\Lambda_\rho}=\mathrm{argmin}_{y\in[0,1]}\{\rho_1(1-y)^2+\rho_0 y^2\}=\rho_1$ is proportional to the $\rho$-probability of $x_t=1$ and $l_t^{\Lambda_\rho}=\mathbf{E}_t(1-\rho(x_t|x_{<t}))^2$. For the $\alpha$-loss $|x-y|^\alpha$ with $\alpha>1$ we get $y_t^{\Lambda_\rho}=(1+\sqrt[\alpha-1]{\rho_0/\rho_1})^{-1}$. For arbitrary finite alphabet $\mathcal{X}$ and vector-valued predictions $\boldsymbol{y}$ the quadratic loss may be generalized to $\ell_{x\boldsymbol{y}}=\frac{1}{2}\boldsymbol{y}^T\mathbf{A}_x\boldsymbol{y}+\boldsymbol{b}_x^T\boldsymbol{y}+c_x$. The *Hellinger loss* can be written for binary outcome in the form $\ell_{xy}=1-\sqrt{|1-x-y|}\in[0,1]$ with $y_t^{\Lambda_\rho}=\rho_1^2/(\rho_0^2+\rho_1^2)$ and $l_t^{\Lambda_\rho}=1-(\mu_0\rho_0+\mu_1\rho_1)/\sqrt{\rho_0^2+\rho_1^2}$. The *logarithmic loss* $\ell_{xy}=-\ln|1-x-y|\in[0,\infty]$ is unbounded. But since the corresponding action is $y_t^{\Lambda_\rho}=\rho_1$ the expected loss is $l_t^{\Lambda_\rho}=-\mathbf{E}_t\ln\rho(x_t|x_{<t})$. Hence $l_t^{\Lambda_\xi}-l_t^{\Lambda_\mu}=d_t$, and the total loss regret $L_n^{\Lambda_\xi}-L_n^{\Lambda_\mu}=D_n\leq\ln w_\mu^{-1}$ is finite anyway and Theorem 3.48 is not needed. Continuous outcome spaces $\mathcal{X}$ are briefly discussed in Section 3.7.5.

### 3.4.4 Proof of Theorem 3.48

The first inequality in Theorem 3.48 was already proven in (3.47). For the second and last inequality, we start, as in Theorem 3.36, by looking for small constants $A>0$ and $B>0$, which satisfy the linear inequality

$$L_n^{\Lambda_\xi} \;\leq\; (A+1)L_n^{\Lambda_\mu}+(B+1)D_n. \tag{3.51}$$

If we could show

$$l_t^{\Lambda_\xi}(x_{<t}) \;\leq\; A' l_t^{\Lambda_\mu}(x_{<t})+B' d_t(x_{<t}), \quad A':=A+1, \quad B':=B+1 \tag{3.52}$$

for all $t\leq n$ and all $x_{<t}$, (3.51) would follow immediately by summation and the definition of $L_n$ and $D_n$. With the abbreviations $m=y_t^{\Lambda_\mu}$ and $s=y_t^{\Lambda_\xi}$

and the abbreviations (3.12) the loss and entropy can then be expressed by $l_t^{\Lambda_\xi}=\sum_i y_i\ell_{is}$, $l_t^{\Lambda_\mu}=\sum_i y_i\ell_{im}$ and $d_t=\sum_i y_i \ln\frac{y_i}{z_i}$. Inserting this into (3.52) we get

$$\sum_{i=1}^N y_i\ell_{is} \;\leq\; A'\sum_{i=1}^N y_i\ell_{im} + B'\sum_{i=1}^N y_i \ln\frac{y_i}{z_i} \tag{3.53}$$

By definition (3.45) of $y_t^{\Lambda_\mu}$ and $y_t^{\Lambda_\xi}$ we have

$$\sum_i y_i\ell_{im} \leq \sum_i y_i\ell_{ij} \quad \text{and} \quad \sum_i z_i\ell_{is} \leq \sum_i z_i\ell_{ij} \quad \text{for all} \quad j. \tag{3.54}$$

Actually, we need the first constraint only for $j=s$ and the second for $j=m$. In Section 3.9 we reduce the problem to the binary $N=2$ case, which we will consider in the following. We take $\sum_{i=0}^1$ instead of $\sum_{i=1}^2$ for convenience.

$$B'\sum_{i=0}^1 y_i \ln\frac{y_i}{z_i} + \sum_{i=0}^1 y_i(A'\ell_{im}-\ell_{is}) \;\stackrel{?}{\geq}\; 0 \tag{3.55}$$

The cases $\ell_{im}>\ell_{is}\forall i$ and $\ell_{is}>\ell_{im}\forall i$ contradict the first/second inequality (3.54). Hence we can assume $\ell_{0m}\geq\ell_{0s}$ and $\ell_{1m}\leq\ell_{1s}$. The symmetric case $\ell_{0m}\leq\ell_{0s}$ and $\ell_{1m}\geq\ell_{1s}$ is proven analogously or can be reduced to the first case by renumbering the indices $(0\leftrightarrow 1)$. Using the abbreviations $a:=\ell_{0m}-\ell_{0s}$, $b:=\ell_{1s}-\ell_{1m}$, $c:=y_1\ell_{1m}+y_0\ell_{0s}$, $y=y_1=1-y_0$ and $z=z_1=1-z_0$, we can write (3.55) as

$$f(y,z) \;:=\; B'[y\ln\tfrac{y}{z}+(1-y)\ln\tfrac{1-y}{1-z}] + A'(1-y)a - yb + Ac \;\stackrel{?}{\geq}\; 0 \tag{3.56}$$

for $zb\leq(1-z)a$ and $0\leq a,b,c,y,z\leq 1$. Constraint (3.54) on $y$ has been dropped since (3.56) will turn out to be true for all $y$. Furthermore, we can assume that $d:=A'(1-y)a-yb\leq 0$ since for $d>0$, $f$ is trivially positive. Multiplying $d$ with a constant $\geq 1$ will decrease $f$. Let us first consider the case $z\leq\frac{1}{2}$. We multiply the $d$ term by $1/b\geq 1$, i.e. replace it with $A'(1-y)\frac{a}{b}-y$. From the constraint on $z$ we know that $\frac{a}{b}\geq\frac{z}{1-z}$. We can decrease $f$ further by replacing $\frac{a}{b}$ by $\frac{z}{1-z}$ and by dropping $Ac$. Hence, (3.56) is proven for $z\leq\frac{1}{2}$ if we can prove

$$B'[y\ln\tfrac{y}{z}+(1-y)\ln\tfrac{1-y}{1-z}] + A'(1-y)\tfrac{z}{1-z} - y \;\stackrel{?}{\geq}\; 0 \quad \text{for} \quad z\leq\tfrac{1}{2}. \tag{3.57}$$

In Section 3.9 we prove that it holds for $B\geq\frac{1}{A}+1$; the case $z\geq\frac{1}{2}$ is treated similarly. We scale $d$ with $1/a\geq 1$, i.e. replace it with $A'(1-y)-y\frac{b}{a}$. From the constraint on $z$ we know that $\frac{b}{a}\leq\frac{1-z}{z}$. We decrease $f$ further by replacing $\frac{b}{a}$ by $\frac{1-z}{z}$ and by dropping $Ac$. Hence (3.56) is proven for $z\geq\frac{1}{2}$ if we can prove

$$B'[y\ln\tfrac{y}{z}+(1-y)\ln\tfrac{1-y}{1-z}] + A'(1-y) - y\tfrac{1-z}{z} \;\stackrel{?}{\geq}\; 0 \quad \text{for} \quad z\geq\tfrac{1}{2}. \tag{3.58}$$

In Section 3.9 we prove that it holds for $B \geq \frac{1}{A}+1$. So, in summary we proved that (3.51) holds for $B \geq \frac{1}{A}+1$. Inserting $B=\frac{1}{A}+1$ into (3.51) and minimizing the r.h.s. w.r.t. $A$ leads to the last bound of Theorem 3.48 with $A=\sqrt{D_n/L_n^{\Lambda_\mu}}$. Actually, inequalities (3.57) and (3.58) also hold for $B \geq \frac{1}{4}A+\frac{1}{A}$, which, by the same minimization argument, proves the slightly tighter bound in Theorem 3.48. Unfortunately, the current proof is very long and complex, and involves some numerical or graphical analysis for determining intersection properties of some higher-order polynomials. This or a simplified proof will be postponed. The cautious reader may check the inequalities (3.57) and (3.58) numerically for $B=\frac{1}{4}A+\frac{1}{A}$. □

### 3.4.5 Convergence of Instantaneous Losses

Since $L_n^{\Lambda_\xi}-L_n^{\Lambda_\mu}$ is not finitely bounded by Theorem 3.48, it cannot be used directly to conclude $l_t^{\Lambda_\xi}-l_t^{\Lambda_\mu} \to 0$. It would follow from $\xi \to \mu$ by continuity if $l_t^{\Lambda_\xi}$ and $l_t^{\Lambda_\mu}$ were continuous functions of $\xi$ and $\mu$. $l_t^{\Lambda_\mu}$ is a continuous piecewise linear concave function of $\mu$, but $l_t^{\Lambda_\xi}$ is, in general, a discontinuous function of $\xi$ (and $\mu$). Fortunately, it is continuous at the one necessary point, $\xi=\mu$. This allows us to bound $l_t^{\Lambda_\xi}-l_t^{\Lambda_\mu}$ in terms of $\xi(x_t|x_{<t})-\mu(x_t|x_{<t})$.

**Theorem 3.59 (Instantaneous loss bound)** Under the same conditions as in Theorem 3.48, the following relations hold for the instantaneous losses $l_t^{\Lambda_\mu}(x_{<t})$ and $l_t^{\Lambda_\xi}(x_{<t})$ at time $t$ of the informed and universal prediction schemes $\Lambda_\mu$ and $\Lambda_\xi$:

$$(i)\ \sum_{t=1}^{n} \mathbf{E}[(l_t^{\Lambda_\xi}(x_{<t})-l_t^{\Lambda_\mu}(x_{<t}))^2] \leq 2D_n \leq 2\ln w_\mu^{-1} < \infty.$$

$$(ii)\ 0 \leq l_t^{\Lambda_\xi}(x_{<t})-l_t^{\Lambda_\mu}(x_{<t}) \leq \sum_{x_t} |\xi(x_t|x_{<t})-\mu(x_t|x_{<t})|$$

$$\leq \sqrt{2d_t(x_{<t})} \to 0 \quad \text{for} \quad t \to \infty \quad \text{w.}\mu\text{.p.1.}$$

$$(iii)\ 0 \leq l_t^{\Lambda_\xi}(x_{<t})-l_t^{\Lambda_\mu}(x_{<t}) \leq 2d_t(x_{<t})+2\sqrt{l_t^{\Lambda_\mu}(x_{<t})\,d_t(x_{<t})} \xrightarrow[w.\mu.p.1]{t\to\infty} 0.$$

Bound $(i)$ implies that the expected number of times $t$ in which $l_t^{\Lambda_\xi}$ exceeds $l_t^{\Lambda_\mu}$ by more than $\varepsilon$ is finite and bounded by $2\varepsilon^{-2}\ln w_\mu^{-1}$, and the probability that the number of these events exceeds $2\varepsilon^{-2}\delta^{-1}\ln w_\mu^{-1}$ is smaller than $\delta$.

**Proof.** $(ii)$ follows from

$$l_t^{\Lambda_\xi}(x_{<t})-l_t^{\Lambda_\mu}(x_{<t}) \equiv \sum_i y_i\ell_{is} - \sum_i y_i\ell_{im} \leq \sum_i (y_i-z_i)(\ell_{is}-\ell_{im})$$

$$\leq \sum_i |y_i-z_i|\cdot|\ell_{is}-\ell_{im}| \leq \sum_i |y_i-z_i| \leq \sqrt{2\sum_i y_i \ln\frac{y_i}{z_i}} \equiv \sqrt{2d_t(x_{<t})}.$$

To arrive at the first inequality we added $\sum_i z_i(\ell_{im}-\ell_{is})$, which is positive due to (3.54). $|\ell_{is}-\ell_{im}|\leq 1$ since $\ell\in[0,1]$. The last inequality follows from Lemma 3.11$a$, and $d_t\to 0$ was proven in Theorem 3.19($ii$).

($i$) follows by squaring ($ii$), taking the expectation $\mathbf{E}$ and sum $\sum_{t=1}^n$, and using (3.18).

($iii$) follows from the proof of Theorem 3.48 by inserting $B=\frac{1}{A}+1=\sqrt{l_t^{\Lambda_\mu}/d_t}+1$ into (3.52). Convergence to zero holds with $\mu$-probability 1, since $l_t^{\Lambda_\mu}\leq 1$ is bounded. The losses $l_t^{\Lambda_\rho}(x_{<t})$ itself need not converge. □

Note, that the inequalities in ($ii$) and ($iii$) hold for all individual sequences. The sum/average is only taken over the current outcome $x_t$, but the history $x_{<t}$ is fixed. Bounds ($ii$) and ($iii$) are in general incomparable, but for large $t$ and for $l_t^{\Lambda_\mu}<\frac{1}{2}$ (especially if $l_t^{\Lambda_\mu}\to 0$) bound ($iii$) is tighter than bound ($ii$).

### 3.4.6 General Loss

Very few restrictions were imposed on the loss $\ell_{x_t y_t}$ in Theorem 3.48, namely that it is static and in the unit interval $[0,1]$. If we look at the proof of Theorem 3.48, we see that the time-independence has not been used at all. The proof is still valid for an individual loss function $\ell^t_{x_t y_t}\in[0,1]$ for each step $t$. The loss might even depend on the actual history $x_{<t}$. The case of a loss $\ell^t_{x_t y_t}(x_{<t})$ bounded to a general interval $[\ell_{min},\ell_{max}]$ can be reduced to the unit interval case by rescaling $\ell$. We introduce a scaled loss $\ell'$

$$0\leq \ell'^t_{x_t y_t}(x_{<t}):=\frac{\ell^t_{x_t y_t}(x_{<t})-\ell_{min}}{\ell_\Delta}\leq 1, \quad \text{where} \quad \ell_\Delta:=\ell_{max}-\ell_{min}.$$

The prediction scheme $\Lambda'_\rho$ based on $\ell'$ is identical to the original prediction scheme $\Lambda_\rho$ based on $\ell$, since argmin in (3.45) is not affected by linear transformation of its argument. From $y_t^{\Lambda'_\rho}=y_t^{\Lambda_\rho}$ it follows that $l_t'^{\Lambda_\rho}=(l_t^{\Lambda_\rho}-\ell_{min})/\ell_\Delta$ and $L_n'^{\Lambda_\rho}=(L_n^{\Lambda_\rho}-\ell_{min})/\ell_\Delta$ ($D'_n\equiv D_n$, since $\ell$ is not involved). Theorem 3.48 is valid for the primed quantities, since $\ell'\in[0,1]$. Inserting $L_n'^{\Lambda_{\mu/\xi}}$ and rearranging terms we get

**Theorem 3.60 (General loss bound)** Let there be sequences $x_1x_2...$ over a finite alphabet $\mathcal{X}$ drawn with probability $\mu(x_{1:n})$ for the first $n$ symbols. A system taking action (or predicting) $y_t \in \mathcal{Y}$ given $x_{<t}$ receives loss $\ell_{x_t y_t}^t(x_{<t}) \in [\ell_{min}, \ell_{min}+\ell_\Delta]$ if $x_t$ is the true $t^{th}$ symbol of the sequence. The $\Lambda_\rho$-system (3.45) acts (or predicts) as to minimize the $\rho$-expected loss. $\Lambda_\xi$ is the universal prediction scheme based on the universal prior $\xi$. $\Lambda_\mu$ is the optimal informed prediction scheme. The total $\mu$-expected losses $L_n^{\Lambda_\xi}$ and $L_n^{\Lambda_\mu}$ of $\Lambda_\xi$ and $\Lambda_\mu$ as defined in (3.46) are bounded in the following way

$$0 \leq L_n^{\Lambda_\xi} - L_n^{\Lambda_\mu} \leq \ell_\Delta D_n + \sqrt{4(L_n^{\Lambda_\mu} - n\ell_{min})\ell_\Delta D_n + \ell_\Delta^2 D_n^2}$$

where $D_n \leq \ln w_\mu^{-1}$ is the relative entropy (3.18), and $w_\mu$ is the weight (3.5) of $\mu$ in $\xi$.

## 3.5 Application to Games of Chance

This section applies the loss bounds to games of chance, defined as a sequence of bets, observations, and rewards. After a brief introduction, we show that if there is a profitable scheme at all, asymptotically the universal $\Lambda_\xi$ scheme will also become profitable. We bound the time needed to reach the winning zone. It is proportional to the relative entropy of $\mu$ and $\xi$ with a factor depending on the profit range and the average profit. We present a numerical example and attempt to give an information-theoretic interpretation of the result.

### 3.5.1 Introduction

Consider investing in the stock market. At time $t$ an amount of money $s_t$ is invested in portfolio $y_t$, where we have access to past knowledge $x_{<t}$ (e.g. charts). After our choice of investment we receive new information $x_t$, and the new portfolio value is $r_t$. The best we can expect is to have a probabilistic model $\mu$ of the behavior of the stock market. The goal is to maximize the net $\mu$-expected profit $p_t = r_t - s_t$. Nobody knows $\mu$, but the assumption of all traders is that there *is* a computable, profitable $\mu$ they try to find or approximate. From Theorem 3.19 we know that Solomonoff-Levin's universal prior $\xi(x_t|x_{<t})$ converges to any computable $\mu(x_t|x_{<t})$ with probability 1. If there is a computable, asymptotically profitable trading scheme at all, the $\Lambda_\xi$ scheme should also be profitable in the long run. To get a practically useful, computable scheme we have to restrict $\mathcal{M}$ to a finite set of computable distributions, e.g. with bounded Levin complexity $Kt$ [LV97, Sec.7.5]. Although convergence of $\xi$ to $\mu$ is pleasing, what we are really interested in is whether $\Lambda_\xi$ is asymptotically profitable and how long it takes to become profitable. This will be explored in the following.

### 3.5.2 Games of Chance

We use Theorem 3.60 to estimate the time needed to reach the winning threshold when using $\Lambda_\xi$ in a game of chance. We assume a game (or a sequence of possibly correlated games) that allows a sequence of bets and observations. In step $t$ we bet, depending on the history $x_{<t}$, a certain amount of money $s_t$, take some action $y_t$, observe outcome $x_t$, and receive reward $r_t$. Our profit, which we want to maximize, is $p_t = r_t - s_t \in [p_{min}, p_{max}]$, where $[p_{min}, p_{max}]$ is the [minimal,maximal] profit per round, and $p_\Delta := p_{max} - p_{min}$ is the profit range. The loss, which we want to minimize, can be defined as the negative profit, $\ell_{x_t y_t} = -p_t$. The probability of outcome $x_t$, possibly depending on the history $x_{<t}$, is $\mu(x_t|x_{<t})$. The total $\mu$-expected profit when using scheme $\Lambda_\rho$ is $P_n^{\Lambda_\rho} = -L_n^{\Lambda_\rho}$. If we knew $\mu$, the optimal strategy to maximize our expected profit is just $\Lambda_\mu$. We assume $P_n^{\Lambda_\mu} > 0$ (otherwise there is no winning strategy at all, since $P_n^{\Lambda_\mu} \geq P_n^{\Lambda_\rho}\, \forall\rho$). Often we are not in the favorable position of knowing $\mu$, but we know (or assume) that $\mu \in \mathcal{M}$ for some $\mathcal{M}$, for instance, that $\mu$ is a computable probability distribution. From Theorem 3.60 we see that the average profit per round $\bar{p}_n^{\Lambda_\xi} := \frac{1}{n} P_n^{\Lambda_\xi}$ of the universal $\Lambda_\xi$ scheme converges to the average profit per round $\bar{p}_n^{\Lambda_\mu} := \frac{1}{n} P_n^{\Lambda_\mu}$ of the optimal informed scheme, i.e. asymptotically we can make the same money even without knowing $\mu$, by just using the universal $\Lambda_\xi$ scheme. Theorem 3.60 allows us to lower-bound the universal profit $P_n^{\Lambda_\xi}$

$$P_n^{\Lambda_\xi} \;\geq\; P_n^{\Lambda_\mu} - p_\Delta D_n - \sqrt{4(np_{max} - P_n^{\Lambda_\mu})p_\Delta D_n + p_\Delta^2 D_n^2}. \tag{3.61}$$

The time needed for $\Lambda_\xi$ to perform well can also be estimated. An interesting quantity is the expected number of rounds needed to reach the winning zone. Using $P_n^{\Lambda_\mu} > 0$ one can show that the r.h.s. of (3.61) is positive if and only if

$$n \;>\; \frac{2p_\Delta(2p_{max} - \bar{p}_n^{\Lambda_\mu})}{(\bar{p}_n^{\Lambda_\mu})^2} \cdot D_n. \tag{3.62}$$

**Theorem 3.63 (Time to win)** Let there be sequences $x_1 x_2 ...$ over a finite alphabet $\mathcal{X}$ drawn with probability $\mu(x_{1:n})$ for the first $n$ symbols. In step $t$ we make a bet, depending on the history $x_{<t}$, take some action $y_t$, and observe outcome $x_t$. Our net profit is $p_t \in [p_{max} - p_\Delta, p_{max}]$. The $\Lambda_\rho$-system (3.45) acts as to maximize the $\rho$-expected profit. $P_n^{\Lambda_\rho}$ is the total and $\bar{p}_n^{\Lambda_\rho} = \frac{1}{n} P_n^{\Lambda_\rho}$ is the average expected profit of the first $n$ rounds. For the universal $\Lambda_\xi$ and for the optimal informed $\Lambda_\mu$ prediction scheme the following holds:

$$(i)\;\; \bar{p}_n^{\Lambda_\xi} \;=\; \bar{p}_n^{\Lambda_\mu} - O(n^{-1/2}) \longrightarrow \bar{p}_n^{\Lambda_\mu} \quad \text{for} \quad n \to \infty,$$

$$(ii)\;\; n > \left(\frac{2p_\Delta}{\bar{p}_n^{\Lambda_\mu}}\right)^2 \cdot k_\mu \;\;\wedge\;\; \bar{p}_n^{\Lambda_\mu} > 0 \;\;\Longrightarrow\;\; \bar{p}_n^{\Lambda_\xi} > 0,$$

where $w_\mu = e^{-k_\mu}$ is the weight (3.5) of $\mu$ in $\xi$.

By dividing (3.61) by $n$ and using $D_n \leq k_\mu$ (3.18) we see that the leading order of $\bar{p}_n^{\Lambda_\xi} - \bar{p}_n^{\Lambda_\mu}$ is bounded by $\sqrt{4p_\Delta p_{max} k_\mu / n}$, which proves $(i)$. The condition in $(ii)$ is actually a weakening of (3.62). $P_n^{\Lambda_\xi}$ is trivially positive for $p_{min} > 0$, since in this wonderful case *all* profits are positive. For negative $p_{min}$ the condition of $(ii)$ implies (3.62), since $p_\Delta > p_{max}$, and (3.62) implies positive (3.61), i.e. $P_n^{\Lambda_\xi} > 0$, which proves $(ii)$.

If a winning strategy $\Lambda$ with $\bar{p}_n^\Lambda > \varepsilon > 0$ exists, then $\Lambda_\xi$ is asymptotically also a winning strategy with the same average profit.

### 3.5.3 Example

Let us consider a game with two dice, one with two black and four white faces, the other with four black and two white faces. The dealer who repeatedly throws the dice uses one or the other die according to some deterministic rule, which correlates the throws (e.g. the first die could be used in round $t$ iff the $t^{th}$ digit of $\pi$ is 7). We can bet on black or white, the stake $s$ is 3\$ in every round, and our return $r$ is 5\$ for every correct prediction.

The profit is $p_t = r\delta_{x_t y_t} - s$. The coloring of the dice and the selection strategy of the dealer unambiguously determine $\mu$. $\mu(x_t|x_{<t})$ is $\frac{1}{3}$ or $\frac{2}{3}$, depending on which die was chosen. One should bet on the more probable outcome ($\gamma = \frac{1}{2}$ in Section 3.4.3). If we knew $\mu$ the expected profit per round would be $\bar{p}_n^{\Lambda_\mu} = p_n^{\Lambda_\mu} = \frac{2}{3}r - s = \frac{1}{3}\$ > 0$. If we do not know $\mu$ we should use Solomonoff-Levin's universal prior with $D_n \leq k_\mu = K(\mu)\cdot\ln 2$, where $K(\mu)$ is the length of the shortest program coding $\mu$ (see Subsection 3.2.9). Then we know that betting on the outcome with higher $\xi$ probability leads asymptotically to the same profit (Theorem 3.63$(i)$), and $\Lambda_\xi$ reaches the winning threshold no later than $n_{thresh} = 900\ln 2\cdot K(\mu)$ (Theorem 3.63$(ii)$) or sharper $n_{thresh} = 330\ln 2\cdot K(\mu)$ from (3.62), where $p_{max} = r - s = 2\$$ and $p_\Delta = r = 5\$$ have been used.

If the die selection strategy reflected in $\mu$ is not too complicated, the $\Lambda_\xi$ prediction system reaches the winning zone after a few thousand rounds. The number of rounds is not really small because the expected profit per round is one order of magnitude smaller than the return. This leads to a constant of two orders of magnitude size in front of $K(\mu)$. Stated otherwise, it is due to the large stochastic noise, which makes it difficult to extract the signal, i.e. the structure of the rule $\mu$ (see next subsection). Furthermore, this is only a bound for the turnaround value of $n_{thresh}$. The true expected turnaround $n$ might be smaller. However, for every game for which there exists a computable winning strategy with $\bar{p}_n^\Lambda > \varepsilon > 0$, $\Lambda_\xi$ is guaranteed to get into the winning zone for some $n \sim K(\mu)$.

### 3.5.4 Information-Theoretic Interpretation

We try to give an intuitive explanation of Theorem 3.63$(ii)$. We know that $\xi(x_t|x_{<t})$ converges to $\mu(x_t|x_{<t})$ for $t\to\infty$. In a sense, $\Lambda_\xi$ learns $\mu$ from past data $x_{<t}$. The information content in $\mu$ relative to $\xi$ is $D_\infty/\ln 2 \leq k_\mu/\ln 2$. One

might think of a Shannon-Fano prefix code of $\nu\in\mathcal{M}$ of length $\lceil k_\nu/\ln 2\rceil$, which exists since the Kraft inequality $\sum_\nu 2^{-\lceil k_\nu/\ln 2\rceil}\leq\sum_\nu w_\nu\leq 1$ is satisfied. So, $k_\mu/\ln 2$ bits have to be learned before $\Lambda_\xi$ can be as good as $\Lambda_\mu$. In the worst case, the only information conveyed by $x_t$ is in form of the received profit $p_t$. Remember that we always know the profit $p_t$ before the next cycle starts.

Assume that the distribution of the profits in the interval $[p_{min},p_{max}]$ is mainly due to noise, and there is only a small informative signal of amplitude $\bar{p}_n^{\Lambda_\mu}$. To reliably determine the sign of a signal of amplitude $\bar{p}_n^{\Lambda_\mu}$, disturbed by noise of amplitude $p_\Delta$, we have to resubmit a bit $O((p_\Delta/\bar{p}_n^{\Lambda_\mu})^2)$ times (this reduces the standard deviation below the signal amplitude $\bar{p}_n^{\Lambda_\mu}$). To learn $\mu$, $k_\mu/\ln 2$ bits have to be transmitted, which requires $n\geq O((p_\Delta/\bar{p}_n^{\Lambda_\mu})^2)\cdot k_\mu/\ln 2$ cycles. This expression coincides with the condition in (*ii*). Identifying the signal amplitude with $\bar{p}_n^{\Lambda_\mu}$ is the weakest part of this consideration, as we have no argument why this should be true. It may be interesting to make the analogy more rigorous, which may also lead to a simpler proof of (*ii*) not based on Theorems 3.48 and 3.60 with its rather complex proof.

## 3.6 Optimality Properties

In this section we discuss the quality of the universal predictor and the bounds. We show that there are $\mathcal{M}$ and $\mu\in\mathcal{M}$ and weights $w_\nu$ such that the derived error bounds are tight. This shows that the loss bounds cannot be improved in general. We show Pareto optimality of $\xi$ in the sense that there is no other predictor which performs at least as well in all environments $\nu\in\mathcal{M}$ and strictly better in at least one. Optimal predictors can always be based on mixture distributions $\xi$. This still leaves open how to choose the weights. We give an Occam's razor argument that the choice $w_\nu=2^{-K(\nu)}$, where $K(\nu)$ is the length of the shortest program describing $\nu$, is optimal.

### 3.6.1 Lower Error Bound

We want to show that there exists a class $\mathcal{M}$ of distributions such that *any* predictor $\Theta$ ignorant of the distribution $\mu\in\mathcal{M}$ from which the observed sequence is sampled must make some minimal additional number of errors as compared to the best informed predictor $\Theta_\mu$. For deterministic environments a lower bound can easily be obtained by a combinatoric argument. Consider a class $\mathcal{M}$ containing $2^n$ binary sequences such that each prefix of length $n$ occurs exactly once. Assume any deterministic predictor $\Theta$ (not knowing the sequence in advance), then for every prediction $x_t^\Theta$ of $\Theta$ at times $t\leq n$ there exists a sequence with opposite symbol $x_t=1-x_t^\Theta$. Hence, $E_\infty^\Theta\geq E_n^\Theta=n=\log_2|\mathcal{M}|$ is a lower worst-case bound for every predictor $\Theta$, (this includes $\Theta_\xi$, of course). This shows that the upper bound $E_\infty^{\Theta_\xi}\leq\log_2|\mathcal{M}|$ for uniform $w$ obtained in the discussion after Theorem 3.36 is sharp. In the general probabilistic case we can show by a similar argument that the upper bound of Theorem 3.36 is sharp for $\Theta_\xi$ and "static" predictors, and sharp within a factor of 2 for general predictors. We do not know whether the factor-2 gap can be closed.

**Theorem 3.64 (Lower error bound)** For every $n$ there is an $\mathcal{M}$ and $\mu\in\mathcal{M}$ and weights $w_\nu$ such that for all $t\leq n$

$$(i)\quad e_t^{\Theta_\xi}-e_t^{\Theta_\mu} = \sqrt{2s_t} \quad\text{and}\quad E_n^{\Theta_\xi}-E_n^{\Theta_\mu} = S_n+\sqrt{4E_n^{\Theta_\mu}S_n+S_n^2},$$

where $E_n^{\Theta_\xi}$ and $E_n^{\Theta_\mu}$ are the total expected number of errors of $\Theta_\xi$ and $\Theta_\mu$, and $s_t$ and $S_n$ are defined in (3.14). More generally, the equalities hold for *any "static"* deterministic predictor $\theta$ for which $y_t^\Theta$ is independent of $x_{<t}$. For every $n$ and *arbitrary* deterministic predictor $\Theta$, there exists an $\mathcal{M}$ and $\mu\in\mathcal{M}$ such that for all $t\leq n$

$$(ii)\quad e_t^{\Theta}-e_t^{\Theta_\mu} \geq \tfrac{1}{2}\sqrt{2s_t(x_{<t})} \quad\text{and}\quad E_n^{\Theta}-E_n^{\Theta_\mu} \geq \tfrac{1}{2}[S_n+\sqrt{4E_n^{\Theta_\mu}S_n+S_n^2}].$$

**Proof.** $(i)$ The proof parallels and generalizes the deterministic case. Consider a class $\mathcal{M}$ of $2^n$ distributions (over binary alphabet) indexed by $a\equiv a_1...a_n\in\{0,1\}^n$. For each $t$ we want a distribution with posterior probability $\frac{1}{2}(1+\varepsilon)$ for $x_t=1$ and one with posterior probability $\frac{1}{2}(1-\varepsilon)$ for $x_t=1$ independent of the past $x_{<t}$ with $0<\varepsilon\leq\frac{1}{2}$. That is

$$\mu_a(x_1...x_n)=\mu_{a_1}(x_1)\cdot...\cdot\mu_{a_n}(x_n), \quad\text{where}\quad \mu_{a_t}(x_t)=\begin{cases}\frac{1}{2}(1+\varepsilon) & \text{for } x_t=a_t\\ \frac{1}{2}(1-\varepsilon) & \text{for } x_t\neq a_t\end{cases}$$

We are not interested in predictions beyond time $n$, but for completeness we may define $\mu_a$ to assign probability 1 to $x_t=1$ for all $t>n$. If $\mu=\mu_a$, the informed scheme $\Theta_\mu$ always predicts the bit that has highest $\mu$-probability, i.e. $y_t^{\Theta_\mu}=a_t$

$$\Longrightarrow\quad e_t^{\Theta_\mu}=1-\mu_{a_t}(y_t^{\Theta_\mu})=\tfrac{1}{2}(1-\varepsilon) \quad\Longrightarrow\quad E_n^{\Theta_\mu}=\tfrac{n}{2}(1-\varepsilon).$$

Since $E_n^{\Theta_\mu}$ is the same for all $a$, we seek to maximize $E_n^\Theta$ for a given predictor $\Theta$ in the following. Assume $\Theta$ predicts $y_t^\Theta$ (independent of history $x_{<t}$). Since we want lower bounds, we seek a worst-case $\mu$. A success $y_t^\Theta=x_t$ has lowest possible probability $\frac{1}{2}(1-\varepsilon)$ if $a_t=1-y_t^\Theta$.

$$\Longrightarrow\quad e_t^{\Theta}=1-\mu_{a_t}(y_t^{\Theta})=\tfrac{1}{2}(1+\varepsilon) \quad\Longrightarrow\quad E_n^{\Theta}=\tfrac{n}{2}(1+\varepsilon).$$

So we have $e_t^\Theta-e_t^{\Theta_\mu}=\varepsilon$ and $E_n^\Theta-E_n^{\Theta_\mu}=n\varepsilon$ for the regrets. We need to eliminate $n$ and $\varepsilon$ in favor of $s_t$, $S_n$, and $E_n^{\Theta_\mu}$. If we assume uniform weights $w_{\mu_a}=2^{-n}$ for all $\mu_a$ we get

$$\xi(x_{1:n}) = \sum_a w_{\mu_a}\mu_a(x_{1:n}) = 2^{-n}\prod_{t=1}^n\sum_{a_t\in\{0,1\}}\mu_{a_t}(x_t) = 2^{-n}\prod_{t=1}^n 1 = 2^{-n},$$

i.e. $\xi$ is an unbiased Bernoulli sequence ($\xi(x_t|x_{<t})=\frac{1}{2}$).

$$\Longrightarrow \quad s_t(x_{<t}) \;=\; \sum_{x_t}(\tfrac{1}{2}-\mu_{a_t}(x_t))^2 \;=\; \tfrac{1}{2}\varepsilon^2 \quad\text{and}\quad S_n = \tfrac{n}{2}\varepsilon^2.$$

So we have $\varepsilon=\sqrt{2s_t}$, which proves the instantaneous regret formula $e_t^\Theta-e_t^{\Theta_\mu}=\sqrt{2s_t}$ for static $\Theta$. Inserting $\varepsilon=\sqrt{\frac{2}{n}S_n}$ into $E_n^{\Theta_\mu}$ and solving w.r.t. $\sqrt{2n}$, we get $\sqrt{2n}=\sqrt{S_n}+\sqrt{4E_n^{\Theta_\mu}+S_n}$. So, we finally get

$$E_n^\Theta - E_n^{\Theta_\mu} \;=\; n\varepsilon \;=\; \sqrt{S_n}\sqrt{2n} \;=\; S_n + \sqrt{4E_n^{\Theta_\mu}S_n + S_n^2},$$

which proves the total regret formula in $(i)$ for static $\Theta$. We can choose[7] $y_t^{\Theta_\xi}\equiv 0$ to be a static predictor. Together this shows $(i)$.

$(ii)$ For non-static predictors, $a_t=1-y_t^\Theta$ in the proof of $(i)$ depends on $x_{<t}$, which is not allowed. For general, but fixed $a_t$ we have $e_t^\Theta(x_{<t})=1-\mu_{a_t}(y_t^\Theta)$. This quantity may assume any value between $\frac{1}{2}(1-\varepsilon)$ and $\frac{1}{2}(1+\varepsilon)$ when averaged over $x_{<t}$ and is, hence, of little direct help. But if we additionally average the result over all environments $\mu_a$, we get

$$< E_n^\Theta >_a \;=< \sum_{t=1}^n \mathbf{E}[e_t^\Theta(x_{<t})] >_a \;=\; \sum_{t=1}^n \mathbf{E}[< e_t^\Theta(x_{<t}) >_a] \;=\; \sum_{t=1}^n \mathbf{E}[\tfrac{1}{2}] = \tfrac{1}{2}n$$

whatever $\Theta$ is chosen: a sort of no-free-lunch theorem [WM97], stating that on *uniform* average all predictors perform equally well/poorly. The expectation of $E_n^\Theta$ w.r.t. $a$ can only be $\frac{1}{2}n$ if $E_n^\Theta\geq\frac{1}{2}n$ for some $a$. Fixing such an $a$ and choosing $\mu=\mu_a$, we get $E_n^\Theta-E_n^{\Theta_\mu}\geq\frac{1}{2}n\varepsilon=\frac{1}{2}[S_n+\sqrt{4E_n^{\Theta_\mu}S_n+S_n^2}]$, and similarly $e_n^\Theta-e_n^{\Theta_\mu}\geq\frac{1}{2}\varepsilon=\frac{1}{2}\sqrt{2s_t(x_{<t})}$. □

Since $d_t/s_t=1+O(\varepsilon^2)$ we have $D_n/S_n\to 1$ for $\varepsilon\to 0$. Hence the error bound of Theorem 3.36 with $S_n$ replaced by $D_n$ is asymptotically tight for $E_n^{\Theta_\mu}/D_n\to\infty$ (which implies $\varepsilon\to 0$). This shows that without restrictions on the loss function that exclude the error loss, the loss bound in Theorem 3.48 can also not be improved. Note that the bounds are tight even when $\mathcal{M}$ is restricted to Markov or i.i.d. environments, since the presented counterexample is i.i.d. Finally, $E_n^\Theta-E_n^{\Theta_\mu}=n\varepsilon=n\sqrt{\frac{2S_n}{n}}\to\sqrt{2nD_n}$, which shows that the bound (3.50) of Merhav and Feder is also asymptotically tight.

A set $\mathcal{M}$ independent of $n$ leading to a good (but not tight) lower bound is $\mathcal{M}=\{\mu_1,\mu_2\}$ with $\mu_{1/2}(1|x_{<t})=\frac{1}{2}\pm\varepsilon_t$ with $\varepsilon_t=\min\{\frac{1}{2},\sqrt{\ln w_{\mu_1}^{-1}}/\sqrt{t\ln t}\}$. For $w_{\mu_1}\ll w_{\mu_2}$ and $n\to\infty$ one can show that $E_n^{\Theta_\xi}-E_n^{\Theta_{\mu_1}}\sim\frac{1}{\ln n}\sqrt{E_n^{\Theta_\mu}\ln w_{\mu_1}^{-1}}$ (Problem 3.6).

Unfortunately, there are many important special cases for which the loss bound (3.48) is not tight. For continuous $\mathcal{Y}$ and logarithmic or quadratic loss

[7] This choice may be made unique by slightly non-uniform $w_{\mu_a}=\prod_{t=1}^n[\frac{1}{2}+(\frac{1}{2}-a_t)\delta]$ with $\delta\ll 1$.

function, for instance, we have seen that the regret $L_\infty^{\Lambda_\xi} - L_\infty^{\Lambda_\mu} \leq \ln w_\mu^{-1} < \infty$ is finite. For arbitrary loss function, but $\mu$ bounded away from certain critical values, the regret is also finite. For instance, consider the special error loss, binary alphabet, and $|\mu(x_t|x_{<t}) - \frac{1}{2}| > \varepsilon$ for all $t$ and $x$; $\Theta_\mu$ predicts 0 if $\mu(0|x_{<t}) > \frac{1}{2}$. If also $\xi(0|x_{<t}) > \frac{1}{2}$, then $\Theta_\xi$ makes the same prediction as $\Theta_\mu$, while for $\xi(0|x_{<t}) < \frac{1}{2}$ the predictions differ. In the latter case $|\xi(0|x_{<t}) - \mu(0|x_{<t})| > \varepsilon$. Conversely for $\mu(0|x_{<t}) < \frac{1}{2}$. So in any case $e_t^{\Theta_\xi} - e_t^{\Theta_\mu} \leq \frac{1}{\varepsilon^2}[\xi(x_t|x_{<t}) - \mu(x_t|x_{<t})]^2$. Using Definition (3.34) and Theorem 3.19($i$) we see that $E_\infty^{\Theta_\xi} - E_\infty^{\Theta_\mu} \leq \frac{1}{\varepsilon^2}\ln w_\mu^{-1} < \infty$ is finite too. Nevertheless, Theorem 3.64 is important as it tells us that bound (3.48) can only be strengthened by making further assumptions on $\ell$ or $\mathcal{M}$.

### 3.6.2 Pareto Optimality of $\xi$

In this subsection we want to establish a different kind of optimality property of $\xi$. Let $\mathcal{F}(\mu,\rho)$ be any of the performance measures of $\rho$ relative to $\mu$ considered in the previous sections (e.g. $s_t$, or $D_n$, or $L_n$, ...). It is easy to find $\rho$ more tailored toward $\mu$ such that $\mathcal{F}(\mu,\rho) < \mathcal{F}(\mu,\xi)$. This improvement may be achieved by increasing $w_\mu$, but probably at the expense of increasing $\mathcal{F}$ for other $\nu$, i.e. $\mathcal{F}(\nu,\rho) > \mathcal{F}(\nu,\xi)$ for some $\nu \in \mathcal{M}$. Since we do not know $\mu$ in advance, we may ask whether there exists a $\rho$ with better or equal performance for *all* $\nu \in \mathcal{M}$ and a strictly better performance for one $\nu \in \mathcal{M}$. This would clearly render $\xi$ suboptimal w.r.t. to $\mathcal{F}$. We show that there is no such $\rho$ for all performance measures studied in this book.

**Definition 3.65 (Pareto optimality)** Let $\mathcal{F}(\mu,\rho)$ be any performance measure of $\rho$ relative to $\mu$. The universal prior $\xi$ is called Pareto optimal w.r.t. $\mathcal{F}$ if there is no $\rho$ with $\mathcal{F}(\nu,\rho) \leq \mathcal{F}(\nu,\xi)$ for all $\nu \in \mathcal{M}$ and strict inequality for at least one $\nu$.

**Theorem 3.66 (Pareto optimal performance measures)** A prior $\xi$ is Pareto optimal w.r.t. the instantaneous and total squared distances $s_t$ and $S_n$ (3.14), entropy distances $d_t$ and $D_n$ (3.16), errors $e_t$ and $E_n$ (3.34), and losses $l_t$ and $L_n$ (3.46).

**Proof.** We first prove Theorem 3.66 for the instantaneous expected loss $l_t$. We need the more general $\rho$-expected instantaneous losses

$$l_{t\rho}^{\Lambda}(x_{<t}) \;:=\; \sum_{x_t} \rho(x_t|x_{<t})\ell_{x_t y_t^\Lambda} \tag{3.67}$$

for a predictor $\Lambda$. We want to arrive at a contradiction by assuming that $\xi$ is not Pareto optimal, i.e. by assuming the existence of a predictor[8] $\Lambda$

[8] According to Definition 3.65 we should look for a $\rho$, but for each deterministic predictor $\Lambda$ there exists a $\rho$ with $\Lambda = \Lambda_\rho$.

with $l_{t\nu}^{\Lambda} \leq l_{t\nu}^{\Lambda_\xi}$ for all $\nu \in \mathcal{M}$ and strict inequality for some $\nu$. Implicit to this assumption is the assumption that $l_{t\nu}^{\Lambda}$ and $l_{t\nu}^{\Lambda_\xi}$ exist; $l_{t\nu}^{\Lambda}$ exists iff $\nu(x_t|x_{<t})$ exists iff $\nu(x_{<t})>0$ iff $w_\nu(x_{<t})>0$.

$$l_{t\xi}^{\Lambda} = \sum_\nu w_\nu(x_{<t}) l_{t\nu}^{\Lambda} < \sum_\nu w_\nu(x_{<t}) l_{t\nu}^{\Lambda_\xi} = l_{t\xi}^{\Lambda_\xi} \leq l_{t\xi}^{\Lambda} .$$

The two equalities follow from inserting (3.7) into (3.67). The strict inequality follows from the assumption and $w_\nu(x_{<t})>0$. The last inequality follows from the fact that $\Lambda_\xi$ minimizes by definition (3.45) the $\xi$-expected loss (similarly to (3.47)). The contradiction $l_{t\xi}^{\Lambda} < l_{t\xi}^{\Lambda}$ proves Pareto optimality of $\xi$ w.r.t. $l_t$.

In the same way we can prove Pareto optimality of $\xi$ w.r.t. the total loss $L_n$ by defining the $\rho$-expected total losses

$$L_{n\rho}^{\Lambda} := \sum_{t=1}^{n} \sum_{x_{<t}} \rho(x_{<t}) l_{t\rho}^{\Lambda}(x_{<t}) = \sum_{t=1}^{n} \sum_{x_{1:t}} \rho(x_{1:t}) \ell_{x_t y_t^{\Lambda}} \tag{3.68}$$

for a predictor $\Lambda$, and by assuming $L_{n\nu}^{\Lambda} \leq L_{n\nu}^{\Lambda_\xi}$ for all $\nu$ and strict inequality for some $\nu$, from which we get the contradiction $L_{n\xi}^{\Lambda} = \sum_\nu w_\nu L_{n\nu}^{\Lambda} < \sum_\nu w_\nu L_{n\nu}^{\Lambda_\xi} = L_{n\xi}^{\Lambda_\xi} \leq L_{n\xi}^{\Lambda}$ with the help of (3.5). The instantaneous and total expected errors $e_t$ and $E_n$ can be considered as special loss functions.

Pareto optimality of $\xi$ w.r.t. $s_t$ (and hence $S_n$) can be understood from geometric insight. A formal proof for $s_t$ goes as follows: With the abbreviations $i = x_t$, $y_{\nu i} = \nu(x_t|x_{<t})$, $z_i = \xi(x_t|x_{<t})$, $r_i = \rho(x_t|x_{<t})$, and $w_\nu = w_\nu(x_{<t}) \geq 0$, we ask for a vector $\boldsymbol{r}$ with $\sum_i (y_{\nu i} - r_i)^2 \leq \sum_i (y_{\nu i} - z_i)^2 \ \forall\nu$. This implies

$$\begin{aligned} 0 &\geq \sum_\nu w_\nu \Big[ \sum_i (y_{\nu i} - r_i)^2 - \sum_i (y_{\nu i} - z_i)^2 \Big] \\ &= \sum_\nu w_\nu \Big[ \sum_i -2y_{\nu i} r_i + r_i^2 + 2y_{\nu i} z_i - z_i^2 \Big] \\ &= \sum_i -2z_i r_i + r_i^2 + 2z_i z_i - z_i^2 = \sum_i (r_i - z_i)^2 \geq 0, \end{aligned}$$

where we have used $\sum_\nu w_\nu = 1$ and $\sum_\nu w_\nu y_{\nu i} = z_i$ (3.7). $0 \geq \sum_i (r_i - z_i)^2 \geq 0$ implies $\boldsymbol{r} = \boldsymbol{z}$, proving Pareto optimality of $\xi$ w.r.t. $s_t$. Similarly for $d_t$, the assumption $\sum_i y_{\nu i} \ln\frac{y_{\nu i}}{r_i} \leq \sum_i y_{\nu i} \ln\frac{y_{\nu i}}{z_i} \ \forall\nu$ implies

$$0 \geq \sum_\nu w_\nu \Big[ \sum_i y_{\nu i} \ln\frac{y_{\nu i}}{r_i} - y_{\nu i} \ln\frac{y_{\nu i}}{z_i} \Big] = \sum_\nu w_\nu \sum_i y_{\nu i} \ln\frac{z_i}{r_i} = \sum_i z_i \ln\frac{z_i}{r_i} \geq 0$$

which implies $\boldsymbol{r} = \boldsymbol{z}$, proving Pareto optimality of $\xi$ w.r.t. $d_t$. The proofs for $S_n$ and $D_n$ are similar. □

We have proven that $\xi$ is Pareto optimal w.r.t. $s_t$, $S_n$, $d_t$ and $D_n$ in a strong sense, that is, there is also no $\rho \neq \xi$ with same performance as $\xi$ in all environments. In the case of $e_t$, $E_n$, $l_t$ and $L_n$, there are other $\rho \neq \xi$

with $\mathcal{F}(\nu,\rho)=\mathcal{F}(\nu,\xi)\forall\nu$, but the actions/predictions they invoke are unique ($y_t^{\Lambda_\rho}=y_t^{\Lambda_\xi}$) (if ties in $\text{argmax}_{y_t}$ are broken in a consistent way), and this is all that counts.

Note that $\xi$ is *not* Pareto optimal w.r.t. to all thinkable performance measures. Counterexamples can be given for $\mathcal{F}(\nu,\xi)=\sum_{x_t}|\nu(x_t|x_{<t})-\xi(x_t|x_{<t})|^\alpha$ for $\alpha\neq 2$ (see Problem 3.5). Nevertheless, for all measures that are relevant from a decision-theoretic point of view, i.e. for all loss functions $l_t$ and $L_n$, $\xi$ has the welcome property of being Pareto optimal.

### 3.6.3 Balanced Pareto Optimality of $\xi$

Pareto optimality should be regarded as a necessary condition for a prediction scheme aiming to be optimal. From a practical point of view, a significant decrease of $\mathcal{F}$ for many $\nu$ may be desirable, even if this causes a small increase of $\mathcal{F}$ for a few other $\nu$. The impossibility of such a "balanced" improvement is a more demanding condition on $\xi$ than pure Pareto optimality. The next theorem shows that $\Lambda_\xi$ is also balanced Pareto optimal. We only consider the performance measure $L_n$ and suppress the index $n$ for convenience.

**Theorem 3.69 (Balanced Pareto optimality w.r.t. $L$)**

$$\Delta_\nu := L_\nu^{\tilde{\Lambda}} - L_\nu^{\Lambda_\xi}, \quad \Delta := \sum_{\nu\in\mathcal{M}} w_\nu \Delta_\nu \quad \Rightarrow \quad \Delta\geq 0.$$

This implies the following: Assume $\tilde{\Lambda}$ has larger loss than $\Lambda_\xi$ on environments $\mathcal{L}$ by a total weighted amount of $\Delta_\mathcal{L}:=\sum_{\lambda\in\mathcal{L}}w_\lambda\Delta_\lambda$. Then $\tilde{\Lambda}$ can have smaller loss on $\eta\in\mathcal{H}:=\mathcal{M}\setminus\mathcal{L}$, but the improvement is bounded by $\Delta_\mathcal{H}:=|\sum_{\eta\in\mathcal{H}}w_\eta\Delta_\eta|\leq\Delta_\mathcal{L}$. In particular $|\Delta_\eta|\leq w_\eta^{-1}\max_{\lambda\in\mathcal{L}}\Delta_\lambda$.

This means that a weighted loss decrease $\Delta_\mathcal{H}$ by using $\tilde{\Lambda}$ instead of $\Lambda_\xi$ is compensated by an at least as large weighted increase $\Delta_\mathcal{L}$ on other environments. If the increase is small, the decrease can also only be small. In the special case of only a single environment with increased loss $\Delta_\lambda$, the decrease is bound by $\Delta_\eta\leq\frac{w_\lambda}{w_\eta}|\Delta_\lambda|$, i.e. an increase by an amount $\Delta_\lambda$ can only cause a decrease by at most the same amount times a factor $\frac{w_\lambda}{w_\eta}$. An increase can only cause a smaller decrease in simpler environments, but can cause a scaled decrease in more complex environments. Finally, note that pure Pareto optimality (3.66) follows from balanced Pareto optimality in the special case of no increase $\Delta_\mathcal{L}\equiv 0$.

**Proof.** $\Delta\geq 0$ follows from $\Delta=\sum_\nu w_\nu[L_\nu^{\tilde{\Lambda}}-L_\nu^{\Lambda_\xi}]=L_\xi^{\tilde{\Lambda}}-L_\xi^{\Lambda_\xi}\geq 0$, where we have used linearity of $L_\rho$ in $\rho$ and $L_\xi^{\Lambda_\xi}\leq L_\xi^{\tilde{\Lambda}}$. The remainder of Theorem 3.69 is obvious from $0\leq\Delta=\Delta_\mathcal{L}-\Delta_\mathcal{H}$ and by bounding the weighted average $\Delta_\eta$ by its maximum. □

The term *Pareto optimal* has been taken from the economics literature, but there is the closely related notion of unimprovable strategies [BM98] or admissible estimators [Fer67] in statistics for parameter estimation, for which results similar to Theorem 3.66 exist. Furthermore, it would be interesting to show under which conditions the class of *all* Bayes mixtures (i.e. with all possible values for the weights) is complete in the sense that *every* Pareto optimal strategy can be based on a Bayes mixture. Pareto optimality is sort of a minimal demand on a prediction scheme aiming to be optimal. A scheme that is not even Pareto optimal cannot be regarded as optimal in any reasonable sense. Pareto optimality of $\xi$ w.r.t. most performance measures emphasizes the distinctiveness of Bayes mixture strategies.

### 3.6.4 On the Optimal Choice of Weights

In the following we indicate the dependency of $\xi$ on $w$ explicitly by writing $\xi_w$. We have shown that the $\Lambda_{\xi_w}$ prediction schemes are (balanced) Pareto optimal, i.e. that *no* prediction scheme $\Lambda$, whether based on a Bayes mixture or not, can be uniformly better. Least assumptions on the environment are made for $\mathcal{M}$ which are as large as possible. In Section 2.4 we have discussed the set $\mathcal{M}$ of all enumerable semimeasures, which we regarded as sufficiently large from a computational point of view (see [Sch02a] for even larger sets, but which are still in the computational realm). Agreeing on this $\mathcal{M}$ still leaves open the question of how to choose the weights (prior beliefs) $w_\nu$, since every $\xi_w$ with $w_\nu > 0\ \forall\nu$ is Pareto optimal and leads asymptotically to optimal predictions.

We have derived bounds for the mean squared sum $S_{n\nu}^{\xi_w} \leq \ln w_\nu^{-1}$ and for the loss regret $L_{n\nu}^{\Lambda_{\xi_w}} - L_{n\nu}^{\Lambda_\nu} \leq 2\ln w_\nu^{-1} + 2\sqrt{\ln w_\nu^{-1} L_{n\nu}^{\Lambda_\nu}}$. All bounds decrease monotonically with increasing $w_\nu$. So it is desirable to assign high weights to all $\nu \in \mathcal{M}$. Due to the (semi)probability constraint $\sum_\nu w_\nu \leq 1$, one has to find a compromise. In the following we argue that in the class of enumerable weight functions with short program there is an optimal compromise, namely $w_\nu = 2^{-K(\nu)}$, which gives Solomonoff-Levin's prior.

Consider the class of enumerable weight functions with short programs, namely $\mathcal{V} := \{v_{(.)} : \mathcal{M} \to I\!R^+ \text{ with } \sum_\nu v_\nu \leq 1 \text{ and } K(v) = O(1)\}$. Let $w_\nu := 2^{-K(\nu)}$ and $v_{(\cdot)} \in \mathcal{V}$. Theorem 2.10(*vii*) says that $K(x) \leq -\log_2 P(x) + K(P) + O(1)$ for all $x$ if $P$ is an enumerable discrete semimeasure. Identifying $P$ with $v$ and $x$ with (the program index describing) $\nu$ we get

$$\ln w_\nu^{-1} \;\leq \ln v_\nu^{-1} + O(1).$$

This means that the bounds for $\xi_w$ depending on $\ln w_\nu^{-1}$ are at most $O(1)$ larger than the bounds for $\xi_v$ depending on $\ln v_\nu^{-1}$. So we lose at most an additive constant of order 1 in the bounds when using $\xi_w$ instead of $\xi_v$. In using Solomonoff-Levin's prior $\xi_w$ we are on the safe side, getting (within $O(1)$) best bounds for *all* environments.

**Theorem 3.70 (Optimality of universal weights)** Within the set $\mathcal{V}$ of enumerable weight functions with short program, the universal weights $w_\nu = 2^{-K(\nu)}$ lead to the smallest loss bounds within an additive (to $\ln w_\mu^{-1}$) constant in all enumerable environments.

Since the above justifies the use of Solomonoff-Levin's prior, and Solomonoff-Levin's prior assigns high probability to an environment if and only if it has low (Kolmogorov) complexity, one may interpret the result as a justification of Occam's razor. But note that this is more of a bootstrap argument, since we used Occam's razor in Section 2.1 to justify the restriction to enumerable semimeasures. We also considered only weight functions $v$ with low complexity $K(v) = O(1)$. What did not enter as an assumption but came out as a result is that the specific universal weights $w_\nu = 2^{-K(\nu)}$ are optimal. See Problem 3.7 for a discussion of the (non)uniqueness of $w_\nu$.

### 3.6.5 Occam's razor versus No Free Lunches

We do not regard Theorem 3.69 as a no-free-lunch (NFL) theorem [WM97]. Since most environments are completely random, a small concession on the loss in each of these completely uninteresting environments provides enough margin $\Delta_\mathcal{H}$ to yield distinguished performance on the few nonrandom (interesting) environments. Indeed, we would interpret the NFL theorems for optimization and search in [WM97] as balanced Pareto optimality results. Interestingly, whereas for prediction only Bayes mixtures are Pareto optimal, for search and optimization every algorithm is Pareto optimal. There is an ongoing battle between believers in Occam's razor and believers in no-free-lunches that cannot be dealt with here [Sto01, SH02].

## 3.7 Miscellaneous

This section generalizes the setting and results obtained so far in various ways. First, we consider multistep/delayed predictions, where the next $h$ / the $h^{th}$-next symbol shall be predicted. We show convergence of $\xi$ to $\mu$ i.m.(s. for bounded $h$). Second, we generalize the setup to continuous probability classes $\mathcal{M} = \{\mu_\theta\}$ consisting of continuously parameterized distributions $\mu_\theta$ with parameter $\theta \in I\!R^d$. Under certain smoothness and regularity conditions a bound for the relative entropy between $\mu$ and $\xi$, which is central for all presented results, can still be derived. The bound depends on the Fisher information of $\mu$ and grows only logarithmically with $n$, the intuitive reason being the necessity to describe $\theta$ to an accuracy $O(n^{-1/2})$. Third, we describe two ways of using the prediction schemes for partial sequence prediction, where not every symbol needs to be predicted. Performing and predicting a sequence of independent experiments and online learning of classification tasks are special cases. Fourth, we compare the universal prediction scheme studied here to the popular predictors based on expert advice (PEA). Although the algorithms, the settings, and the proofs are quite different, the PEA bounds and the error bound derived here have the same structure. Finally, we outline possible extensions of the presented theory and results, including infinite alphabets, more active systems influencing the environment, learning aspects, a unification with PEA, and the minimal description length principle.

### 3.7.1 Multistep Predictions

**Introduction.** In multistep prediction we want to predict $x_{t:n}$ from $x_{<t}$. For instance, every day a weather forecaster in the morning of day $t$ predicts the weather for the next three days $t$, $t+1$, and $t+2=n$. Up to now we have considered prediction problems with a lookahead of one time-step only: Given $x_{<t}$, predict $x_t$. Greedy minimization of the expected loss $l_t(x_{<t})$ at time $t$ was optimal. Looking farther ahead ($>t$) was not necessary, because the prediction/decision/action $y_t$ has no influence on the environment $\mu$. For acting agents, described in detail in later chapters, multistep lookahead is necessary for optimal actions. Another application of multistep predictions is 'delayed sequence prediction', in which not the next, but next-to-next or $h^{th}$-next symbol shall be predicted.

**Notation and basic relations.** We are interested in multistep posteriors $\rho(x_{t:n}|x_{<t})=\rho(x_{1:n})/\rho(x_{<t})$, which generalize the one-step posteriors ($n=t$) considered so far. We abbreviate $\rho_{t:n}^{\cdots}:=\rho(x_{t:n}^{\cdots}|x_{<t})$, where $^{\cdots}$ are any superscripts (e.g. empty or $'$). We define the conditional probability vector $\boldsymbol{\rho}_{t:n}:=\rho_{t:n}(\cdot|x_{<t})\in I\!R^N$, where $N=|\mathcal{X}|^{n-t+1}$ and the $i^{th}$ component of vector $\boldsymbol{\rho}_{t:n}$ is $\rho_{t:n}(i|x_{<t})$ with identification $\{1,...,N\}\ni i\,\widehat{=}\,x_{t:n}\in\mathcal{X}^{n-t+1}$. Let $f\in\{a,b,d,h,s\}$ be any of the distances defined in (3.10), i.e. $f(\boldsymbol{y},\boldsymbol{z})=\sum_{i=1}^N\hat{f}(y_i,z_i)$ with $\hat{a}(y,z)=|y-z|$, $\hat{b}(y,z)=y|\ln\frac{y}{z}|$, $\hat{d}(y,z)=y\ln\frac{y}{z}$, $\hat{h}(y,z)=(\sqrt{y}-\sqrt{z})^2$, $\hat{s}(y,z)=(y-z)^2$. We define $f_{t:n}(x_{<t}):=f(\boldsymbol{\mu}_{t:n},\boldsymbol{\xi}_{t:n})$, generalizing (3.13)–(3.17). The definitions $f_t(x_{<t}):=f_{t:t}(x_{<t})$ and $F_n:=\sum_{t=1}^n\mathbf{E}[f_t]$, $F\in\{A,B,D,H,S\}$ are consistent with (3.13)–(3.17). Lemma 3.11 ($b-d\leq a\leq\sqrt{2d}$, $h\leq d$, $s\leq d$) implies $b_{t:n}-d_{t:n}\leq a_{t:n}\leq\sqrt{2d_{t:n}}$, $h_{t:n}\leq d_{t:n}$, $s_{t:n}\leq d_{t:n}$. We define $\mathbf{E}_{t:k}[f(x_{1:k})]:=\sum'_{x_{t:k}}\mu_{t:k}f(x_{1:k})$, cf. (3.4). For the relative entropy we have $D_n=d_{1:n}$ and $\mathbf{E}_{t:k}[d_{k+1:n}]=d_{t:n}-d_{t:k}$ for $t\leq k<n$, which implies $\mathbf{E}[d_{t:n}]=D_n-D_{t-1}=\sum_{k=t}^n\mathbf{E}[d_k]\geq 0$, and $d_{t:n}$ is monotone increasing in $n$.

**Convergence i.m.s. for bounded horizon.** Henceforth, we no longer need the Hellinger distance, and we use $h$ for the horizon. Assume we want to predict the next $h$ symbols, i.e. $n=n_t=t+h-1$. We want to determine how fast $\xi'_{t:n_t}\equiv\xi(x'_{t:t+h-1}|x_{<t})$ converges to $\mu'_{t:n_t}\equiv\mu(x'_{t:t+h-1}|x_{<t})$. To prove convergence i.m.s. we have to bound the expectation sum of $s_{t:n_t}$ or $a_{t:n_t}^2\equiv(\sum_{x'_{t:n_t}}|\xi'_{t:n_t}-\mu'_{t:n_t}|)^2$:

$$\tfrac{1}{2}\sum_{t=1}^{\infty}\mathbf{E}[a_{t:n_t}^2] \leq \sum_{t=1}^{\infty}\mathbf{E}[d_{t:n_t}] = \sum_{t=1}^{\infty}\sum_{k=t}^{n_t}\mathbf{E}[d_k] \leq h\sum_{k=1}^{\infty}\mathbf{E}[d_k] = h\cdot D_\infty \leq h\cdot\ln w_\mu^{-1} < \infty. \tag{3.71}$$

In the second inequality we have used that the number of times $d_k \geq 0$ occurs for some $k$ in the double sum is $\min\{h,k\} \leq h$. The bound implies $a_{t:n_t} \to 0$ i.m.s., which implies $\xi'_{t:n_t} \xrightarrow{t\to\infty} \mu'_{t:n_t}$ i.m.s. by dropping the sum in the definition of $a_{t:n_t}$. The bound loosens by a factor of $h$ for $h$-step prediction as compared to 1-step prediction. The same bound holds for bounded horizon $h_t := n_t - t + 1 \leq h < \infty\,\forall t$ (increase $\sum_{k=t}^{n_t}$ to $\sum_{k=t}^{t+h-1}$), i.e.

$$\xi(x'_{t:n_t}|x_{<t}) \to \mu(x'_{t:n_t}|x_{<t}) \text{ i.m.s. for } t\to\infty \text{ if } h_t := n_t - t + 1 \leq h < \infty\,\forall t.$$

**Delayed sequence prediction.** A delayed feedback, where at time $t$, $x$ is only known up to time $t-h$ for some delay $h$, is common in many practical problems. This is equivalent to predicting $x_{n_t}$ from $x_{<t}$ with $n_t = t+h-1$. The probability of $x'_{n_t}$, given $x_{<t}$, is

$$\rho(x'_{n_t}|x_{<t}) := \sum_{x'_{t:n_t-1}} \rho'_{t:n_t} = \sum_{x_{t:n_t-1}} \rho(x_{<n_t}x'_{n_t})/\rho(x_{<t}).$$

Using $\displaystyle\sum_{x'_{n_t}} |\xi(x'_{n_t}|x_{<t}) - \mu(x'_{n_t}|x_{<t})| \leq \sum_{x'_{t:n_t}} |\xi'_{t:n_t} - \mu'_{t:n_t}| \equiv a_{t:n_t}$

and bound (3.71) we get

$$\sum_{t=1}^{\infty}\mathbf{E}\Big(\sum_{x'_{n_t}} |\xi(x'_{n_t}|x_{<t}) - \mu(x'_{n_t}|x_{<t})|\Big)^2 \leq \sum_{t=1}^{\infty}\mathbf{E}[a_{t:n_t}^2] \leq 2h\cdot\ln w_\mu^{-1} < \infty,$$

which implies $\xi(x'_{n_t}|x_{<t}) \xrightarrow{t\to\infty} \mu(x'_{n_t}|x_{<t})$ i.m.s. The loss bounds of Theorem 3.59($i$,$ii$) also generalize to the delayed case. Loss bounds similar to Theorem 3.59($iii$) and Theorem 3.48 should also be derivable.

**Convergence i.m. for arbitrary horizon.** Convergence i.m.s. does generally not hold for unbounded horizon $h_t$ (see Problem 3.15). Remarkably, convergence i.m. holds nevertheless: For any limit path $n \geq t \to \infty$ we have

$$\lim_{t,n\to\infty}\mathbf{E}[d_{t:n}] = \lim_{t,n\to\infty}[D_n - D_{t-1}] = \lim_{n\to\infty} D_n - \lim_{t\to\infty} D_{t-1} = D_\infty - D_\infty = 0.$$

So for any $n_t = t + h_t - 1$ we have $\frac{1}{2}\mathbf{E}[a_{t:n_t}^2] \leq \mathbf{E}[d_{t:n_t}] \xrightarrow{t\to\infty} 0$, which implies

$$\xi(x'_{t:n_t}|x_{<t}) \underset{i.m.}{\xrightarrow{t\to\infty}} \mu(x'_{t:n_t}|x_{<t}) \quad\text{and}\quad \xi(x'_{n_t}|x_{<t}) \underset{i.m.}{\xrightarrow{t\to\infty}} \mu(x'_{n_t}|x_{<t}) \quad\text{for any } h_t.$$

Convergence i.m. is weaker than convergence w.p.1 and i.m.s., and is potentially slow. We expect cases where convergence is very slow when $h_t$ grows very fast. We do not know whether convergence is reasonably fast for slowly growing horizons, e.g. for $h_t = \log t$ (or $h_t = t$).

### 3.7.2 Continuous Probability Classes $\mathcal{M}$

We have considered thus far countable probability classes $\mathcal{M}$, which makes sense from a computational point of view as emphasized in Section 3.2.9. On the other hand, in statistical parameter estimation one often has a continuous hypothesis class (e.g. a Bernoulli($\theta$) process with unknown $\theta\in[0,1]$). Let

$$\mathcal{M} \;:=\; \{\mu_\theta : \theta\in\Theta\subseteq I\!R^d\}$$

be a family of probability distributions parameterized by a $d$-dimensional continuous parameter $\theta$. Let $\mu\equiv\mu_{\theta_0}\in\mathcal{M}$ be the true generating distribution and $\theta_0$ be in the interior of the compact set $\Theta$. We may restrict $\mathcal{M}$ to a countable dense subset like $\{\mu_\theta\}$ with computable (or rational) $\theta$. If $\theta_0$ is itself a computable real (or rational) vector then Theorem 3.60 applies. From a practical point of view, the assumption of a computable $\theta_0$ is not so serious. It is more from a traditional analysis point of view that one would like quantities and results depending smoothly on $\theta$, and not depending in a weird fashion on the computational complexity of $\theta$. For instance, the weight $w(\theta)$ is often a continuous probability density

$$\xi(x_{1:n}) \;:=\; \int_\Theta d\theta\, w(\theta)\cdot\mu_\theta(x_{1:n}), \qquad \int_\Theta d\theta\, w(\theta) \;=\; 1, \qquad w(\theta) \;\geq\; 0. \quad (3.72)$$

The most important property of $\xi$ used in this book is $\xi(x_{1:n})\geq w_\nu\cdot\nu(x_{1:n})$ which was obtained from (3.5) by dropping the sum over $\nu$. The analogous construction here is to restrict the integral over $\Theta$ to a small vicinity $N_\delta$ of $\theta$. For sufficiently smooth $\mu_\theta$ and $w(\theta)$ we expect $\xi(x_{1:n})\gtrsim|N_{\delta_n}|\cdot w(\theta)\cdot\mu_\theta(x_{1:n})$, where $|N_{\delta_n}|$ is the volume of $N_{\delta_n}$. This in turn leads to $D_n\lesssim\ln w_\mu^{-1}+\ln|N_{\delta_n}|^{-1}$, where $w_\mu:=w(\theta_0)$. $N_{\delta_n}$ should be the largest possible region in which $\ln\mu_\theta$ is approximately flat on average. The averaged instantaneous, mean, and total curvature matrices of $\ln\mu$ are

$$j_t(x_{<t}) \;:=\; \mathbf{E}_t[\nabla_\theta \ln\mu_\theta(x_t|x_{<t})\nabla_\theta^T \ln\mu_\theta(x_t|x_{<t})]_{|\theta=\theta_0}, \qquad \bar{\jmath}_n \;:=\; \tfrac{1}{n}J_n,$$

$$J_n \;:=\; \sum_{t=1}^n \mathbf{E}[j_t(x_{<t})] \;=\; \mathbf{E}[\nabla_\theta \ln\mu_\theta(x_{1:n})\nabla_\theta^T \ln\mu_\theta(x_{1:n})]_{|\theta=\theta_0}. \quad (3.73)$$

They are the Fisher information of $\mu$ and may be viewed as measures of the parametric complexity of $\mu_\theta$ at $\theta=\theta_0$. The last equality can be shown by using the fact that the $\mu$-expected value of $\nabla\ln\mu\cdot\nabla^T\ln\mu$ coincides with $-\nabla\nabla^T\ln\mu$ (since $\mathcal{X}$ is finite) and a similar line of reasoning as in (3.18) for $D_n$.

**Theorem 3.74 (Continuous entropy bound)** Let $\mu_\theta$ be twice continuously differentiable at $\theta_0 \in \Theta \subseteq I\!R^d$ and $w(\theta)$ be continuous and positive at $\theta_0$. Furthermore we assume that the inverse of the mean Fisher information matrix $(\bar{\jmath}_n)^{-1}$ exists, is bounded for $n\to\infty$, and is uniformly (in $n$) continuous at $\theta_0$. Then the relative entropy $D_n$ between $\mu \equiv \mu_{\theta_0}$ and $\xi$ defined in (3.72) can be bounded by

$$D_n \;:=\; \mathbf{E}\ln\frac{\mu(x_{1:n})}{\xi(x_{1:n})} \;\leq\; \ln w_\mu^{-1} + \frac{d}{2}\ln\frac{n}{2\pi} + \frac{1}{2}\ln\det\bar{\jmath}_n + o(1) \;=:\; k_\mu,$$

where $w_\mu \equiv w(\theta_0)$ is the weight density (3.72) of $\mu$ in $\xi$, and $o(1)$ tends to zero for $n\to\infty$.

For independent and identically distributed distributions $\mu_\theta(x_{1:n}) = \mu_\theta(x_1)\cdot\ldots\cdot\mu_\theta(x_n)\;\forall\theta$ this bound was proven in [CB90, Thm.2.3]. In this case $J^{[CB90]}(\theta_0) \equiv \bar{\jmath}_n \equiv \jmath_n$ independent of $n$. For stationary ($k^{th}$-order) Markov processes $\bar{\jmath}_n$ is also constant. The proof generalizes to arbitrary $\mu_\theta$ by replacing $J^{[CB90]}(\theta_0)$ with $\bar{\jmath}_n$ everywhere in their proof. For the proof to go through, the vicinity $N_{\delta_n} := \{\theta : ||\theta-\theta_0||_{\bar{\jmath}_n} \leq \delta_n\}$ of $\theta_0$ must contract to a point set $\{\theta_0\}$ for $n\to\infty$ and $\delta_n \to 0$. $\bar{\jmath}_n$ is always positive semi-definite, as can be seen from the definition. The boundedness condition of $\bar{\jmath}_n^{-1}$ implies a strictly positive lower bound independent of $n$ on the eigenvalues of $\bar{\jmath}_n$ for all sufficiently large $n$, which ensures $N_{\delta_n} \to \{\theta_0\}$. The uniform continuity of $\bar{\jmath}_n$ ensures that the remainder $o(1)$ from the Taylor expansion of $D_n$ is independent of $n$. Note that twice continuous differentiability of $D_n$ at $\theta_0$ [CB90, Con.2] follows for finite $\mathcal{X}$ from twice continuous differentiability of $\mu_\theta$. Under some additional technical conditions one can even prove an equality $D_n = \ln w_\mu^{-1} + \frac{d}{2}\ln\frac{n}{2\pi e} + \frac{1}{2}\ln\det\bar{\jmath}_n + o(1)$ for the i.i.d. case [CB90, (1.4)], which is probably also valid for general $\mu$.

The $\ln w_\mu^{-1}$ part in the bound is the same as for countable $\mathcal{M}$. The $\frac{d}{2}\ln\frac{n}{2\pi}$ contribution can be understood as follows: Consider $\theta \in [0,1)$ and restrict the continuous $\mathcal{M}$ to $\theta$ that are finite binary fractions. Assign a weight $w(\theta) \approx 2^{-l}$ to a $\theta$ with binary representation of length $l$; $D_n \lesssim l\cdot\ln 2$ in this case. But what if $\theta$ is not a finite binary fraction? A continuous parameter can typically be estimated with accuracy $O(n^{-1/2})$ after $n$ observations. The data do not allow to distinguish a $\tilde{\theta}$ from the true $\theta$ if $|\tilde{\theta}-\theta| < O(n^{-1/2})$. There is such a $\tilde{\theta}$ with binary representation of length $l = \log_2 O(\sqrt{n})$. Hence we expect $D_n \lesssim \frac{1}{2}\ln n + O(1)$, or $\frac{d}{2}\ln n + O(1)$ for a $d$-dimensional parameter space. In general, the $O(1)$ term depends on the parametric complexity of $\mu_\theta$ and is explicated by the third $\frac{1}{2}\ln\det\bar{\jmath}_n$ term in Theorem 3.74. See [CB90, p454] for an alternative explanation. Note that a uniform weight $w(\theta) = \frac{1}{|\Theta|}$ does not lead to a uniform bound, unlike the discrete case. A uniform bound is obtained for Bernando's (or in the scalar case Jeffreys') reference prior $w(\theta) \sim \sqrt{\det\bar{\jmath}_\infty(\theta)}$ if $\jmath_\infty$ exists [Ris96].

For finite alphabet $\mathcal{X}$ we consider throughout this book, $j_t^{-1}<\infty$ independent of $t$ and $x_{<t}$ in case of i.i.d. sequences. More generally, the conditions of Theorem 3.74 are satisfied for the practically very important class of stationary ($k^{th}$-order) finite-state Markov processes ($k=0$ is i.i.d.).

Theorem 3.74 shows that Theorems 3.19–3.60 are also applicable to the case of continuously parameterized probability classes. Theorem 3.74 is also valid for a mixture of the discrete and continuous cases, $\xi=\sum_a\int d\theta\, w^a(\theta)\,\mu_\theta^a$ with $\sum_a\int d\theta\, w^a(\theta)=1$.

### 3.7.3 Further Applications

**Partial sequence prediction.** There are at least two ways to treat partial sequence prediction. By this we mean that not every symbol of the sequence needs to be predicted, say, given sequences of the form $z_1x_1...z_nx_n$, we want to predict the $x$'s only. The first way is to keep the $\Lambda_\rho$ prediction schemes of the last sections mainly as they are and to use a time-dependent loss function that assigns zero loss $\ell_{zy}^t\equiv 0$ at the $z$ positions. Any dummy prediction $y$ is then consistent with (3.45). The losses for predicting $x$ are generally nonzero. This solution is satisfactory as long as the $z$'s are drawn from a probability distribution. The second and preferable way does not rely on a probability distribution over the $z$. We replace all distributions $\rho(x_{1:n})$ ($\rho=\mu,\,\nu,\,\xi$) everywhere by distributions $\rho(x_{1:n}|z_{1:n})$ conditioned on $z_{1:n}$. The $z_{1:n}$ conditions cause nowhere problems as they can essentially be thought of as fixed (or as oracles or spectators). So the bounds in Theorems 3.19–3.74 also hold in this case for all individual $z$'s. Applications are:

**Independent experiments and classification (CF).** A typical experimental situation is a sequence of independent (i.i.d) experiments, predictions and observations. At time $t$ one arranges an experiment $z_t$ (or observes data $z_t$), then tries to make a prediction, and finally observes the true outcome $x_t$. Often one has a parameterized class of models (hypothesis space) $\mu_\theta(x_t|z_t)$ and wants to infer the true $\theta$ in order to make improved predictions. This is a special case of partial sequence prediction, where the hypothesis space $\mathcal{M}=\{\mu_\theta(x_{1:n}|z_{1:n})=\mu_\theta(x_1|z_1)\cdot...\cdot\mu_\theta(x_n|z_n)\}$ consists of i.i.d. distributions, but note that $\xi$ is not i.i.d. This is the same setting as for online learning of classification tasks, where a $z\in\mathcal{Z}$ should be classified as an $x\in\mathcal{X}$ (cf. Problem 3.12). The previous paragraph reduced this setting to sequence prediction.

### 3.7.4 Prediction with Expert Advice

There are two schools of universal sequence prediction: We considered expected performance bounds for Bayesian prediction based on mixtures of environments, as is common in information theory and statistics [MF98]. The other approach uses predictors based on expert advice (PEA) algorithms with worst-case loss bounds in the spirit of Littlestone, Warmuth, Vovk and others. The two schools usually do not refer to each other much. We briefly describe PEA and compare both approaches. For a more comprehensive comparison see [MF98]. In the following we focus on topics not covered in [MF98]. PEA was invented in [LW89, LW94] and [Vov92] and further developed in [CB97, HKW98, KW99] and by many others. Many variations known by many names (prediction/learning with expert advice, weighted majority/average, aggregating strategy, boosting, hedge algorithm, ...) have since been invented. Early works in this direction are [Daw84, Ris89]. See [Vov01] for a review and further references. We describe the setting and basic idea of PEA for binary alphabet. Consider a finite binary sequence $x_1x_2...x_n\in\{0,1\}^n$ and a finite set $\mathcal{E}$ of experts $e\in\mathcal{E}$ making predictions $x_t^e$ in the unit interval [0,1] based on past observations $x_1x_2...x_{t-1}$. The loss of expert $e$ in step $t$ is defined as $|x_t-x_t^e|$. In the case of binary predictions $x_t^e\in\{0,1\}$, $|x_t-x_t^e|$ coincides with our error measure (3.34). The PEA algorithm $p_{\beta n}$ combines the predictions of all experts. It forms its own prediction[9] $x_t^p\in[0,1]$ according to some weighted average of the expert's predictions $x_t^e$. There are certain update rules for the weights depending on some parameter $\beta$. Various bounds for the total loss $L_p(\boldsymbol{x}):=\sum_{t=1}^n|x_t-x_t^p|$ of PEA in terms of the total loss $L_\varepsilon(\boldsymbol{x}):=\sum_{t=1}^n|x_t-x_t^\varepsilon|$ of the best expert $\varepsilon\in\mathcal{E}$ have been proven. It is possible to fine-tune $\beta$ and to eliminate the necessity of knowing $n$ in advance. The first bound of this kind was obtained in [CB97]:

$$L_p(\boldsymbol{x}) \;\leq\; L_\varepsilon(\boldsymbol{x})+2.8\ln|\mathcal{E}|+4\sqrt{L_\varepsilon(\boldsymbol{x})\ln|\mathcal{E}|}. \tag{3.75}$$

The constants 2.8 and 4 were improved in [ACBG02, YEYS04]. The last bound in Theorem 3.36 with $S_n\leq D_n\leq\ln|\mathcal{M}|$ for uniform weights and with $E_n^{\Theta_\mu}$ increased to $E_n^\Theta$ reads

$$E_n^{\Theta_\xi} \;\leq\; E_n^\Theta+2\ln|\mathcal{M}|+2\sqrt{E_n^\Theta\ln|\mathcal{M}|}.$$

It has a quite similar structure to (3.75), although the algorithms, the settings, the proofs, and the interpretation are quite different. Whereas PEA performs well in any environment, but only relative to a given set of experts $\mathcal{E}$, our $\Theta_\xi$ predictor competes with the best possible $\Theta_\mu$ predictor (and hence with any other $\Theta$ predictor), but only in expectation and for a given set of environments $\mathcal{M}$. PEA depends on the set of experts, $\Theta_\xi$ depends on the set of environments $\mathcal{M}$. The basic $p_{\beta n}$ algorithm was extended in different directions: incorporation of different initial weights ($|\mathcal{E}|\rightsquigarrow w_\nu^{-1}$) [LW89, Vov92], more general loss functions [HKW98], continuous-valued outcomes [HKW98], and multidimensional predictions [KW99] (but not yet for the absolute loss). The work [Yam98] lies somewhat in between PEA and this book; "PEA"

[9] The original PEA version [LW89] had discrete deterministic prediction $x_t^p\in\{0,1\}$ with (necessarily) twice as many errors as the best expert and now is only of historical interest.

techniques are used to prove expected loss bounds, but only for sequences of independent symbols/experiments and limited classes of loss functions. Finally, note that the predictions of PEA are continuous. This is appropriate for weather forecasters, who announce the probability of rain, but the *decision* to wear sunglasses or to take an umbrella is binary, and the suffered loss depends on this binary decision, not on the probability estimate. It is possible to convert the continuous prediction of PEA into a probabilistic binary prediction by predicting 1 with probability $x_t^p \in [0,1]$. The probability of making an error is then $|x_t - x_t^p|$. Note that the expectation is taken over the probabilistic prediction, whereas for the deterministic $\Theta_\xi$ algorithm the expectation is taken over the environmental distribution $\mu$. The multidimensional case [KW99] could then be interpreted as a (probabilistic) prediction of symbols over an alphabet $\mathcal{X} = \{0,1\}^d$, but error bounds for the absolute loss have yet to be proven. In [FS97] the regret is bounded by $\ln|\mathcal{E}| + \sqrt{2\tilde{L}\ln|\mathcal{E}|}$ for arbitrary unit loss function and alphabet, where $\tilde{L}$ is an upper bound on $L_\varepsilon$, which has to be known in advance. It is possible to generalize PEA and bound (3.75) to arbitrary alphabet and weights and to general loss functions with probabilistic interpretation [HP04].

### 3.7.5 Outlook

In the following we discuss several directions in which the findings of this book may be extended.

**Infinite alphabet.** In many cases the basic prediction unit is not a letter, but a number (for inducing number sequences), or a word (for completing sentences), or a real number or vector (for physical measurements). The prediction may either be generalized to a block-by-block prediction of symbols, or, more suitably, the finite alphabet $\mathcal{X}$ could be generalized to countable (numbers, words) or continuous (real or vector) alphabets. The presented theorems are independent of the size of $\mathcal{X}$ and hence should generalize to countably infinite alphabets by appropriately taking the limit $|\mathcal{X}| \to \infty$ and to continuous alphabets by a denseness or separability argument. Since the proofs are also independent of the size of $\mathcal{X}$, we may directly replace all finite sums over $\mathcal{X}$ by infinite sums or integrals and carefully check the validity of each operation. We expect all theorems to remain valid in full generality, except for minor technical existence and convergence constraints.

An infinite prediction space $\mathcal{Y}$ was no problem at all as long as we assumed the existence of $y_t^{\Lambda_\rho} \in \mathcal{Y}$ (3.45). In case $y_t^{\Lambda_\rho} \in \mathcal{Y}$ does not exist one may define $y_t^{\Lambda_\rho} \in \mathcal{Y}$ in a way to achieve a loss at most $\varepsilon_t = o(t^{-1})$ larger than the infimum loss. We expect a small finite correction of the order of $\varepsilon = \sum_{t=1}^\infty \varepsilon_t < \infty$ in the loss bounds somehow.

**More active systems.** Prediction means guessing the future, but not influencing it. A small step in the direction to more active systems was to allow the

$\Lambda$ system to act and to receive a loss $\ell_{x_t y_t}$ depending on the action $y_t$ and the outcome $x_t$. The probability $\mu$ is still independent of the action, and the loss function $\ell^t$ has to be known in advance. This ensures that the greedy strategy (3.45) is optimal. The loss function may be generalized to depend not only on the history $x_{<t}$, but also on the historic actions $y_{<t}$ with $\mu$ still independent of the action. It would be interesting to know whether the scheme $\Lambda$ and/or the loss bounds generalize to this case. The full model of an acting agent influencing the environment is developed in the next chapter, but non-asymptotic loss bounds have yet to be proven.

**Miscellaneous.** Another direction is to investigate the learning aspect of universal prediction. Many prediction schemes explicitly learn and exploit a model of the environment. Learning and exploitation are melted together in the framework of universal Bayesian prediction. A separation of these two aspects in the spirit of hypothesis learning with MDL [VL00] could lead to new insights. Also, the separation of noise from useful data, usually an important issue [GTV01], did not play a role here. The attempt at an information-theoretic interpretation of Theorem 3.63 may be made more rigorous in this or another way. In the end, this may lead to a simpler proof of Theorem 3.63 and maybe even for the loss bounds. A unified picture of the loss bounds obtained here and the loss bounds for predictors based on expert advice (PEA) could also be fruitful. Yamanishi [Yam98] used PEA methods to prove expected loss bounds for Bayesian prediction, so maybe the proof technique presented here could be used *vice versa* to prove more general loss bounds for PEA. Maximum-Likelihood predictors may also be studied. Since $2^{-K(x)}$ (or some of its variants) is a close approximation of $\xi_U$, it is generally believed that predictions based on $K$ are as good as predictions based on $\xi_U$. Convergence and loss bounds for predictors based on $K$ would prove this conjecture, but it is easy to see that $K$ completely fails for predictive purposes [Hut03d]. Also, more promising variants like the monotone complexity $Km$ and universal two-part MDL, both extremely close to $\xi_U$, fail in certain situations (see Problems 2.8 and 3.17 or [Hut03d, PH04a]). Finally, the reader is invited to apply the $\Lambda_\xi$ predictor to his favorite induction problem by choosing a suitable $\mathcal{M}$ with computable $\xi$.

## 3.8 Summary

We compared universal predictions based on Bayes mixtures $\xi$ to the infeasible informed predictor based on the unknown true generating distribution $\mu$. We showed that the universal posterior $\xi$ converges to $\mu$ and that $\xi/\mu \to 1$. Our main focus was on a decision-theoretic setting, where each prediction $y_t \in \mathcal{X}$ (or more generally action $y_t \in \mathcal{Y}$) results in a loss $\ell_{x_t y_t}$ if $x_t$ is the true next symbol of the sequence. We showed that the $\Lambda_\xi$ predictor suffers only slightly more loss than the $\Lambda_\mu$ predictor. We also showed that the derived error and loss bounds cannot be improved in general, i.e. without making extra assumptions on $\ell$, $\mu$, $\mathcal{M}$, or $w_\nu$. Within a factor of 2 this is also true for any $\mu$-independent predictor. We demonstrated Pareto optimality of $\xi$ in the sense that there is no other predictor that performs better or equal in all environments $\nu\in\mathcal{M}$ and strictly better in at least one. Optimal predictors can (in most cases) be based on mixture distributions $\xi$. Finally, we gave an Occam's razor argument that Solomonoff-Levin's prior with weights $w_\nu=2^{-K(\nu)}$ is optimal, where $K(\nu)$ is the Kolmogorov complexity of $\nu$. Of course, optimality always depends on the setup, the assumptions, and the chosen criteria. For instance, the universal predictor was not always Pareto optimal, but at least for many popular, and for all decision-theoretic performance measures it was. Bayes predictors are also not necessarily optimal under worst-case criteria [CBL01]. We also derived a bound for the relative entropy between $\xi$ and $\mu$ in the case of a continuously parameterized family of environments, which allowed us to generalize the loss bounds to continuous $\mathcal{M}$. Furthermore, we discussed the duality between the Bayes mixture and expert-mixture approaches and results, classification tasks, games of chances, infinite alphabet, active systems influencing the environment, and others.

## 3.9 Technical Proofs

### 3.9.1 How to Deal with $\mu=0$

Some expressions (like conditional or inverse probabilities) are undefined for zero $\mu$. We thought of the following solutions:

**Avoid the problem.** We may restrict ourselves to $\mu(x_{1:n})>0\,\forall x_{1:n}$.
+ The treatment in this book is then rigorous, and a zero $\mu$ can be approximated to an arbitrary precision. From a practical point of view, this is a completely satisfactory approach.
– Theoretically unsatisfactory, because deterministic environments (for which $\mu(x_t|x_{<t})=0$ for all but one $x_t$) are of special interest, and not just esoteric limits.

**Take the limit.** Develop all theorems for $\mu^{(i)}>0$ and finally perform the limit $\mu^{(i)}\stackrel{i\to\infty}{\longrightarrow}\mu$, where $\mu$ might be zero for some strings. For instance, $\mu^{(\varepsilon)}(x_{1:n}):=(1-\varepsilon)\mu(x_{1:n})+\varepsilon/2^n$ and $\varepsilon\to 0$ will do.
+ Rigorous treatment with the advantage not having to deal with the problem until the end. If all spaces are finite, then interchange of finite sums or maxs with $\lim_{i\to\infty}$ is safe.
– Problematic for infinite spaces (e.g. alphabet, time, ...), since limits may not be interchangeable.

**Face the problem.** A way of facing the problem, which is different from Section 3.2.1, is to restrict the set of strings to one with nonzero $\mu$-probability. Define the critical set $Z:=\bigcup_{x\in\mathcal{X}^*:\mu(x)=0}\Gamma_x$, where $\Gamma_x:=\{\omega:\omega_{1:\ell(x)}=x\}$

is defined as the cylinder set containing all infinite sequences $\omega$ starting with $x$.

+ Since $Z$ is a countable (for countable alphabet) union of cylinder sets $\Gamma_x$ of measure zero, $Z$ itself is measurable with $\mu$-measure zero. So all theorems proven with $\mu$-probability 1 on $\Gamma_\epsilon\setminus Z$ still hold on $\Gamma_\epsilon$ with $\mu$-probability 1, since $\mu(Z)=0$.
– All sums over $x$ have to be restricted appropriately. More seriously, other measures on $\Gamma_\epsilon$, especially $\xi$, deteriorate to semimeasures on $\Gamma_\epsilon\setminus Z$ (see Section 2.4 and Problem 3.1).

**Ignore the problem.** Address other more severe or interesting problems first. Why waste time fooling with exceptions when everything seems to work well anyway and there are more important problems to solve.

+ Time-efficient "physicist" approach (like: always exchange limits and integrals until you get into trouble).
– The approach is risky and a mathematician would turn in his grave.

We usually faced the problem, but decided to avoid/ignore these subtleties in the main text (see Problem 3.8). We will also not explicate every subtlety in the following proofs. Subtleties regarding $y,z=0/1$ have been checked but will be passed over. $0\ln\frac{0}{z_i}:=0$ even for $z_i=0$. Positive means $\geq 0$. The probability constraints in (3.76) on $y$ and $z$ are assumed to hold throughout this section. Finally, $z>0$ if $y>0$.

### 3.9.2 Entropy Inequalities (Lemma 3.11)

We show that

$$\frac{1}{2}\sum_{i=1}^N f(y_i-z_i) \leq f\Big(\sqrt{\frac{1}{2}\sum_{i=1}^N y_i\ln\frac{y_i}{z_i}}\Big) \quad \text{for} \quad y_i\geq 0, \quad z_i\geq 0, \quad \sum_{i=1}^N y_i=1=\sum_{i=1}^N z_i \tag{3.76}$$

for any convex and even ($f(x)=f(-x)$) function with $f(0)\leq 0$. For $f(x)=x^2$ we get inequality (3.11$s$), and for $f(x)=|x|$ we get inequality (3.11$a$). To prove (3.76) we partition (for the moment arbitrarily) $i\in\{1,...,N\}=G^+\cup G^-$, $G^+\cap G^-=\{\}$, and define $y^\pm:=\sum_{i\in G^\pm} y_i$ and $z^\pm:=\sum_{i\in G^\pm} z_i$. It is well known that the relative entropy is positive, i.e.

$$\sum_{i\in G^\pm} p_i\ln\frac{p_i}{q_i} \geq 0 \quad \text{for} \quad p_i\geq 0, \quad q_i\geq 0, \quad \sum_{i\in G^\pm} p_i = 1 = \sum_{i\in G^\pm} q_i. \tag{3.77}$$

Note that there are four probability distributions ($p_i$ and $q_i$ for $i\in G^+$ and $i\in G^-$). For $i\in G^\pm$, $p_i:=y_i/y^\pm$ and $q_i:=z_i/z^\pm$ satisfy the conditions on $p$ and $q$. Inserting this into (3.77) and rearranging terms, we get

$$\sum_{i\in G^\pm} y_i\ln\frac{y_i}{z_i} \geq y^\pm\ln\frac{y^\pm}{z^\pm}.$$

If we sum over $\pm$ and define $y \equiv y^+ = 1-y^-$ and $z \equiv z^+ = 1-z^-$, we get

$$\sum_{i=1}^N y_i \ln\frac{y_i}{z_i} \;\geq\; \sum_\pm y^\pm \ln\frac{y^\pm}{z^\pm} \;=\; y\ln\frac{y}{z} + (1-y)\ln\frac{1-y}{1-z} \;\geq\; 2(y-z)^2. \quad (3.78)$$

The last inequality is elementary and well known. For the special choice $G^\pm := \{i : y_i \gtrless z_i\}$, we can upper-bound $\sum_i f(y_i - z_i)$ as follows

$$\sum_{i\in G^\pm} f(y_i - z_i) \stackrel{(a)}{=} \sum_{i\in G^\pm} f(|y_i - z_i|) \stackrel{(b)}{\leq} f(\sum_{i\in G^\pm} |y_i - z_i|) \stackrel{(c)}{=} f(|\sum_{i\in G^\pm} y_i - z_i|)$$

$$\stackrel{(d)}{=} f(|y^\pm - z^\pm|) \stackrel{(e)}{=} f(|y-z|) \stackrel{(f)}{=} f(\sqrt{(y-z)^2}) \stackrel{(g)}{\leq} f(\sqrt{\frac{1}{2}\sum_{i=1}^N y_i \ln\frac{y_i}{z_i}}) \quad (3.79)$$

$(a)$ follows from the symmetry of $f$. $(b)$ follows from the convexity of $f$ and from $f(0)\leq 0$: Inserting $y=0$ and $x=a+b$ in the convexity definition $\alpha f(x) + (1-\alpha)f(y) \geq f(\alpha x + (1-\alpha)y)$ leads to $\alpha f(a+b) + (1-\alpha)f(0) \geq f(\alpha(a+b))$. Inserting $\alpha = \frac{a}{a+b}$ and $\alpha = \frac{b}{a+b}$ and adding both inequalities gives $f(a+b) + f(0) \geq f(a) + f(b)$ for $a,b \geq 0$. Using $f(0)\leq 0$ we get $f(\sum_i x_i) \geq \sum_i f(x_i)$ for $x_i \geq 0$ by induction. $(c)$ is true, since all $y_i - z_i$ are positive/negative for $i\in G^\pm$ due to the special choice of $G^\pm$. $(d)$ and $(e)$ follow from the definition of $y^{(\pm)}$ and $z^{(\pm)}$, and $(f)$ is obvious. $(g)$ follows from (3.78) and the monotonicity of $\sqrt{\;}$ and $f$ for positive arguments: Inserting $b=y=-x$ and $\alpha=\frac{1}{2}$ into the convexity definition and using the symmetry of $f$ we get $f(b)\geq f(0)$. Inserting this into $f(a+b)+f(0)\geq f(a)+f(b)$ we get $f(a+b)\geq f(a)$, which proves that $f$ is monotone increasing for positive arguments $(a,b\geq 0)$. Inequality (3.76) follows by summation of (3.79) over $\pm$ and noting that $f(\sqrt{\;})$ is independent of $\pm$. This proves Lemma 3.11$f$.

Inserting $f(x) = x^2$ yields Lemma 3.11$s$; inserting $f(x) = |x|$ yields Lemma 3.11$a$. Lemma 3.11$b$ follows from

$$\sum_{i=1}^N y_i \left|\ln\frac{y_i}{z_i}\right| - \sum_i y_i \ln\frac{y_i}{z_i} \;=\; -2\sum_{i\in G^-} y_i \ln\frac{y_i}{z_i} \;\leq\; 2\sum_{i\in G^-} z_i - y_i \;=\; \sum_{i=1}^N |y_i - z_i|,$$

where we have used $-\ln x \leq \frac{1}{x} - 1$. Lemma 3.11$h$ is proven differently. For arbitrary $y\geq 0$ and $z\geq 0$ we define

$$f(y,z) := y\ln\frac{y}{z} - (\sqrt{y}-\sqrt{z})^2 + z - y = 2yg(\sqrt{z/y}) \text{ with } g(t) := -\ln t + t - 1 \geq 0.$$

This shows $f\geq 0$, and hence $\sum_i f(y_i,z_i)\geq 0$, which implies

$$\sum_i y_i \ln\frac{y_i}{z_i} - \sum_i (\sqrt{y_i} - \sqrt{z_i})^2 \geq \sum_i y_i - \sum_i z_i = 1 - 1 = 0.$$

This proves Lemma 3.11$h$. □

### 3.9.3 Error Inequality (Theorem 3.36)

Here we give a direct proof of the second bound in Theorem 3.36. Again, we try to find small constants $A$ and $B$ that satisfy the linear inequality

$$E_n^{\Theta_\xi} \le (A+1)E_n^{\Theta_\mu} + (B+1)S_n. \tag{3.80}$$

If we could show

$$e_t^{\Theta_\xi}(x_{<t}) \le (A+1)e_t^{\Theta_\mu}(x_{<t}) + (B+1)s_t(x_{<t}) \tag{3.81}$$

for all $t \le n$ and all $x_{<t}$, (3.80) would follow immediately by summation and the definition of $E_n$ and $S_n$. With the abbreviations (3.12) and the abbreviations $m = x_t^{\Theta_\mu}$ and $s = x_t^{\Theta_\xi}$, the various error functions can then be expressed by $e_t^{\Theta_\xi} = 1 - y_s$, $e_t^{\Theta_\mu} = 1 - y_m$ and $s_t = \sum_i (y_i - z_i)^2$. Inserting this into (3.81) we get

$$1 - y_s \le (A+1)(1 - y_m) + (B+1)\sum_{i=1}^N (y_i - z_i)^2. \tag{3.82}$$

By definition of $x_t^{\Theta_\mu}$ and $x_t^{\Theta_\xi}$ we have $y_m \ge y_i$ and $z_s \ge z_i$ for all $i$. We prove a sequence of inequalities which show that

$$(B+1)\sum_{i=1}^N (y_i - z_i)^2 + (A+1)(1 - y_m) - (1 - y_s) \ge \ldots \tag{3.83}$$

is positive for suitable $A \ge 0$ and $B \ge 0$, which proves (3.82). For $m = s$ (3.83) is obviously positive. So we will assume $m \ne s$ in the following. From the square we keep only contributions from $i = m$ and $i = s$.

$$\ldots \ge (B+1)[(y_m - z_m)^2 + (y_s - z_s)^2] + (A+1)(1 - y_m) - (1 - y_s) \ge \ldots$$

By definition of $y$, $z$, $\mathcal{M}$ and $s$ we have the constraints $y_m + y_s \le 1$, $z_m + z_s \le 1$, $y_m \ge y_s \ge 0$ and $z_s \ge z_m \ge 0$. From the latter two it is easy to see that the square terms (as a function of $z_m$ and $z_s$) are minimized by $z_m = z_s = \frac{1}{2}(y_m + y_s)$. Furthermore, we define $x := y_m - y_s$ and eliminate $y_s$.

$$\ldots \ge (B+1)\tfrac{1}{2}x^2 + A(1 - y_m) - x \ge \ldots \tag{3.84}$$

The constraint on $y_m + y_s \le 1$ translates into $y_m \le \frac{x+1}{2}$, hence (3.84) is minimized by $y_m = \frac{x+1}{2}$.

$$\ldots \ge \tfrac{1}{2}[(B+1)x^2 - (A+2)x + A] \ge \ldots \tag{3.85}$$

(3.85) is quadratic in $x$ and minimized by $x^* = \frac{A+2}{2(B+1)}$. Inserting $x^*$ gives

$$\ldots \ge \frac{4AB - A^2 - 4}{8(B+1)} \ge 0 \quad \text{for} \quad B \ge \tfrac{1}{4}A + \tfrac{1}{A}, \quad A > 0, \quad (\Rightarrow B \ge 1). \tag{3.86}$$

Inequality (3.80) therefore holds for any $A>0$, provided we insert $B=\frac{1}{4}A+\frac{1}{A}$. Thus we might minimize the r.h.s. of (3.80) w.r.t. $A$, leading to the upper bound

$$E_n^{\Theta_\xi} \leq E_n^{\Theta_\mu}+S_n+\sqrt{4E_n^{\Theta_\mu}S_n+S_n^2} \qquad \text{for} \qquad A^2=\frac{S_n}{E_n^{\Theta_\mu}+\frac{1}{4}S_n},$$

which completes the proof of Theorem 3.36. □

### 3.9.4 Binary Loss Inequality for $z\leq\frac{1}{2}$ (3.57)

With the definition

$$f(y,z) := B'\cdot\left[y\ln\frac{y}{z}+(1-y)\ln\frac{1-y}{1-z}\right]+A'\cdot(1-y)\frac{z}{1-z}-y, \qquad z\leq\tfrac{1}{2}, \tag{3.87}$$

we show $f(y,z)\geq 0$ for suitable $A'\equiv A+1$ and $B'\equiv B+1$. We do this by showing that $f\geq 0$ at all extremal values and "at" boundaries. Keeping $y$ fixed, $f\to+\infty$ for $z\to 0$, if we choose $B'>0$. For the boundary $z=\frac{1}{2}$ we lower-bound the relative entropy by the sum over squares (Lemma 3.11$s$)

$$f(y,\tfrac{1}{2}) \geq 2B'(y-\tfrac{1}{2})^2+A'(1-y)-y.$$

The r.h.s. is quadratic in $y$ with minimum at $y^*=\frac{A'+2B'+1}{4B'}$, which implies

$$f(y,\tfrac{1}{2}) \;\geq\; f(y^*,\tfrac{1}{2}) \;\geq\; \frac{4AB-A^2-4}{8(B+1)} \;\geq 0 \quad \text{for} \quad B\geq\tfrac{1}{4}A+\tfrac{1}{A}, \quad A>0$$

(which implies $B\geq 1$). Furthermore, for $A\geq 4$ and $B\geq 1$ we have $f(y,\frac{1}{2})\geq 2(1-y)(3-2y)\geq 0$. Hence $f(y,\frac{1}{2})\geq 0$ for $B\geq\frac{1}{A}+1$, since for $A\geq 4$ it implies $B\geq 1$, and for $A\leq 4$ it implies $B\geq\frac{1}{4}A+\frac{1}{A}$.

The extremal condition $\partial f/\partial z=0$ (keeping $y$ fixed) leads to

$$y \;=\; y^* \;:=\; z\cdot\frac{B'(1-z)+A'}{B'(1-z)+A'z}.$$

Inserting $y^*$ into the definition of $f$, and again replacing the relative entropy by the sum over squares (Lemma 3.11$s$), we get

$$f(y^*,z) \;\geq\; 2B'(y^*-z)^2+A'(1-y^*)\tfrac{z}{1-z}-y^* \;=\; \tfrac{z(1-z)}{(B'(1-z)+A'z)^2}\cdot g(z),$$

$$g(z) \;:=\; 2B'A'^2z(1-z)+[(A'-1)B'(1-z)-A'](B'+A'\tfrac{z}{1-z}).$$

We have reduced the problem to showing $g\geq 0$. If the bracket [...] is positive, then $g$ is positive. If the bracket is negative, we can decrease $g$ by increasing $\frac{z}{1-z}\leq 1$ in $(B'+A'\frac{z}{1-z})$ to 1. The resulting expression is now quadratic in $z$

with minima at the boundary values $z=0$ and $z=\frac{1}{2}$. It is therefore sufficient to check

$$g(0) \geq (AB-1)(A+B+2) \geq 0 \quad \text{and} \quad g(\tfrac{1}{2}) \geq \tfrac{1}{2}(AB-1)(2A+B+3) \geq 0$$

which is true for $B\geq\frac{1}{A}$. In summary, we have proved (3.87) for $B\geq\frac{1}{A}+1$ and $A>0$. □

### 3.9.5 Binary Loss Inequality for $z\geq\frac{1}{2}$ (3.58)

With the definition

$$f(y,z) \;:=\; B'\cdot\left[y\ln\frac{y}{z}+(1-y)\ln\frac{1-y}{1-z}\right]+A'\cdot(1-y)-y\frac{1-z}{z}, \qquad z\geq\tfrac{1}{2} \tag{3.88}$$

we show $f(y,z)\geq 0$ for suitable $A'\equiv A+1>1$ and $B'\equiv B+1>2$, similarly as in the last subsection by proving that $f\geq 0$ at all extremal values and "at" boundaries. Keeping $y$ fixed, $f\to+\infty$ for $z\to 1$. The boundary $z=\frac{1}{2}$ was already checked in the last paragraph. The extremal condition $\partial f/\partial z=0$ (keeping $y$ fixed) leads to

$$y \;=\; y^* \;:=\; z\cdot\frac{B'z}{(B'+1)z-1}.$$

Inserting $y^*$ into the definition of $f$ and replacing the relative entropy by the sum over squares (Lemma 3.11$s$), we get

$$f(y^*,z) \;\geq\; 2B'(y^*-z)^2+A'(1-y^*)-y^*\tfrac{1-z}{z} \;=\; \tfrac{z(1-z)}{((B'+1)z-1)^2}\cdot g(z),$$
$$g(z) \;:=\; [(A'-1)B'z-A'+2z(1-z)](B'+1-\tfrac{1}{z})+2(1-z)^2.$$

We have reduced the problem to showing $g\geq 0$. Since $(B'+1-\frac{1}{z})\geq 0$ it is sufficient to show that the bracket is positive. We solve $[...]\geq 0$ w.r.t. $B$ and get

$$B\geq\frac{1-2z(1-z)}{z}\cdot\frac{1}{A}+\frac{1-z}{z}.$$

For $B\geq\frac{1}{A}+1$ this is satisfied for all $\frac{1}{2}\leq z\leq 1$. In summary, we have proved (3.88) for $B\geq\frac{1}{A}+1$ and $A>0$. □

### 3.9.6 General Loss Inequality (3.53)

We reduce

$$f(\boldsymbol{y},\boldsymbol{z}) := B'\sum_{i=1}^N y_i\ln\frac{y_i}{z_i}+A'\sum_{i=1}^N y_i\ell_{im}-\sum_{i=1}^N y_i\ell_{is} \;\geq\; 0 \tag{3.89}$$

$$\text{for} \quad \sum_{i=1}^{N} z_i d_i \geq 0, \quad d_i := \ell_{im} - \ell_{is} \tag{3.90}$$

to the binary $N=2$ case. We do this by keeping $\boldsymbol{y}$ fixed and showing that $f$ as a function of $\boldsymbol{z}$ is positive at all extrema in the interior of the simplex $\Delta:=\{\boldsymbol{z}:\sum_i z_i=1, z_i\geq 0\}$ of the domain of $\boldsymbol{z}$ and "at" all boundaries. First, the boundaries $z_i\to 0$ are safe as $f\to\infty$ for $B'>0$. Variation of $f$ w.r.t. to $\boldsymbol{z}$ leads to a minimum at $\boldsymbol{z}=\boldsymbol{y}$. If $\sum_i y_i d_i\geq 0$, we have

$$f(\boldsymbol{y},\boldsymbol{y}) = \sum_i y_i(A'\ell_{im}-\ell_{is}) \geq \sum_i y_i(\ell_{im}-\ell_{is}) = \sum_i y_i d_i \geq 0.$$

In the first inequality we used $A'>1$. If $\sum_i y_i d_i<0$, $\boldsymbol{z}=\boldsymbol{y}$ is outside the valid domain due to constraint (3.90), and the valid minima are attained at the boundary $\Delta\cap P$ with $P:=\{\boldsymbol{z}:\sum_i z_i d_i=0\}$. We implement the constraints with the help of Lagrange multipliers and extremize

$$L(\boldsymbol{y},\boldsymbol{z}) := f(\boldsymbol{y},\boldsymbol{z}) + B'\lambda\sum z_i + B'\mu\sum z_i d_i.$$

$\partial L/\partial z_i=0$ leads to $y_i=y_i^*:=z_i(\lambda+\mu d_i)$. Summing this equation over $i$, we obtain $\lambda=1$. $\mu$ is a function of $\boldsymbol{y}$ for which a formal expression might be given. If we eliminate $y_i$ in favor of $z_i$, we get

$$f(\boldsymbol{y}^*,\boldsymbol{z}) = \sum_i c_i z_i \quad \text{with} \quad c_i := (1+\mu d_i)(B'\ln(1+\mu d_i) + A'\ell_{im} - \ell_{is}).$$

In principle $\mu$ is a function of $\boldsymbol{y}$, but we can treat $\mu$ directly as an independent variable, since $\boldsymbol{y}$ has been eliminated.

The next step is to determine the extrema of the function $f=\sum c_i z_i$ for $\boldsymbol{z}\in\Delta\cap P$. For clarity we state the line of reasoning for $N=3$ first. In this case $\Delta$ is a triangle. As $f$ is linear in $\boldsymbol{z}$ it assumes its extrema at the vertices of the triangle, where all $z_i=0$ except one. But we have to take into account a further constraint $\boldsymbol{z}\in P$. The plane $P$ intersects triangle $\Delta$ in a finite line (for $\Delta\cap P=\{\}$ the only boundaries are $z_i\to 0$, which have already been treated). Again, as $f$ is linear, it assumes its extrema at the ends of the line, i.e. at edges of the triangle $\Delta$ on which all but two $z_i$ are zero. Similarly for $N>3$, the extrema of $f$, restricted to the polytope $\Delta\cap P$, are assumed in the corners of $\Delta\cap P$, which lie on the edges of simplex $\Delta$, where all but two $z_i$ are zero. We conclude that a necessary condition for a minimum of $f$ at the boundary is that at most two $z_i$ are nonzero. But this implies that all but two $y_i$ are zero. If we had eliminated $\boldsymbol{z}$ in favor of $\boldsymbol{y}$, we could not have made the analogous conclusion because $y_i=0$ does not necessarily imply $z_i=0$. We have effectively reduced the problem of showing $f(\boldsymbol{y}^*,\boldsymbol{z})\geq 0$ to the case $N=2$. We can go back one step further and prove (3.89) for $N=2$, which implies $f(\boldsymbol{y}^*,\boldsymbol{z})\geq 0$ for $N=2$. A proof of (3.89) for $N=2$ implies, by the arguments given above, that it holds for all $N$. This is what we set out to show here. □

The $N=2$ case was proven in Section 3.4 and the two previous subsections.

## 3.10 History & References

There are good introductions and surveys of Solomonoff sequence prediction [LV92a, LV97], inductive inference in general [AS83, Sol97, MF98], competitive online statistics [Vov01], and reasoning under uncertainty [Grü98]. The latter also contains a more serious discussion of the case $\mu \not\in \mathcal{M}$ in a related context. The convergence $\xi \to \mu$ of Theorem 3.19(*iii*) was first proven in [BD62] in a more general framework with the help of martingales, but the martingale proof does not provide a speed of convergence. Solomonoff's contribution was to focus on the set of all computable distributions [Sol64, Eq.(13)] and to prove Theorem 3.19(*i*) for binary alphabet, which shows that convergence (*iii*) of $\xi$ to $\mu$ is rapid [Sol78]. The generalization of (*i*) to arbitrary finite alphabet was probably first shown by the author in [Hut01a], but may have occurred earlier somewhere in the statistics literature. Convergence (*v*) of the ratio $\xi/\mu$ to 1 w.$\mu$.p.1 was first shown by Gács with the help of martingales [LV97], again not allowing one to estimate the speed of convergence. The elementary proof of $\xi/\mu \to 1$ i.m.s., i.e. of (*iv*) and (*v*) is from [Hut03a], showing that convergence is rapid. (*vi*) directly follows from Lemma 3.11(*a*). Lemma 3.11(*a*) is due to Pinsker [Pin64] and Csiszàr [Csi67], and can be found in [CT91, Lem.12.6.1]. A proof of Lemma 3.11(*h*) can be found in [BM98, p178]. Lemma 3.11(*b*) is also known [Bar00]. Most other results in this chapter are from [Hut03a, Hut03c].

## 3.11 Problems

**3.1 (Semimeasures)** [C30u/C40o] All results in this chapter were obtained for probability measures $\mu$, and $\xi$ and $w_\nu$, i.e. $\sum_{x_{1:t}} \xi(x_{1:t}) = \sum_{x_{1:t}} \mu(x_{1:t}) = \sum_\nu w_\nu = 1$. On the other hand, the primary class $\mathcal{M}$ of interest in this book is the class of all enumerable semimeasures and $\sum_\nu w_\nu \leq 1$, (see Section 2.4). In general, each of the following four items could be semi (<) or not (=): ($\xi$, $\mu$, $\mathcal{M}$, $w_\nu$), where $\mathcal{M}$ is semi if some elements are semi.

Which of the $2^4$ combinations make sense? (Hint: 6 of the 16). Show that the entropy inequalities (Lemma 3.11) hold for (<,=,<,<), but not for (<,<,<,<). Nevertheless, show that $\xi \to \mu$ (Theorem 3.19 *iii*) for (<,<,<,<) with maximal $\mu$ semi-probability, i.e. fails with $\mu$ semi-probability 0. Generalize all other theorems in this chapter as far as possible to the semi case.

**3.2 (Dominance of the Speed prior)** [C40oi] Which (semi)measures are multiplicatively dominated by the Speed prior defined in [Sch02b]? Show that computable deterministic environments are not dominated by the Speed prior, but for quickly computable deterministic environments domination holds with a slowly decreasing "constant". Does the Speed prior dominate quickly computable truly probabilistic environments for some suitable definition of 'quick' and 'truly'? Is there an easily characterizable class of (all?) dominated probabilistic environments?

**3.3 (Comparing two mixtures)** [C05u/C40o] Consider two mixtures $\xi$ and $\xi'$ over $\mathcal{M}$. Show that $\sum_{t=1}^\infty \mathbf{E}[\sum_{x_t}(\xi(x_t|x_{<t})-\xi'(x_t|x_{<t}))^2] \leq 2[\ln w_\mu^{-1} + \ln w_\mu'^{-1}] < \infty$, i.e. $\xi(x_t|x_{<t}) \to \xi'(x_t|x_{<t})$ for $t\to\infty$ with $\mu$-probability 1 for all $\mu\in\mathcal{M}$ (cf. Theorem 3.19$i$). Furthermore, show that $L_n^{\Lambda_\xi} - L_n^{\Lambda'_\xi} \leq O(\sqrt{L_n^{\Lambda_\mu}})$ (cf. Theorem 3.48). Is the stronger result $L_\infty^{\Lambda_\xi} - L_\infty^{\Lambda'_\xi} < \infty$ also true?

**3.4 (Convergence and loss bounds with high probability)** [C30oi] Show that $\mathbf{P}[\sum_{t=1}^n(\mu(x_t|x_{<t}) - \xi(x_t|x_{<t}))^2 \geq \frac{1}{\varepsilon}\ln w_\mu^{-1}] \leq \varepsilon$ and $\mathbf{P}[\sum_{t=1}^n(l_t^{\Lambda_\xi} - l_t^{\Lambda_\mu})^2 \geq \frac{2}{\varepsilon}\ln w_\mu^{-1}] \leq \varepsilon$, where $\mathbf{P}$ denotes $\mu$-probability. Use Theorem 3.19($i$) and Theorem 3.59($i$) and Markov's inequality. Is it possible to prove similar high-probability bounds for the ratio $l_t^{\Lambda_\xi}/l_t^{\Lambda_\mu}$, possibly exploiting Theorem 3.48 and Corollary 3.49($iii$)? High-probability bounds on $l_t^\Lambda(x_{<t})$ still involve an expectation over $x_t$ (see definition of $l_t^\Lambda$). Is it possible to prove high-probability bounds on the difference or ratio of $\ell_{x_t y_t^{\Lambda_\xi}}$ and $\ell_{x_t y_t^{\Lambda_\mu}}$, which do not involve any expectations?

**3.5 (Pareto optimality)** [C30u] Show that $\xi$ is not Pareto optimal w.r.t. the $\alpha$-norm $\mathcal{F}(\nu,\xi) = ||\nu-\xi||_\alpha = \sqrt[\alpha]{\sum_{x_t}|\nu(x_t|x_{<t})-\xi(x_t|x_{<t})|^\alpha}$ if $\alpha\neq 2$. Further, Pareto optimality of $\xi$ w.r.t. $\mathcal{F}_1$ *and* $\mathcal{F}_2$ is neither a necessary nor a sufficient condition for $\xi$ being Pareto optimal w.r.t. their sum $\mathcal{F}_1+\mathcal{F}_2$. Finally, if $\xi$ is Pareto optimal w.r.t. $\mathcal{F}$, then $\xi$ is also Pareto optimal w.r.t. any monotone increasing function of $\mathcal{F}$ (e.g. $\mathcal{F}^\alpha$, $\alpha\geq 0$).

Hint: Intuition on this problem can be gained by considering probability vectors $\boldsymbol{x},\boldsymbol{y},\boldsymbol{z}\in\Delta\subset I\!R^3$, where $\Delta$ is the two-dimensional probability triangle, and $\boldsymbol{z} = w\boldsymbol{x}+(1-w)\boldsymbol{y}$ is a mixture of $\boldsymbol{x}$ and $\boldsymbol{y}$. Consider the sets $M_{\boldsymbol{x}} := \{\boldsymbol{r} : \mathcal{F}(\boldsymbol{x},\boldsymbol{r}) \leq \mathcal{F}(\boldsymbol{x},\boldsymbol{z})\}$ and analogously $M_{\boldsymbol{y}}$. $M_{\boldsymbol{x}}\cap M_{\boldsymbol{y}}$ is not empty; it contains $\boldsymbol{z}$. If $M_{\boldsymbol{x}}\cap M_{\boldsymbol{y}}$ has an interior, then $\boldsymbol{z}$ is not Pareto optimal. Visualize the one-dimensional boundaries of the two-dimensional areas $M_{\boldsymbol{x}}$ and $M_{\boldsymbol{y}}$ qualitatively for the various performance measures $\mathcal{F}$. Now consider mixtures of three vectors in $I\!R^4$. This should give you enough intuition to prove Pareto optimality and to construct counter examples.

**3.6 (Lower error bound)** [C30u] It is possible to derive good (but not tight) lower error bounds with a fixed ($n$ independent) set $\mathcal{M}$ and weights, as opposed to the $n$ dependent set $\mathcal{M}$ chosen in the proof of Theorem 3.64. For instance, choose $\mathcal{M} = \{\mu_1,\mu_2\}$ with $\mu_{1/2}(1|x_{<t}) = \frac{1}{2}\pm\varepsilon_t$ with $\varepsilon_t = \min\{\frac{1}{2},\sqrt{\ln w_{\mu_1}^{-1}}/\sqrt{t\ln t}\}$. For $w_{\mu_1} \ll w_{\mu_2}$ and $n\to\infty$ show that $E_n^{\Theta_\xi} - E_n^{\Theta_\mu} \sim \frac{1}{\ln n}\sqrt{E_n^{\Theta_\mu} D_n}$. Is it possible to derive a tight(er) lower bound for different, but $n$ independent $\mathcal{M}$?

**3.7 ((Non)uniqueness of universal weights)** [C25ui] Section 3.6.4 showed that the universal weights $w_\nu = 2^{-K(\nu)}$ are optimal in a sense precisely stated in Theorem 3.70. Show that this choice for $w_\nu$ is not unique (not even within a constant factor). For instance, for $v_\nu = O(1)$ for $\nu=\xi_w$ and $v_\nu$ arbitrary (e.g. 0) for all other $\nu$, the obvious dominance $\xi_\nu \geq v_\nu \nu$ can be improved to $\xi_\nu \overset{\times}{\geq} w_\nu \nu$. Indeed, formally every choice of weights $v_\nu > 0\,\forall\nu$ leads within a multiplicative constant to the same universal distribution, but this constant is not necessarily of "acceptable" size. Suitably define "acceptable size" by considering the implications for the loss bounds. Construct (counter)examples and necessary/sufficient conditions for weights to be acceptable.

**3.8 (Deal with zero $\mu$)** [C35uo] Verify that all results in this chapter remain valid even if $\mu$ is allowed to take value zero for some arguments. Take the limit or face the problem as discussed in Section 3.9.1. Note, that when facing the problem, $\xi$ deteriorates to a semimeasure (see Section 2.4 and Problem 3.1).

**3.9 (Relations between random convergence criteria)** [C30sm] Prove the relations in Lemma 3.9 between the various convergence criteria given in Definition 3.8. Show that no other implications hold by constructing example random sequences. More precisely, implications are strict with reverse being wrong, and disconnected criteria (in the transitive hull) are incomparable. Show that convergence i.m.s. implies that $z_t$ deviates from $z_*$ by more than $\varepsilon$ only finitely many times and give a bound on the number.

**3.10 (Individual $\xi \to \mu$ convergence)** [C45om] In Problem 2.3 the open question whether $\xi_U(x_t|x_{<t})$ converges to $\mu(x_t|x_{<t})$ (in ratio or difference sense) individually for all Martin-Löf random sequences was posed (short $\xi_U \overset{\text{M.L.}}{\longrightarrow} \mu$). Theorem 3.22 shows that $\xi_U \overset{\text{M.L.}}{\longrightarrow} \mu$ cannot be decided from $\xi_U$ being a mixture distribution (3.5) or from the dominance property (3.6) alone. $\xi \notin \mathcal{M}$ for the classes used in Theorem 3.22. Construct $\xi_\mathcal{M} \in \mathcal{M}$ and prove a theorem analogous to Theorem 3.22 for these $\mathcal{M}$'s (Start with two-element classes $\mathcal{M}$, then enlarge $\mathcal{M}$ as far as possible). Hence $\xi_U \in \mathcal{M}_U$ is also not sufficient to resolve $\xi_U \overset{\text{M.L.}}{\longrightarrow} \mu$. Convert Theorem 3.19(*ii*/*iv*) to potential $\mu$.M.L.-randomness tests, but show that they are not effective. Try also to generalize Vovk's result [Vov87] to nonrecursive distributions. Where is the problem? With these insights, try again to solve Problem 2.3.

**3.11 (Speed of $\xi \to \mu$ convergence)** [C35o] Theorem 3.19(*i*) shows that $\sum_{t=1}^\infty s_t < \infty$. If $s_t$ were monotone decreasing ($s_{t+1} \leq s_t$) this would imply that $s_t$ tends to zero faster than $1/t$, i.e. $s_t = o(1/t)$. Show (or refute) that this monotonicity is generally wrong for some class $\mathcal{M}$ of measures. For the *semi*measure $\xi = \xi_U$, instantaneous convergence is extremely slow (see Problem 2.7), although $s_t$ converges fast to zero in an average sense. Provide necessary and/or sufficient conditions on $\mathcal{M}$ such that $s_t = o(1/t)$.

**3.12 (Learnability of the universal Turing machine)** [C05u] Consider the problem of learning a function $f : \mathcal{Z} \to \mathcal{X}$. A sequence of sample pairs $(z_1,x_1)$, $(z_2,x_2)$, ..., $(z_{n-1},x_{n-1})$ with $x_i = f(z_i)$ is given. The task is to predict $x_n = f(z_n)$ from $z_n$. This setup is a special case of the one described in Section 3.7.3. Show that if $f$ is a recursive function, then the $\Theta_\xi$ predictor makes at most a finite number of prediction errors, more precisely, at most $2\ln 2\cdot K(f)+O(1)$ errors. Consider now the universal (partial) function $f(z):=U(z)$, where $U$ is some universal Turing machine. Show that $U$ is learnable in the sense that after finite time $n$, $\Theta_\xi$ correctly predicts $x_i=U(z_i)$ for all $i>n$ as long as all $z_i$ are in the domain of $f$. What happens if for some $z_i$, $U$ does not halt?

**3.13 (Posterization)** [C30ui] Show that many properties of Kolmogorov complexity, Solomonoff's prior, and (policies based on) Bayes mixtures remain valid after "posterization". By posterization we mean replacing $L_n$, $w_\nu$, $K(\nu)$, $\nu(x_{1:n})$, etc., by the posteriors $L_{kn}:=\sum_{t=k}^n\mathbf{E}[\ell_{x_ty_t}|x_{<k}]$, $w_\nu(x_{<k})$, $K(\nu|x_{<k})$, $\nu(x_{k:n}|x_{<k})$, etc. Show that, strangely enough, for $\mathcal{M}=\mathcal{M}_U$ and $w_\nu=2^{-K(\nu)}$ it is not true that $w_\nu(x_{<k})\stackrel{\times}{=}2^{-K(\nu|x_{<k})}$ (not even $\log_2 w_\nu(x_{<k})=-K(\nu|x_{<k})+O(\log)$ holds). The important $\geq$ direction fails. Is the other direction $\leq$ true? So bounds, of e.g. $L_{kn}^{\Lambda_\xi}$, in terms of $\ln w_\mu(x_{<k})$ cannot be converted to bounds in terms of $K(\mu|x_{<k})$ unlike the $k=1$ case. But if we go one step back, we see that a bound on $\ln[\mu(x_{k:n}|x_{<k})/\xi(x_{k:n}|x_{<k})]$ is sufficient to bound $L_{kn}^{\Lambda_\xi}$. Use Problem 2.6(*iii*) to bound this expression by $\ln 2\cdot\tilde{K}(\mu|x_{<k})+O(?)$. The more information the history $x_{<k}$ contains about the environment $\mu$, the smaller the bounds get.

**3.14 (Probabilistic error bounds)** [C35s/C35o] Instead of making (deterministic) predictions that minimize the $\rho$-expected loss or error, we may define probabilistic $\rho$-predictors that predict $x_t$ from $x_{<t}$ with probability $\rho(x_t|x_{<t})$, and compare the performance of the $\xi$-predictor with the $\mu$- or other $\rho$-predictors. For simplicity, consider binary sequences (drawn from $\mu$) and the error loss. Let $e_t^\rho(x_{<t}):=\mathbf{E}_t[1-\rho(x_t|x_{<t})]$ be the error probability in the $t^{th}$ prediction, and $E_n^\rho:=\sum_{t=1}^n\mathbf{E}[e_t^\rho(x_{<t})]$ be the $\mu$-expected total number of errors in the first $n$ predictions. Prove the following error relations between universal ($\rho=\xi$), informed ($\rho=\mu$), and general ($\rho$) predictors:

$$\begin{array}{llll}
(i) & |E_n^\xi-E_n^\mu| & \leq \tfrac{1}{2}A_n & \leq D_n+\sqrt{2E_n^\mu D_n} \\
(ii) & E_n^\xi & \geq \tfrac{1}{2}[S_n+E_n^\mu] & \\
(iii) & E_n^\xi & \geq E_n^\mu+D_n-\sqrt{2E_n^\mu D_n} & \geq D_n \quad \text{for} \quad E_n^\mu\geq 2D_n \\
(iv) & E_n^\mu & \leq 2E_n^\rho, \quad e_n^\mu\leq 2e_n^\rho & \text{for any } \rho \\
(v) & E_n^\xi & \leq 2E_n^\rho+D_n+\sqrt{4E_n^\rho D_n} & \text{for any } \rho,
\end{array}$$

where $A_n$ and $S_n\leq D_n\leq\ln w_\mu^{-1}$ are defined in (3.13)–(3.16), and $w_\mu$ is the weight (3.5) of $\mu$ in $\xi$ ((*i*)–(*v*) were proven in [Hut01c]). This shows that the $\xi$-predictor performs well as compared to the $\mu$-predictor ($E_n^\xi-E_n^\mu=O(\sqrt{E_n^\mu})$), but does not exclude the possibility that it makes twice as many errors as other (better) predictors (there is a factor 2 in $E_n^\xi/E_n^\rho\leq 2+O((E_n^\rho)^{-1/2})$). Give an example for $E_n^\xi/E_n^{\Theta_\mu}\geq 2$, showing that the factor 2 in (*iv*) and (*v*) cannot be improved in general. Generalize (*i*)–(*v*) to nonbinary alphabet and arbitrary loss functions.

**3.15 (Posterior convergence for unbounded horizon)** [C15ui/C30o] Show that for unbounded horizon $h_t \to \infty$, there exist $\mathcal{M}$, $\mu$, and $w_\nu$ such that $\sum_{t=1}^\infty \mathbf{E}[d_{t:t+h_t-1}] = \infty$ (whereas $\sum_{t=1}^\infty \mathbf{E}[d_t] < \infty$). Show that condition $\sup_t h_t = \infty$ is not sufficient. Show that convergence can be (very) slow when $h_t$ grows (very) fast. To answer this question one has to define '(very) slows/fast', e.g. as logarithmical/exponentially increasing, or slower/faster than any unbounded computable function. Is convergence reasonably fast for slowly growing horizon, e.g. for $h_t = \log t$? What about $h_t = t$?

**3.16 (Distance bounds)** [C25u] Consider the distance measures $a$, $d$, $h$, and $s$ defined in (3.10) for probabilities $y_i \geq 0$ with $\sum_i y_i = 1$ and semi-probabilities $z_i \geq 0$ with $\sum_i z_i \leq 1$; $1 \leq i \leq N$. Show that the bounds $s \leq d$, $a^2 \leq 2d$, $h \leq d$, $h \leq a$, $s \leq 4h$, $s \leq a^2$, $a^2 \leq Ns$ hold and are tight in the sense that the ratio of the l.h.s. to the r.h.s. is bounded by 1 and can get (arbitrarily close to) 1. Show that $s \leq a^2$ and $s \leq 4h$ can be improved to $s \leq \frac{1}{2} a^2$ and $s \leq 2h$ if $(z_i)$ is restricted to a probability $\sum_i z_i = 1$.

**3.17 (Prediction by minimum description length)** [C40s] Instead of the mixture distribution $\xi = \sum_\nu w_\nu \nu$, consider the maximum a posteriori estimator $\varrho(x) := \max\{w_\nu \nu(x) : \nu \in \mathcal{M}\}$ or equivalently the two-part minimum description length (MDL) estimator $\varrho := \text{argmin}_{\nu\in\mathcal{M}}\{\log_2 \nu(x)^{-1} + \log_2 w_\nu^{-1}\}$, where as before, $\mathcal{M}$ is a countable set of (semi)measures and $\sum_{\nu\in\mathcal{M}} w_\nu \leq 1$. Show that $\sum_{t=1}^\infty \mathbf{E}[\sum_{x_t'}(\mu(x_t'|x_{<t}) - \varrho_{(norm)}(x_t'|x_{<t}))^2] \overset{\times}{\leq} w_\mu^{-1}$, where $\varrho(x_t|x_{<t}) := \varrho(x_{1:t})/\varrho(x_{<t})$ and $\varrho_{norm}(x_t|x_{<t}) := \varrho(x_{1:t})/\sum_{x_t} \varrho(x_{1:t})$. Show that these bounds are (within a multiplicative constant) tight (even for i.i.d. environments). This shows that MDL converges i.m.s. but convergence speed can be exponentially worse than for $\xi$ (cf. Theorem 3.19($i$)).

Hint: Show that $\xi - \rho$ is a semimeasure, although $\rho$ is not. Bound the relative entropy "$D_n$" between $\mu$ and $\varrho_{norm}$ and the absolute distance "$A_n$" between $\varrho_{norm}$ and $\varrho$. For the lower bound consider the deterministic environments $1^i 0^\infty$ (or Bernoulli$(\theta_i = \frac{1}{2} + 2^{-i})$) for $1 \leq i \leq N$ with uniform weights. See [PH04a, PH04b] for solutions.

**3.18 (Reduced prediction of aspects)** [C20u] Assume we want to predict only certain reduced aspects $x_t'$ of $x_t \in \mathcal{X}_t$, e.g. $x_t$ are detailed meteorological data of day $t$, but we are only interested whether it rains ($x_t' = 0$) or there is sunshine ($x_t' = 1$). Formally, let $x_t' = f(x_t) \in \mathcal{X}'$, and loss $\ell_{x_t y_t} = \ell'_{x_t' y_t}$ depend on $x_t$ through $x_t'$ only. Assume $x_{1:\infty}$ is sampled from $\mu$. Let $\mathcal{M}' := \{\nu' : \nu \in \mathcal{M}\}$ and $w_\nu' := w_\nu$, where $\nu'(x_{1:n}') := \sum_{x_{1:n}: f(x_t) = x_t' \forall 1 \leq t \leq n} \nu(x_{1:n})$. Show that $x_{1:\infty}'$ is sampled from $\mu'$ and regret bound (3.48) also holds for the reduced (primed) quantities, although in general $\Lambda_{\mu/\xi} \neq \Lambda'_{\mu/\xi}$. Furthermore, show that the bound can significantly improve, e.g. for $\mathcal{M} = \mathcal{M}_U$, $w_\nu = 2^{-K(\nu)}$, $K(f) = O(1)$, if $K(\mu') \ll K(\mu)$. This does *not* mean that to predict in the reduced space $\mathcal{X}'$ is to be preferred, since what matters is not a bound on the regret, but the true loss $L_n^{\Lambda_\xi} \leftrightarrow L_n^{\Lambda_\xi'}$. Give examples for which $L_n^{\Lambda_{\mu/\xi}} = 0$ but $L_n^{\Lambda'_{\mu/\xi}} \geq \frac{n-1}{2}$, i.e. it sometimes does much harm to go to the reduced space (the regret improves simply because $\Lambda_\mu$ deteriorates). Show that always $L_n^{\Lambda'_\mu} \geq L_n^{\Lambda_\mu}$, i.e. reduction never helps if $\mu$ is known, but $L_n^{\Lambda'_\xi} \leq 1 \ll n-1 \leq L_n^{\Lambda_\xi}$ is possible, i.e. reduction may help or harm if $\mu$ is unknown. Is $L_n^{\Lambda'_\xi} \lesseqgtr L_n^{\Lambda_\xi}$ possible for $\mathcal{M} = \mathcal{M}_U$ (cf. Problem 2.9)?

Hint: All examples can be chosen of the form $\mathcal{X} = \mathcal{X}' \times \mathcal{X}''$ with $\mathcal{X}' = \mathcal{X}'' = \mathbb{B}$ ($x_t = x'_t x''_t$) and $\ell'$ being the error loss. (a) $\mu(x_{1:n}) = \mu'(x'_{1:n}) \cdot \mu''(x''_{1:n})$ with $K(\mu') = O(1)$ and $K(\mu'') \stackrel{+}{=} K(\mu)$ large, (b) $x'_t = x''_{t-1}$ with $x''$ being Bernoulli($\frac{1}{2}$), (c) $\mathcal{M} = \{\nu_0, ..., \nu_n\}$ with $\nu_0((01)^\infty) = 1$, $\nu_i((00)^{i-1}(10)) = 1$ for $1 \leq i \leq n$, $w_i = 2^{-i-1}$, $\mu = \nu_n$.

Napoleon: *How is it that, although you say so much about the Universe, you say nothing about its Creator?*
Laplace: *No, Sire, I had no need of that hypothesis.*
Lagrange: *Ah, but it is such a good hypothesis: it explains so many things!*
Laplace: *Indeed, Sire, Monsieur Lagrange has, with his usual sagacity, put his finger on the precise difficulty with the hypothesis: it explains everything, but predicts nothing.*
— Conversation between Laplace and Lagrange mediated by Napoleon

Alan Turing
(1912–1954)

# 4 Agents in Known Probabilistic Environments



The general framework for AI might be viewed as the design and study of intelligent agents [RN95]. An agent is a cybernetic system with some internal state, which acts with output $y_k$ on some environment in cycle $k$, perceives some input $x_k$ from the environment and updates its internal state. Then the next cycle follows. We split the input $x_k$ into a regular part $o_k$ and a reward $r_k$, often called reinforcement feedback. From time to time the environment provides nonzero reward to the agent. The task of the agent is to maximize its utility, defined as the sum of future rewards. A probabilistic environment can be described by the conditional probability $\mu$ for the inputs $x_1...x_n$ to the agent under the condition that the agent outputs $y_1...y_n$. Most, if not all, environments are of this type. We give formal expressions for the outputs of the agent that maximize the total $\mu$-expected reward sum, called value. This model is called the AI$\mu$ model. As every AI problem can be brought into this form, the problem of maximizing utility is hence being formally solved, if $\mu$ is known. Furthermore, we study some special aspects of the AI$\mu$ model. We introduce factorizable probability distributions describing environments with independent episodes. They occur in several problem classes studied in Chapter 6 and are a special case of more general separable probability distributions defined in Section 5.3. We also clarify the connection to the Bellman equations of sequential decision theory and discuss similarities and differences. We discuss minor parameters of our model, including (the size of) the input and output spaces $\mathcal{X}$ and $\mathcal{Y}$ and the lifetime of the agent, and their universal choice, which we have in mind. There is nothing remarkable in this chapter; it is the essence of sequential decision theory [NM44, Bel57, BT96, SB98], presented in a very general form. Notation and formulas needed in later sections are simply developed. There are two major remaining problems: the problem of the unknown true probability distribution $\mu$, which is solved in Chapter 5, and computational aspects, which are addressed in Chapter 7.

## 4.1 The AI$\mu$ Model in Functional Form

### 4.1.1 The Cybernetic Agent Model

A good way to start thinking about intelligent systems is to consider more generally cybernetic systems, in AI usually called agents. This avoids having to struggle with the meaning of intelligence from the very beginning. A cybernetic system is a control circuit with input $x$ and output $y$ and an internal state. From an external input and the internal state the agent calculates deterministically or stochastically an output. This output (action) modifies the environment and leads to a new input (perception). This continues ad infinitum or for a finite number of cycles.

**Definition 4.1 (The agent model)** An agent is a system that interacts with an environment in cycles $k=1,2,3,...$. In cycle $k$ the action (output) $y_k \in \mathcal{Y}$ of the agent is determined by a policy $p$ that depends on the I/O history $y_1x_1...y_{k-1}x_{k-1}$. The environment reacts to this action, and leads to a new perception (input) $x_k \in \mathcal{X}$ determined by a deterministic function $q$ or probability distribution $\mu$, which depends on the history $y_1x_1...y_{k-1}x_{k-1}y_k$. Then the next cycle $k+1$ starts.

There is significant overlap between control theory studied by engineers and agent theory studied in AI, but both fields differ in notation and emphasis. Table 4.2 compares notation and emphasis in both fields. Only the interchange of input↔output can cause confusion. With few exceptions we use the notion common in AI.

**Table 4.2 (Notation and emphasis in AI versus control theory)**
The upper part ($\widehat{=}$) of the table compares notation used in AI or reinforcement learning to notation used in control theory. The lower part ($\Leftrightarrow$) compares the objectives of both fields.

| artificial intelligence | | control theory |
|---:|:---:|:---|
| agent | $\widehat{=}$ | controller |
| environment | $\widehat{=}$ | system |
| policy | $\widehat{=}$ | control=policy |
| transition matrix | $\widehat{=}$ | transition matrix? |
| input=perception | $\widehat{=}$ | output+reward |
| observation | $\widehat{=}$ | output |
| action=output | $\widehat{=}$ | input |
| (instantaneous) reward | $\widehat{=}$ | immediate or one-period cost |
| cumulative reward=value | $\widehat{=}$ | expected (total) cost(-to-go) |
| model learning | $\widehat{=}$ | system identification |
| exploitation | $\widehat{=}$? | (optimal?) stochastic control? |
| reactive agent | $\widehat{=}$ | closed-loop control |
| prewired agent? | $\widehat{=}$ | open-loop control |
| Markov decision process | $\widehat{=}$ | controlled Markov chain |
| belief state | $\widehat{=}$ | information state |
| Bellman equation | $\widehat{=}$ | Bellman equation (Ricatti eq.?) |
| reinforcement learning | $\widehat{=}$ | sequential decision theory |
| reinforcement learning | $\widehat{=}$ | adaptive control |
| ? | $\widehat{=}$ | consistent control |
| ? | $\widehat{=}$ | self-tuning control |
| ? | $\widehat{=}$ | self-optimizing control |
| exploration↔exploitation problem | $\widehat{=}$ | estimation↔control problem |
| qualitative solution | $\Leftrightarrow$ | high precision |
| complex environment | $\Leftrightarrow$ | simple machine |
| temporal difference learning | $\Leftrightarrow$ | value/policy iteration |
| temporal difference learning? | $\Leftrightarrow$ | dynamic programming |

As mentioned, we need some reward assignment to the cybernetic system. The input $x$ is divided into two parts, the standard input $o$ and some reward input $r$. If input and output are represented by strings, a deterministic cybernetic system can be modeled by a Turing machine $p$, where $p$ is called the policy of the agent, which determines the (re)action to a perception. If the environment is also computable it might be modeled by a Turing machine $q$ as well. The interaction of the agent with the environment can be illustrated as follows:

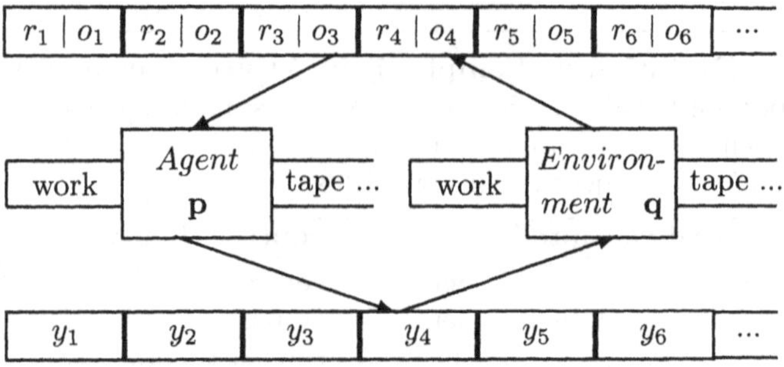


Both $p$ as well as $q$ have unidirectional input and output tapes and bidirectional work tapes. What entangles the agent with the environment is the fact that the upper tape serves as input tape for $p$, as well as output tape for $q$, and that the lower tape serves as output tape for $p$ as well as input tape for $q$. Further, the reading head must always be left of the writing head, i.e. the symbols must first be written before they are read. Both $p$ and $q$ have their own mutually inaccessible work tapes containing their own "secrets". The heads move in the following way. In the $k^{th}$ cycle $p$ writes $y_k$, $q$ reads $y_k$, $q$ writes $x_k \equiv r_k o_k$, $p$ reads $x_k \equiv r_k o_k$, followed by the $(k+1)^{th}$ cycle and so on. The whole process starts with the first cycle, with all heads on tape start and work tapes being empty. We call Turing machines behaving in this way *chronological Turing machines*. Before continuing, some notations on strings and probability distributions are appropriate.

### 4.1.2 Strings

We denote strings over the alphabet $\mathcal{X}$ by $s = x_1 x_2 ... x_n$, with $x_k \in \mathcal{X}$, where $\mathcal{X}$ is alternatively interpreted as a nonempty subset of $I\!N$ or itself as a prefix-free set of binary strings. The length of $s$ is $\ell(s) = \ell(x_1) + ... + \ell(x_n)$. Analogous definitions hold for $y_k \in \mathcal{Y}$. We call $x_k$ the $k^{th}$ input word and $y_k$ the $k^{th}$ output word (rather than letter). The string $s = y_1 x_1 ... y_n x_n$ represents the input/output in chronological order. Due to the prefix property of the $x_k$ and $y_k$, $s$ can be uniquely separated into its words. The words appearing in strings are always in chronological order. We further introduce the following abbreviations: $\epsilon$ is the empty string, $x_{n:m} := x_n x_{n+1} ... x_{m-1} x_m$ for $n \leq m$ and $\epsilon$ for $n > m$. $x_{<n} := x_1 ... x_{n-1}$. Analogously for $y$. Further, $y\!x_n := y_n x_n$, $y\!x_{n:m} := y_n x_n ... y_m x_m$, and so on.

### 4.1.3 AI Model for Known Deterministic Environment

Let us define for the chronological Turing machine $p$ a partial function also named $p : \mathcal{X}^* \to \mathcal{Y}^*$ with $y_{1:k} = p(x_{<k})$, where $y_{1:k}$ is the output of Turing machine $p$ on input $x_{<k}$ in cycle $k$, i.e. where $p$ has read up to $x_{k-1}$ but

no further.[1] In an analogous way, we define $q:\mathcal{Y}^*\to\mathcal{X}^*$ with $x_{1:k}=q(y_{1:k})$. Conversely, for every partial recursive chronological function we can define a corresponding chronological Turing machine. Each (agent,environment) pair $(p,q)$ produces a unique I/O sequence $\omega^{pq}:=y_1^{pq}x_1^{pq}y_2^{pq}x_2^{pq}...$. When we look at the definitions of $p$ and $q$ we see a nice symmetry between the cybernetic system and the environment. Until now, not much intelligence is in our agent. Now the credit assignment comes into the game and removes the symmetry somewhat. We split the input $x_k\in\mathcal{X}:=\mathcal{R}\times\mathcal{O}$ into a regular part $o_k\in\mathcal{O}$ and a reward $r_k\in\mathcal{R}\subset\mathbb{R}$. We define $x_k\equiv r_k o_k$ and $r_k\equiv r(x_k)$. The goal of the agent should be to maximize received rewards. This is called reinforcement learning. The reason for the asymmetry is that eventually we (humans) will be the environment with which the agent will communicate and *we* want to dictate what is good and what is wrong, not the other way round. This one-way learning, the agent learns from the environment, and not conversely, neither prevents the agent from becoming more intelligent than the environment, nor does it prevent the environment learning from the agent because the environment can itself interpret the outputs $y_k$ as a regular and a reward part. The environment is just not forced to learn, whereas the agent is. In cases where we restrict the reward to two values $r\in\mathcal{R}=\mathbb{B}:=\{0,1\}$, $r=1$ is interpreted as a positive feedback, called *good* or *correct*, and $r=0$ a negative feedback, called *bad* or *error*. Further, let us restrict for a while the lifetime (number of cycles) $m$ of the agent to a large but finite value. Let

$$V_{km}^{pq} := \sum_{i=k}^{m} r(x_i^{pq})$$

be the future total reward (called future utility), the agent $p$ receives from the environment $q$ in the cycles $k$ to $m$. It is now natural to call the agent $p^*$, which maximizes $V_{1m}$ (called total utility), the *best* one:[2]

$$p^* := \arg\max_p V_{1m}^{pq} \quad\Rightarrow\quad V_{km}^{p^*q} \geq V_{km}^{pq} \quad \forall p: y_{<k}^{pq}=y_{<k}^{p^*q}. \tag{4.3}$$

For $k=1$ the condition on $p$ is nil. For $k>1$ it states that $p$ shall be consistent with $p^*$ in the sense that they have the same history. If $\mathcal{X}$, $\mathcal{Y}$ and $m$ are finite, the number of different behaviors of the agent, i.e. the search space, is finite. Therefore, because we have assumed that $q$ is known, $p^*$ can effectively be determined by pre-analyzing all behaviors. The main reason for restricting to finite $m$ was not to ensure computability of $p^*$ but that the limit $m\to\infty$ might not exist. The ease with which we defined and computed the optimal policy $p^*$ is not remarkable. Just the (unrealistic) assumption of a completely known deterministic environment $q$ has trivialized everything.

[1] Note that a possible additional dependence of $p$ on $y_{<k}$ as mentioned in Definition 4.1 can be eliminated by recursive substitution; see below. Similarly for $q$.

[2] $\text{argmax}_p V(p)$ is the $p$ that maximizes $V(\cdot)$. If there is more than one maximum we might choose the lexicographically smallest one for definiteness.

### 4.1.4 AI Model for Known Prior Probability

Let us now weaken our assumptions by replacing the environment $q$ with a probability distribution $\mu(q)$ over chronological functions. Here $\mu$ might be interpreted in two ways. Either the environment itself behaves stochastically defined by $\mu$ or the true environment is deterministic, but we only have subjective (probabilistic) information of which is the true environment. Combinations of both cases are also possible. We assume here that $\mu$ is known and describes the true stochastic behavior of the environment. The case of unknown $\mu$ with the agent having some beliefs about the environment lies at the heart of the AI$\xi$ model described in Chapter 5.

The *best* agent is now the one that maximizes the *expected* utility (called value function) $V_\mu^p \equiv V_{1m}^{p\mu} := \sum_q \mu(q) V_{1m}^{pq}$. This defines the AI$\mu$ model.

**Definition 4.4 (The AI$\mu$ model)** The AI$\mu$ model is the agent with policy $p^\mu$ that maximizes the $\mu$-expected total reward $r_1 + ... + r_m$, i.e. $p^* \equiv p^\mu := \text{argmax}_p V_\mu^p$. Its value is $V_\mu^* := V_\mu^{p^\mu}$.

We need the concept of a *value function* in a slightly more general form.

**Definition 4.5 (The $\mu$/true/generating value function)** The agent's perception $x$ consists of a regular observation $o \in \mathcal{O}$ and a reward $r \in \mathcal{R} \subset I\!R$. In cycle $k$ the *value* $V_{km}^{p\mu}(y\!x_{<k})$ is defined as the $\mu$-expectation of the future reward sum $r_k + ... + r_m$ with actions generated by policy $p$, and fixed history $y\!x_{<k}$. We say that $V_{km}^{p\mu}(y\!x_{<k})$ is the (future) *value* of policy $p$ in environment $\mu$ given history $y\!x_{<k}$, or shorter, the $\mu$ or true or generating value of $p$ given $y\!x_{<k}$. $V_\mu^p := V_{1m}^{p\mu}$ is the (total) value of $p$.

We now give a more formal definition for $V_{km}^{p\mu}$. Let us assume we are in cycle $k$ with history $\dot{y}\dot{x}_1 ... \dot{y}\dot{x}_{k-1}$ and ask for the *best* output $y_k$. Further, let $\dot{Q}_k := \{q : q(\dot{y}_{<k}) = \dot{x}_{<k}\}$ be the set of all environments producing the above history. We say that $q \in \dot{Q}_k$ is *consistent* with history $\dot{y}\dot{x}_{<k}$. The expected reward for the next $m-k+1$ cycles (given the above history) is called the value of policy $p$ and is given by a conditional probability:

$$V_{km}^{p\mu}(\dot{y}\dot{x}_{<k}) \;:=\; \frac{\sum_{q \in \dot{Q}_k} \mu(q) V_{km}^{pq}}{\sum_{q \in \dot{Q}_k} \mu(q)}. \tag{4.6}$$

Policy $p$ and environment $\mu$ do not determine history $\dot{y}\dot{x}_{<k}$, unlike the deterministic case because the history is no longer deterministically determined by $p$ and $q$, but depends on $p$ and $\mu$ *and* on the outcome of a stochastic process. Every new cycle adds new information ($\dot{x}_i$) to the agent. This is indicated by the dots over the symbols. In cycle $k$ we have to maximize the expected future rewards, taking into account the information in the history $\dot{y}\dot{x}_{<k}$. This

information is not already present in $p$ and $q/\mu$ at the agent's start, unlike in the deterministic case.

Furthermore, we want to generalize the finite lifetime $m$ to a dynamic (computable) farsightedness $h_k \equiv m_k - k + 1 \geq 1$, called horizon. For $m_k = m$ we have our original finite lifetime; for $h_k = h$ the agent maximizes in every cycle the next $h$ expected rewards. A discussion of the choices for $m_k$ is delayed to Section 5.7. The next $h_k$ rewards are maximized by

$$p_k^* \;:=\; \arg\max_{p\in\dot{P}_k} V_{km_k}^{p\mu}(\dot{y}\dot{x}_{<k}),$$

where $\dot{P}_k := \{p : \exists y_k : p(\dot{x}_{<k}) = \dot{y}_{<k}y_k\}$ is the set of policies *consistent* with the current history. Note that $p_k^*$ depends on $k$ and is used only in step $k$ to determine $\dot{y}_k$ by $p_k^*(\dot{x}_{<k}|\dot{y}_{<k}) = \dot{y}_{<k}\dot{y}_k$. After writing $\dot{y}_k$ the environment replies with $\dot{x}_k$ with (conditional) probability $\mu(\dot{Q}_{k+1})/\mu(\dot{Q}_k)$. This probabilistic outcome provides new information to the agent. The cycle $k+1$ starts with determining $\dot{y}_{k+1}$ from $p_{k+1}^*$ (which can differ from $p_k^*$ for dynamic $m_k$) and so on. Note that $p_k^*$ implicitly also depends on $\dot{y}_{<k}$ because $\dot{P}_k$ and $\dot{Q}_k$ do so. But recursively inserting $p_{k-1}^*$ and so on, we can define

$$p^*(\dot{x}_{<k}) \;:=\; p_k^*(\dot{x}_{<k}|p_{k-1}^*(\dot{x}_{<k-1}|...p_1^*)). \tag{4.7}$$

It is a chronological function and is computable if $\mathcal{X}$, $\mathcal{Y}$ and $m_k$ are finite and $\mu$ is computable. For constant $m$ one can show that the policy (4.7) coincides with the AI$\mu$ model (Definition 4.4). This also proves

$$V_{km}^{*\mu}(\dot{y}\dot{x}_{<k}) \geq V_{km}^{p\mu}(\dot{y}\dot{x}_{<k}) \quad \forall p \in \dot{P}_k, \quad \text{i.e. for all } p \text{ consistent with } \dot{y}\dot{x}_{<k}, \tag{4.8}$$

similarly to (4.3) (see Problem 4.1). For $k=1$ this is obvious. We also call (4.7) AI$\mu$ model. For deterministic[3] $\mu$ this model reduces to the deterministic case discussed in the last subsection.

It is important to maximize the sum of future rewards and not, for instance, to be greedy and only maximize the next reward, as is done e.g. in sequence prediction. For example, let the environment be a sequence of chess games, and each cycle corresponds to one move. Only at the end of each game is a positive reward $r=1$ given to the agent if it won the game (and made no illegal moves). For the agent, maximizing all future rewards means trying to win as many games in as short as possible time (and avoiding illegal moves). The same performance is reached if we choose $h_k$ much larger than the typical game lengths. Maximization of only the next reward would be a very bad chess playing agent. Even if we would make our reward $r$ finer, e.g. by evaluating the number of chessmen, the agent would play very bad chess for $h_k=1$, indeed.

The AI$\mu$ model still depends on $\mu$ and $m_k$; $m_k$ is addressed in Section 5.7. To get our final universal AI model the idea is to replace $\mu$ by the universal probability $\xi$, defined later. This is motivated by the fact that $\xi$ converges to $\mu$

[3] We call a probability distribution deterministic if it assumes values 0 and 1 only.

in a certain sense for any $\mu$. With $\xi$ instead of $\mu$ our model no longer depends on any parameters, so it is truly universal. It remains to show that it behaves intelligently. But let us continue step by step. In the following we develop an alternative but equivalent formulations of the AI$\mu$ model. Whereas the functional form presented above is more suitable for theoretical considerations, especially for the development of a time-bounded version in Section 7.2, the iterative and recursive formulation of the next Section will be more appropriate for the explicit calculations in most of the other sections.

## 4.2 The AI$\mu$ Model in Recursive and Iterative Form

### 4.2.1 Probability Distributions

We use Greek letters for probability distributions, and underline their arguments to indicate that they are probability arguments. Let $\rho_n(\underline{x}_1...\underline{x}_n)$ be the probability that an (infinite) string starts with $x_1...x_n$. We drop the index on $\rho$ if it is clear from its arguments:

$$\sum_{x_n\in\mathcal{X}}\rho(\underline{x}_{1:n}) \equiv \sum_{x_n}\rho_n(\underline{x}_{1:n}) = \rho_{n-1}(\underline{x}_{<n}) \equiv \rho(\underline{x}_{<n}), \quad \rho(\epsilon) \equiv \rho_0(\epsilon) = 1. \quad (4.9)$$

We also need conditional probabilities. We prefer a notation that preserves the chronological order of the words, in contrast to the standard notation $\rho(\cdot|\cdot)$, which flips it. We extend the definition of $\rho$ to the conditional case with the following convention for its arguments: An underlined argument $\underline{x}_k$ is a probability variable, and other non-underlined arguments $x_k$ represent conditions. With this convention, the conditional probability has the form $\rho(x_{<n}\underline{x}_n) = \rho(\underline{x}_{1:n})/\rho(\underline{x}_{<n})$. The equation states that the probability that a string $x_1...x_{n-1}$ is followed by $x_n$ is equal to the probability of $x_1...x_n*$ divided by the probability of $x_1...x_{n-1}*$. We use $x*$ as an abbreviation for 'strings starting with $x$'.

The introduced notation is also suitable for defining the conditional probability $\rho(y_1\underline{x}_1...y_n\underline{x}_n)$ that the environment reacts with $x_1...x_n$ under the condition that the output of the agent is $y_1...y_n$. The environment is chronological, i.e. input $x_i$ depends on $yx_{<i}y_i$ only. In the probabilistic case this means that $\rho(\underline{yx}_{<k}y_k):=\sum_{x_k}\rho(\underline{yx}_{1:k})$ is independent of $y_k$, hence a tailing $y_k$ in the arguments of $\rho$ can be dropped. Probability distributions with this property will be called *chronological*. The $y$ are always conditions, i.e. are never underlined, whereas additional conditioning for the $x$ can be obtained with the chain rule

$$\rho(yx_{<n}\underline{yx}_n) = \rho(\underline{yx}_{1:n})/\rho(\underline{yx}_{<n}) \quad \text{and} \quad (4.10)$$

$$\rho(\underline{yx}_{1:n}) = \rho(\underline{yx}_1)\cdot\rho(yx_1\underline{yx}_2)\cdot...\cdot\rho(yx_{<n}\underline{yx}_n). \quad (4.11)$$

The second equation is the first equation applied $n$ times.

### 4.2.2 Explicit Form of the AI$\mu$ Model

Let us define the AI$\mu$ model $p^*$ in a different way. In the next subsection we will show that the $p^*$ model defined here is identical to the functional definition of $p^*$ given in the last section.

Let $\mu(\dot{y}\dot{x}_{<k}y\underline{x}_k)$ be the true probability of input $x_k$ in cycle $k$, given the history $\dot{y}\dot{x}_{<k}y_k$; $\mu(y\underline{x}_{1:k})$ is the true chronological prior probability that the environment reacts with $x_{1:k}$ if provided with actions $y_{1:k}$ from the agent. We assume the cybernetic model depicted on page 128 (see also book cover) to be valid. Next we define the value $V^{*\mu}_{k+1,m}(\dot{y}\dot{x}_{1:k})$ to be the $\mu$-expected reward sum $r_{k+1}+...+r_m$ in cycles $k+1$ to $m$ with outputs $y_i$ generated by agent $p^*$ that maximizes the expected reward sum, and responses $x_i$ from the environment, drawn according to $\mu$. Adding $r(x_k)\equiv r_k$ we get the reward including cycle $k$. The probability of $x_k$, given $\dot{y}\dot{x}_{<k}y_k$, is given by the conditional probability $\mu(\dot{y}\dot{x}_{<k}y\underline{x}_k)$. So the expected reward sum in cycles $k$ to $m$ given $\dot{y}\dot{x}_{<k}y_k$ is

$$V^{*\mu}_{km}(\dot{y}\dot{x}_{<k}y_k) \;:=\; \sum_{x_k}[r(x_k)+V^{*\mu}_{k+1,m}(\dot{y}\dot{x}_{1:k})]\cdot\mu(\dot{y}\dot{x}_{<k}y\underline{x}_k). \qquad (4.12)$$

Now we ask how $p^*$ chooses $y_k$: It should choose $y_k$ so as to maximize the future rewards. So the expected reward in cycles $k$ to $m$ given $\dot{y}\dot{x}_{<k}$ and $y_k$ chosen by $p^*$ is $V^{*\mu}_{km}(\dot{y}\dot{x}_{<k}):=\max_{y_k}V^{*\mu}_{km}(\dot{y}\dot{x}_{<k}y_k)$ (see Figure 4.13).

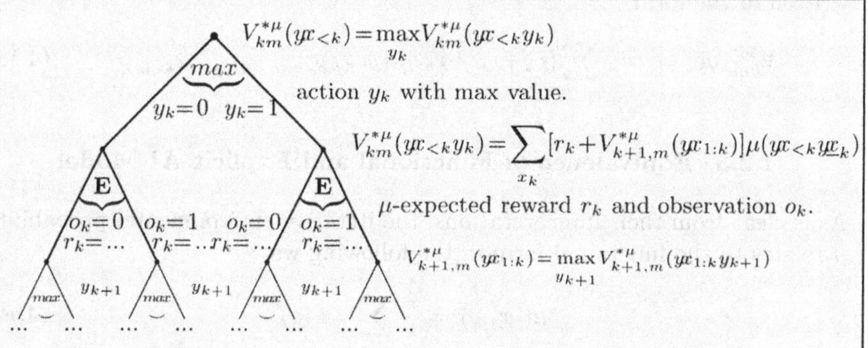


**Figure 4.13 (Expectimax tree/algorithm for $\mathcal{O}=\mathcal{Y}=\mathbb{B}$)**

Together with the induction start

$$V^{*\mu}_{m+1,m}(\dot{y}\dot{x}_{1:m}) \;:=\; 0, \qquad (4.14)$$

$V^{*\mu}_{km}$ is completely defined. We might summarize one cycle into the formula

$$V^{*\mu}_{km}(\dot{y}\dot{x}_{<k}) \;=\; \max_{y_k}\sum_{x_k}[r(x_k)+V^{*\mu}_{k+1,m}(\dot{y}\dot{x}_{1:k})]\cdot\mu(\dot{y}\dot{x}_{<k}y\underline{x}_k). \qquad (4.15)$$

We introduce a dynamic (computable) farsightedness $h_k\equiv m_k-k+1\geq 1$, called horizon. For $m_k=m$, where $m$ is the lifetime of the agent, we achieve optimal behavior, for limited farsightedness $h_k = h$ ($m = m_k = h+k-1$), the agent maximizes in every cycle the next $h$ expected rewards. A discussion of the choices for $m_k$ is delayed to Section 5.7. If $m_k$ is our horizon function of $p^*$ and $\dot{y}\dot{x}_{<k}$ is the actual history in cycle $k$, the output $\dot{y}_k$ of the agent is explicitly given by

$$\dot{y}_k = \arg\max_{y_k} V^{*\mu}_{km_k}(\dot{y}\dot{x}_{<k}y_k), \tag{4.16}$$

which in turn defines the policy $p^*$. Then the environment responds $\dot{x}_k$ with probability $\mu(\dot{y}\dot{x}_{<k}\dot{y}\dot{\underline{x}}_k)$. Then cycle $k+1$ starts. We might unfold the recursion (4.15) further and give $\dot{y}_k$ nonrecursively as

$$\dot{y}_k \equiv \dot{y}^\mu_k := \arg\max_{y_k}\sum_{x_k}\max_{y_{k+1}}\sum_{x_{k+1}}...\max_{y_{m_k}}\sum_{x_{m_k}}(r(x_k)+...+r(x_{m_k}))\cdot\mu(\dot{y}\dot{x}_{<k}\underline{yx}_{k:m_k}) \tag{4.17}$$

This has a direct interpretation: The probability of inputs $x_{k:m_k}$ in cycle $k$ when the agent outputs $y_{k:m_k}$ with actual history $\dot{y}\dot{x}_{<k}$ is $\mu(\dot{y}\dot{x}_{<k}\underline{yx}_{k:m_k})$. The future reward in this case is $r(x_k)+...+r(x_{m_k})$. The best expected reward is obtained by averaging over the $x_i$ ($\sum_{x_i}$) and maximizing over the $y_i$. This has to be done in chronological order to correctly incorporate the dependencies of $x_i$ and $y_i$ on the history. This is essentially the expectimax algorithm/tree [Mic66, RN95]. The AI$\mu$ model is *optimal* in the sense that no other policy leads to higher expected reward. The value for a general policy $p$ can be written in the form

$$V^{p\mu}_{km}(yx_{<k}) := \sum_{x_{1:m}}(r_k+ ... +r_m)\mu(yx_{<k}\underline{yx}_{k:m})_{|y_{1:m}=p(x_{<m})} \tag{4.18}$$

### 4.2.3 Equivalence of Functional and Explicit AI Model

As is clear from their interpretations, the iterative environmental probability $\mu$ relates to the functional form in the following way:

$$\mu(\underline{yx}_{1:k}) = \sum_{q:q(y_{1:k})=x_{1:k}} \mu(q) \tag{4.19}$$

**Theorem 4.20 (Equivalence of functional and explicit AI model)** The actions of the functional AI model (4.7) coincide with the actions of the iterative (=recursive) AI model (4.15)–(4.17) with environments identified by (4.19).

**Proof.** We prove the equivalence of (4.7) and (4.16) only for $k=2$ and $m_2=3$. The proof of the general case is completely analogous, except that the notation becomes quite messy.

Let us first evaluate (4.6) for fixed $\dot{y}_1\dot{x}_1$ and some $p\in\dot{P}_2$, i.e. $p(\dot{x}_1)=\dot{y}_1y_2$ for some $y_2$. If the next input to the agent is $x_2$, $p$ will respond with $p(\dot{x}_1x_2)=\dot{y}_1y_2y_3$ for some $y_3$ depending on $x_2$. We write $y_3(x_2)$ in the following. Dependency on dotted words like $\dot{x}_1$ is not shown, as the dotted words are fixed. The numerator of (4.6) simplifies to

$$\sum_{q\in\dot{Q}_2}\mu(q)V_{23}^{pq} = \sum_{q:q(\dot{y}_1)=\dot{x}_1}\mu(q)V_{23}^{pq} = \sum_{x_2x_3}(r(x_2)+r(x_3))\sum_{q:q(\dot{y}_1y_2y_3(x_2))=\dot{x}_1x_2x_3}\mu(q)$$
$$= \sum_{x_2x_3}(r(x_2)+r(x_3))\cdot\mu(\dot{y}_1\underline{\dot{x}}_1y_2\underline{x}_2y_3(x_2)\underline{x}_3).$$

In the first equality we inserted the definition of $\dot{Q}_2$. In the second equality we split the sum over $q$ by first summing over $q$ with fixed $x_2x_3$. This allows us to pull $V_{23}=r(x_2)+r(x_3)$ out of the inner sum. Then we sum over $x_2x_3$. Further, we have inserted $p$, i.e. replaced $p$ by $y_2$ and $y_3(\cdot)$. In the last equality we used (4.19). The denominator reduces to

$$\sum_{q\in\dot{Q}_2}\mu(q) = \sum_{q:q(\dot{y}_1)=\dot{x}_1}\mu(q) = \mu(\dot{y}_1\underline{\dot{x}}_1).$$

For the quotient we get

$$V_{23}^{p\mu}(\dot{y}_1\dot{x}_1) = \sum_{x_2x_3}(r(x_2)+r(x_3))\cdot\mu(\dot{y}_1\dot{x}_1y_2\underline{x}_2y_3(x_2)\underline{x}_3).$$

We have seen that the relevant behavior of $p\in\dot{P}_2$ in cycles 2 and 3 is completely determined by $y_2$ and the function $y_3(\cdot)$.

$$\max_{p\in\dot{P}_2}V_{23}^{p\mu}(\dot{y}_1\dot{x}_1) = \max_{y_2}\max_{y_3(\cdot)}\sum_{x_2x_3}(r(x_2)+r(x_3))\cdot\mu(\dot{y}_1\dot{x}_1y_2\underline{x}_2y_3(x_2)\underline{x}_3)$$
$$= \max_{y_2}\sum_{x_2}\max_{y_3}\sum_{x_3}(r(x_2)+r(x_3))\cdot\mu(\dot{y}_1\dot{x}_1y_2\underline{x}_2y_3\underline{x}_3)$$

In the last equality we have used the fact that the functional minimization over $y_3(\cdot)$ reduces to a simple minimization over the word $y_3$ when interchanging with the sum over its arguments ($\max_{y_3(\cdot)}\sum_{x_2}\equiv\sum_{x_2}\max_{y_3}$). In the functional case $\dot{y}_2$ is therefore determined by

$$\dot{y}_2 = \arg\max_{y_2}\sum_{x_2}\max_{y_3}\sum_{x_3}(r(x_2)+r(x_3))\cdot\mu(\dot{y}_1\dot{x}_1y_2\underline{x}_2y_3\underline{x}_3).$$

This is identical to the iterative definition (4.17) with $k=2$ and $m_2=3$. □

## 4.3 Special Aspects of the AI$\mu$ Model

### 4.3.1 Factorizable Environments

Up to now we have made no restrictions on the form of the prior probability $\mu$ apart from being a chronological probability distribution. On the other hand, we will see that, in order to prove rigorous reward bounds, the prior probability must satisfy some separability condition to be defined later. Here we introduce a very strong form of separability, when $\mu$ factorizes into products. We start with a factorization into two factors. Let us assume that $\mu$ is of the form

$$\mu(y\!x_{1:n}) \;=\; \mu_1(y\!x_{<l})\cdot\mu_2(y\!x_{l:n}) \tag{4.21}$$

for some fixed $l$ and sufficiently large $n \geq m_k$. For this $\mu$ the output $\dot{y}_k$ in cycle $k$ of the AI$\mu$ agent (4.17) for $k \geq l$ depends on $\dot{y}\!\dot{x}_{l:k-1}$ and $\mu_2$ only and is independent of $\dot{y}\!\dot{x}_{<l}$ and $\mu_1$. This is easily seen when inserting

$$\mu(\dot{y}\!\dot{x}_{<k}y\!x_{k:m_k}) = \underbrace{\mu_1(\dot{y}\!\dot{x}_{<l})}_{\equiv 1}\cdot\mu_2(\dot{y}\!\dot{x}_{l:k-1}y\!x_{k:m_k}) \tag{4.22}$$

into (4.17). For $k<l$ the output $\dot{y}_k$ depends on $\dot{y}\!\dot{x}_{<k}$ (this is trivial) and $\mu_1$ only (trivial if $m_k<l$) and is independent of $\mu_2$. The nontrivial case, where the horizon $m_k \geq l$ reaches into the region $\mu_2$, can be proven as follows (we abbreviate $m:=m_k$ in the following). Inserting (4.21) into the definition of $V_{lm}^{*\mu}(y\!x_{<l})$, the factor $\mu_1$ is 1, as in (4.22). We abbreviate $V_{lm}^{*\mu}:=V_{lm}^{*\mu}(y\!x_{<l})$, as it is independent of its argument. One can decompose

$$V_{km}^{*\mu}(y\!x_{<k}) \;=\; V_{k,l-1}^{*\mu}(y\!x_{<k}) \;+\; V_{lm}^{*\mu}. \tag{4.23}$$

For $k=l$ this is true because the first term on the r.h.s. is zero. For $k<l$ we prove the decomposition by induction from $k+1$ to $k$.

$$\begin{aligned} V_{km}^{*\mu}(y\!x_{<k}) &= \max_{y_k}\sum_{x_k}[r(x_k)+V_{k+1,l-1}^{*\mu}(y\!x_{1:k})+V_{lm}^{*\mu}]\cdot\mu_1(y\!x_{<k}y\!x_k) \\ &= \max_{y_k}\left[\sum_{x_k}(r(x_k)+V_{k+1,l-1}^{*\mu}(y\!x_{<k}))\cdot\mu_1(y\!x_{<k}y\!x_k)+V_{lm}^{*\mu}\right] \\ &= V_{k,l-1}^{*\mu}(y\!x_{<k})+V_{lm}^{*\mu}. \end{aligned}$$

Inserting (4.23), valid for $k+1$ by induction hypothesis, into (4.15) gives the first equality. In the second equality we have performed the $x_k$ sum for the $V_{lm}^{*\mu}\cdot\mu_1$ term, which is now independent of $y_k$. It can therefore be pulled out of $\max_{y_k}$. In the last equality we again used the definition (4.15). This completes the induction step and proves (4.23) for $k<l$. Output $\dot{y}_k$ can now be represented as

$$\dot{y}_k \;=\; \arg\max_{y_k} V_{km}^{*\mu}(\dot{y}\!\dot{x}_{<k}y_k) \;=\; \arg\max_{y_k} V_{k,l-1}^{*\mu}(\dot{y}\!\dot{x}_{<k}y_k), \tag{4.24}$$

where (4.16) and (4.23) and the fact that an additive constant $V_{lm}^{*\mu}$ does not change $\text{argmax}_{y_k}$ have been used. $V_{k,l-1}^{*\mu}(\dot{y}\!\dot{x}_{<k}y_k)$ and hence $\dot{y}_k$ is independent of $\mu_2$ for $k<l$. Note that $\dot{y}_k$ is also independent of the choice of $m$, as long as $m \geq l$.

In the general case of an (infinite) sequence of consecutive episodes one can show an analogous result:

**Theorem 4.25 (Factorizable environments $\mu$)** Assume that the cycles are grouped into independent episodes $r=1,2,3,...$, where each episode $r$ consists of the cycles $k=n_r+1,...,n_{r+1}$ for some $0=n_0<n_1<...<n_s=n$:

$$\mu(y\!x_{1:n}) = \prod_{r=0}^{s-1} \mu_r(y\!x_{n_r+1:n_{r+1}}) \tag{4.26}$$

(In the simplest case, when all episodes have the same length $l$ then $n_r=r\cdot l$) Then, $\dot{y}_k$ depends on $\mu_r$ and $x$ and $y$ of episode $r$ only, with $r$ such that $n_r<k\leq n_{r+1}$.

$$\dot{y}_k = \arg\max_{y_k}\sum_{x_k}...\max_{y_t}\sum_{x_t}(r(x_k)+...+r(x_t))\cdot\mu_r(\dot{y}\!\dot{x}_{n_r+1:k-1}y\!x_{k:t}) \tag{4.27}$$

with $t:=\min\{m_k,n_{r+1}\}$. The different episodes are completely independent in the sense that the inputs $x_k$ of different episodes are statistically independent and depend only on $y_k$ of the same episode. The outputs $y_k$ depend on the $x$ and $y$ of the corresponding episode $r$ only, and are independent of the actual I/O of the other episodes.

If all episodes have a length of at most $l$, i.e. $n_{r+1}-n_r\leq l$, and if we choose the horizon $h_k$ to be at least $l$, then $m_k\geq k+l-1\geq n_r+l\geq n_{r+1}$, and hence $t=n_{r+1}$ independent of $m_k$. This means that for factorizable $\mu$ there is no problem in taking the limit $m_k\to\infty$. Maybe this limit can also be performed in the more general case of a sufficiently separable $\mu$. The (problem of the) choice of $m_k$ will be discussed in more detail in Section 5.7.

Although factorizable $\mu$ are too restrictive to cover all AI problems, they often occur in practice in the form of repeated problem solving, and hence, are worthy of study. For example, if the agent has to play games like chess repeatedly, or has to minimize different functions, the different games/functions might be completely independent, i.e. the environmental probability factorizes, where each factor corresponds to a game/function minimization. For details, see Section 6.3 on strategic games and Section 6.4 on function minimization.

Further, for factorizable $\mu$ it is probably easier to derive suitable reward bounds for the universal AI$\xi$ model defined in the next chapter, than for the separable cases which will be introduced in Section 5.3. This could be a first step toward a definition and proof for the general case of separable problems. One goal of this subsection was to show, that the notion of a factorizable $\mu$ could be the first step toward a definition and analysis of the general case of separable $\mu$.

### 4.3.2 Constants and Limits

We have in mind a universal agent with complex interactions that is at least as intelligent and complex as a human being. One might think of an agent whose input $y_k$ comes from a digital video camera, and the output $x_k$ is some image to a monitor[4], only for the rewards we might restrict to the most primitive binary ones, i.e. $r_k \in I\!B$. So we think of the following constant sizes:

$$\begin{array}{cccccccc} 1 & \ll & \langle\ell(y_k x_k)\rangle & \ll & k & \leq m & \ll & |\mathcal{Y}\times\mathcal{X}| \\ 1 & \ll & 2^{16} & \ll & 2^{24} & \leq 2^{32} & \ll & 2^{65536} \end{array}$$

The first two limits say that the actual number $k$ of inputs/outputs should be reasonably large compared to the typical length $\langle\ell\rangle$ of the input/output words, which itself should be rather sizeable. The last limit expresses the fact that the total lifetime $m$ (number of I/O cycles) of the agent is far too small to allow every possible input to occur, or to try every possible output, or to make use of identically repeated inputs or outputs. We do not expect any useful outputs for $k \lesssim \langle\ell\rangle$. More interesting than the lengths of the inputs is the complexity $K(x_1...x_k)$ of all inputs until now (see Definition 2.9). The environment is usually not "perfect". The agent could either interact with a imperfect human or tackle a nondeterministic world (due to quantum mechanics or chaos)[5]. In either case, the sequence contains some noise, leading to $K(x_1...x_k) \propto \langle\ell\rangle \cdot k$. The complexity of the probability distribution of the input sequence is something different. We assume that this noisy world operates according to some simple computable rules: $K(\mu_k) \ll \langle\ell\rangle \cdot k$, i.e. the rules of the world can be highly compressed. We may allow environments in which new aspects appear for $k \to \infty$, causing an unbounded $K(\mu_k)$.

In the following we never use these limits, except when explicitly stated. In some simpler models and examples the size of the constants will even violate these limits (e.g. $\ell(x_k) = \ell(y_k) = 1$), but it is the limits above that the reader should bear in mind. We are only interested in theorems that do not degenerate under the above limits. In order to avoid cumbersome convergence and existence considerations we make the following assumptions throughout this book:

**Assumption 4.28 (Finiteness)** We assume that

- the input/perception space $\mathcal{X}$ is finite,
- the output/action space $\mathcal{Y}$ is finite,
- the rewards are nonnegative and bounded, i.e. $r_k \in \mathcal{R} \subseteq [0, r_{max}]$,
- the horizon $m$ is finite.

---

[4] Humans can only simulate a screen as output device by drawing pictures.

[5] Whether there exist truly stochastic processes at all is a difficult question. At least the quantum indeterminacy comes very close to it. See Section 8.6.2 for a detailed discussion.

Finite $\mathcal{X}$ and bounded $\mathcal{R}$ (each separately) ensure existence of $\mu$-expectations but are sometimes needed together. Finite $\mathcal{Y}$ ensures that $\text{argmax}_{y_k\in\mathcal{Y}}[...]$ exists, i.e. that maxima are attained, while finite $m$ avoids various technical and philosophical problems (Section 5.7), and positive rewards are needed for the time-bounded AIXI*tl* model (Section 7.2). Many theorems can be generalized by relaxing some or all of the above finiteness assumptions.

### 4.3.3 Sequential Decision Theory

In the following we clarify the connection of (4.15) and (4.16) to the Bellman equations [Bel57, BT96] of sequential decision theory and discuss similarities and differences. We consider a Markov decision process (MDP) where, with probability $M_{ij}^a$, the agent under consideration should reach (environmental) state $j\in\mathcal{S}$ when taking action $a\in\mathcal{A}$ in (the current) state $i\in\mathcal{S}$. If the agent receives reward $R(i)$, the optimal policy $p^*$, maximizing expected utility (defined as the sum of future rewards), and the utility $U(i)$ of policy $p^*$ are

$$p^*(i) = \arg\max_a \sum_j M_{ij}^a U(j), \qquad U(i) = R(i) + \max_a \sum_j M_{ij}^a U(j). \quad (4.29)$$

See [RN95, BT96] for details and further references. Let us identify

$$\mathcal{S} = (\mathcal{Y}\times\mathcal{X})^*, \quad \mathcal{A} = \mathcal{Y}, \quad a = y_k, \quad M_{ij}^a = \mu(y\!x_{<k}y\!\underline{x}_k), \quad p^*(i) = y_k,$$

$$i = y\!x_{<k}, \quad R(i) = r(x_{k-1}), \quad U(i) = V_{k-1,m}^*(y\!x_{<k}) = r(x_{k-1}) + V_{km}^*(y\!x_{<k}),$$

$$j = y\!x_{1:k}, \quad R(j) = r(x_k), \qquad U(j) = V_{km}^*(y\!x_{1:k}) = r(x_k) + V_{k+1,m}^*(y\!x_{1:k}),$$

where we further set $M_{ij}^a = 0$ if $i$ is not a starting substring of $j$ or if $a\neq y_k$. This ensures the sum over $j$ in (4.29) to reduce to a sum over $x_k$. If we set $m_k = m$ and insert (4.12) into (4.16), it is easy to see that (4.29) coincides with (4.15) and (4.16).

Note that despite of this formal equivalence, we were forced to use the complete history $y\!x_{<k}$ as environmental state $i$. The AI$\mu$ model assumes neither stationarity, nor Markov property nor complete accessibility of the environment, as any assumption would restrict the applicability of AI$\mu$. The consequence is that every state occurs at most once in the lifetime of the agent. Every moment in the universe is unique! Even if the state space could be identified with the input space $\mathcal{X}$, inputs would usually not occur twice by the assumption $k\ll|\mathcal{X}|$, made in the last subsection. Further, there is no obvious universal similarity relation on $(\mathcal{X}\times\mathcal{Y})^*$ allowing an effective reduction of the size of the state space. Although many algorithms (like value and policy iteration) have problems in solving (4.29) for huge or infinite state spaces in practice, there is no principle problem in determining $p^*$ and $U$, as long as $\mu$ is known and computable, and $|\mathcal{X}|$, $|\mathcal{Y}|$ and $m$ are finite.

Things drastically change if $\mu$ is unknown. Reinforcement learning algorithms [KLM96, SB98] are commonly used in this case to learn the unknown

$\mu$ or directly its value. They succeed if the state space is either small or has effectively been made small by so-called generalization techniques. In any case, the solutions are either ad hoc, or work in restricted domains only, or have serious problems with state space exploration versus exploitation, or are prone to diverge, or have nonoptimal learning rate. There is no universal and optimal solution to this problem so far. In the next chapter we present a new model and argue that it formally solves all these problems in an optimal way. It is not concerned with learning $\mu$ directly. All we do is to replace the true prior probability $\mu$ by a universal probability $\xi$, which is shown to converge to $\mu$ in a sense.

## 4.4 Problems

**4.1 (Value dominance $\rho$)** [C20s] In (4.5) $V_{km}^{p\mu}(y\!x_{<k})$ was defined as the $\mu$-expected future reward sum $r_k+...+r_m$ with actions generated by policy $p$, and fixed history $y\!x_{<k}$. The optimal policy $p^\mu$ was defined as the one with maximal total $\mu$-value, i.e. $p^\mu := \text{argmax}_p V_{1m}^{p\mu}(\epsilon)$. The corresponding value function is $V_{km}^{*\mu}(y\!x_{<k}) := V_{km}^{p^\mu\mu}(y\!x_{<k})$. Obviously, $V_{1m}^{*\mu}(\epsilon) \geq V_{1m}^{p\mu}(\epsilon)\,\forall p$. Show that this implies $V_{km}^{*\mu}(y\!x_{<k}) \geq V_{km}^{p\mu}(y\!x_{<k})$ for all $p$ consistent with $y\!x_{<k}$. The derivation of a result of this form goes hand in hand with the derivation of Bellman's equations [Ber95a].

**4.2 (Probabilistic policies)** [C15usi] In this chapter we only gave formal definitions of the value function for deterministic policies, but allowed probabilistic environments. Generalize the definition of the value function (4.18) to probabilistic policies $\pi$ in the following way:

$$\begin{aligned} V_\mu^\pi &= \textstyle\sum_{y\!x_{1:m}} (r_1+\,...\,+r_m)\,\mu(x_m|y\!x_{<m}y_m)\pi(y_m|y\!x_{<m})...\mu(x_1|y_1)\pi(y_1) \\ &= \textstyle\sum_{y\!x_{1:m}} (r_k+\,...\,+r_m)\,\mu(x_{1:m}|y_{1:m})\pi(y_{1:m}|x_{<m}) \\ &=: \textstyle\sum_{y\!x_{1:m}} (r_k+\,...\,+r_m)\,\rho(y_{1:m}x_{1:m}) \end{aligned}$$

and similarly for $V_{km}^{\pi\mu}(y\!x_{<k})$. We used here the conventional notation $\mu(\cdot|\cdot)$ for conditional probabilities to emphasize the following oddity. The last equality seems to say something like $p(x|y)p(y|x)=p(x\&y)$, which would contradict Bayes' law $p(x|y)p(y)=p(x\&y)$. Show that everything goes right here. What important property (defined in the main text) is used to arrive at the above expressions? Show that $\max_\pi V_\mu^\pi = \max_p V_\mu^p$, and that among optimal policies there is always a deterministic one. Since we are mainly interested in optimal policies, the restriction to deterministic policies is not serious. Generalize the theorems in this and later chapters involving $V_\mu^p$ to $V_\mu^\pi$. A serious treatment of probabilistic policies can be found in game theory, where optimal policies can also be truly probabilistic [OR94].

*The Three Laws of Robotics:*

1. *A robot may not injure a human being, or, through inaction, allow a human being to come to harm.*
2. *A robot must obey the orders given it by human beings except where such orders would conflict with the First Law.*
3. *A robot must protect its own existence as long as such protection does not conflict with the First or Second Law.*

— Isaac Asimov

Isaac Asimov
(1920–1992)

# 5 The Universal Algorithmic Agent AIXI

Active systems, like game playing (SG) and optimization (FM), cannot be reduced to induction systems. The *main idea in this book* is to generalize universal induction to the general agent model described in Chapter 4. For this, we generalize Solomonoff's universal prior $M$ to include actions as conditions and replace $\mu$ by $M$ in the rational agent model, resulting in the AIXI model. In this way the problem that the true prior probability $\mu$ is usually unknown is solved. Convergence of $M \to \mu$ will be shown, indicating that the AIXI model could behave optimally in any computable but unknown environment with reinforcement feedback.

The main focus of Chapter 5 is to investigate what we can expect from a universally optimal agent and to clarify the meanings of *universal*, *optimal*, etc. Similarly to the induction case, it is convenient to consider a general mixture distribution $\xi$ of a weighted sum of distributions $\nu \in \mathcal{M}$, where $\mathcal{M}$ is any class of distributions including the true environment $\mu$. We show that the Bayes optimal policy $p^\xi$ based on the mixture $\xi$ is self-optimizing in the sense that the average value converges asymptotically for all $\mu \in \mathcal{M}$ to the optimal value achieved by the (infeasible) Bayes optimal policy $p^\mu$ which knows $\mu$ in advance. We show that the necessary condition that $\mathcal{M}$ admits self-optimizing policies at all is also sufficient. No other structural assumptions are made on $\mathcal{M}$.

One can show that bandits, i.i.d. processes, classification tasks, certain classes of POMDPs, ($k^{th}$-order) ergodic MDPs, factorizable environments, repeated games, and prediction problems admit self-optimizing policies. Unfortunately, the class $\mathcal{M}_U$ of all enumerable semimeasures does *not* admit self-optimizing policies. This forces us to lower our expectation about universally optimal agents and to introduce other (weaker) performance measures. Finally, we show that $p^\xi$ is Pareto optimal in the sense that there is no other policy yielding higher or equal value in *all* environments $\nu \in \mathcal{M}$ and a strictly higher value in at least one. Pareto optimality holds for any choice of $\mathcal{M}$ (including $\mathcal{M}_U$).

## 5.1 The Universal AIXI Model

### 5.1.1 Definition of the AIXI Model

We have developed enough formalism to suggest our universal AIXI model. All we have to do is to suitably generalize the universal semimeasure $M$ from Section 2.4 and replace the true but unknown prior probability $\mu^{\text{AI}}$ in the AI$\mu$ model by this generalized $M^{\text{AI}}$. In what sense this AIXI model is universal will be discussed subsequently.

In the functional formulation we define the universal probability $M^{\text{AI}}$ of an environment $q$ simply as $2^{-\ell(q)}$:

$$M(q) := 2^{-\ell(q)}$$

The definition could not be easier![1,2] Collecting the formulas of Section 4.1 and replacing $\mu(q)$ by $M(q)$ we get the definition of the AIXI agent in functional form. Given the history $\dot{y}\dot{x}_{<k}$, the functional AIXI agent outputs

$$\dot{y}_k := \arg\max_{y_k} \max_{p:p(\dot{x}_{<k})=\dot{y}_{<k}y_k} \sum_{q:q(\dot{y}_{<k})=\dot{x}_{<k}} 2^{-\ell(q)} \cdot V_{km_k}^{pq} \tag{5.1}$$

in cycle $k$, where $V_{km_k}^{pq}$ is the total reward of cycles $k$ to $m_k$ when agent $p$ interacts with environment $q$. We have dropped the denominator $\sum_q \mu(q)$ from (4.6) as it is independent of the $p \in \dot{P}_k$, and a constant multiplicative factor does not change $\text{argmax}_{y_k}$.

For the iterative formulation, the universal probability $M$ can be obtained by inserting the functional $M(q)$ into (4.19):

$$M(\underline{yx}_{1:k}) = \sum_{q:q(y_{1:k})=x_{1:k}} 2^{-\ell(q)} \tag{5.2}$$

Replacing $\mu$ by $M$ in (4.17) the iterative AIXI agent outputs

$$\dot{y}_k \equiv \dot{y}_k^M := \arg\max_{y_k}\sum_{x_k}\max_{y_{k+1}}\sum_{x_{k+1}} ... \max_{y_{m_k}}\sum_{x_{m_k}}(r(x_k)+ ... +r(x_{m_k}))M(\dot{y}\dot{x}_{<k}\underline{yx}_{k:m_k}) \tag{5.3}$$

in cycle $k$ given the history $\dot{y}\dot{x}_{<k}$.

One subtlety has been passed over. Like in the the sequence prediction (SP) case, $M$ is not a probability distribution but satisfies only the weaker inequalities

$$\sum_{x_n} M(\underline{yx}_{1:n}) \leq M(\underline{yx}_{<n}) \quad \text{and} \quad M(\epsilon) \leq 1. \tag{5.4}$$

Note that the sum on the l.h.s. is *not* independent of $y_n$, unlike for chronological probability distributions. Nevertheless, it is bounded by something (the r.h.s) that is independent of $y_n$. This is because the sum in (5.2) runs over (partial recursive) chronological functions only, and the functions $q$ that satisfy $q(y_{1:n}) = x_{<n}*$ (with $* \in \mathcal{X}$) are a subset of the functions satisfying $q(y_{<n}) = x_{<n}$. We will in general call functions satisfying (5.4) *chronological semimeasures*. The important point is that the conditional probabilities (4.10) are $\leq 1$, like for true probability distributions.

---

[1] It is not necessary to use $2^{-K(q)}$ or something similar as some readers may expect at this point, because for every program $q$ there exists a functionally equivalent program $\tilde{q}$ with $K(q) \stackrel{+}{=} \ell(\tilde{q})$.

[2] Here and later we identify objects with their coding relative to some fixed Turing machine $U$. For example, if $q$ is a function $K(q) := K(\langle q \rangle)$ with $\langle q \rangle$ being a binary coding of $q$ such that $U(\langle q \rangle, y) = q(y)$. Reversely, if $q$ already is a binary string we define $q(y) := U(q, y)$.

The equivalence of the functional and iterative AI model proven in Section 4.2 is true for every chronological semimeasure $\rho$, in particular for $M$; hence we can talk about *the* AIXI model in this respect.[3] It (slightly) depends on the choice of the universal Turing machine. $\ell(\langle q\rangle)$ is defined only up to an additive constant. The AIXI model also depends on the choice of $\mathcal{X}=\mathcal{R}\times\mathcal{O}$ and $\mathcal{Y}$, but we do not expect any bias when the spaces are chosen sufficiently simple, e.g. all strings of length $2^{16}$. Choosing $I\!N$ as the word space would be ideal, but whether the maxima (suprema) exist in this case, must be shown beforehand. The only nontrivial dependence is on the horizon function $m_k$, which will be discussed in Section 5.7. So, apart from $m_k$ and unimportant details, the AIXI agent is uniquely defined by (5.1) or (5.3). It does not depend on any assumption about the environment, apart from being generated by some computable (but unknown!) probability distribution.

### 5.1.2 Universality of $M^{\text{AI}}$ and $\xi^{\text{AI}}$

In which sense the AIXI model is optimal will be clarified later. In this and the next subsection we show that $M^{\text{AI}}$ defined in (5.2) is universal and converges to $\mu^{\text{AI}}$ analogous to the SP case (2.27) and (2.25). The proofs are generalizations from the SP case. The $y$ are pure spectators and cause no difficulties in the generalization. In (2.21) $U(p)=x*$ produces strings starting with $x$, whereas in (5.2) we can demand $q$ to output exactly $n$ words $x_{1:n}$ as $q$ "knows" $n$ from the number of input words $y_1...y_n$.

As in the SP case (2.26), there is an alternative definition of $M$ that coincides with (5.2) within a multiplicative constant of order one:

$$\xi(y\!\!x_{1:n}) \;:=\; \sum_\rho 2^{-K(\rho)}\rho(y\!\!x_{1:n}) \;\stackrel{\times}{=}\; M(y\!\!x_{1:n}), \tag{5.5}$$

where the sum runs over all enumerable chronological semimeasures. The $2^{-K(\rho)}$ weighted sum over probabilistic environments $\rho$ coincides with the sum over $2^{-\ell(q)}$ weighted deterministic environments $q$, as will be proven below. In Section 5.10 we show that an enumeration of all enumerable functions can be converted into an enumeration of enumerable chronological semimeasures $\rho$. $K(\rho)$ is co-enumerable, and therefore $\xi$ defined in (5.5) is itself enumerable. The representation (5.2) is also enumerable. As $\sum_\rho 2^{-K(\rho)}\leq 1$ and the $\rho$'s satisfy (5.4), $\xi$ is a chronological semimeasure as well. If we pick one $\rho$ in (5.5) we get the universality property "for free"

[3] Iterative AIXI is *not* equivalent to recursive AIXI. Solomonoff normalization (2.30) rescues equivalence, but destroys enumerability of the universal value $V_{km}^{p\xi}$. Better is to formally make all partial functions $q$ in (5.2) and elsewhere total by defining $o_k$ somehow and $r_k=0$ if originally undefined, or equivalently replacing $M(...\underline{0})$ by $1-\sum_{r_k\neq 0}M(...\underline{r}_k)$. Then $M$ becomes a (proper) measure, functional and iterative *and* recursive AIXI are equivalent, and $V_{km}^{p\xi}$ is enumerable, although the modified $M$ is not.

$$\xi(\underline{yx}_{1:n}) \;\geq\; 2^{-K(\rho)}\rho(\underline{yx}_{1:n}). \tag{5.6}$$

$\xi$ is a universal element in the sense of (5.6) in the set of all enumerable chronological semimeasures.

To prove universality of $M$ we have to prove (5.5). We note that for every enumerable chronological semimeasure $\rho$ there exists a Turing machine $T$ with

$$\rho(\underline{yx}_{1:n}) \;=\; \sum_{q:T(q,y_{1:n})=x_{1:n}} 2^{-\ell(q)} \qquad \text{and} \qquad \ell(T) \stackrel{+}{=} K(\rho). \tag{5.7}$$

We will not prove (5.7) here (see Problem 5.3). Given $T$, the universality of $M$ follows from

$$\begin{aligned} M(\underline{yx}_{1:n}) \;&=\; \sum_{q:U(q,y_{1:n})=x_{1:n}} 2^{-\ell(q)} \;\geq\; \sum_{q':U(Tq',y_{1:n})=x_{1:n}} 2^{-\ell(Tq')} \\ &=\; 2^{-\ell(T)} \sum_{q':T(q',y_{1:n})=x_{1:n}} 2^{-\ell(q')} \;\stackrel{\times}{=}\; 2^{-K(\rho)}\rho(\underline{yx}_{1:n}). \end{aligned}$$

The first equality and (5.2) are identical by definition. In the inequality we have restricted the sum over all $q$ to $q$ of the form $q{=}Tq'$. The third relation is true as running $U$ on $Tz$ is a simulation of $T$ on $z$. The last equality follows from (5.7). All enumerable universal chronological semimeasures coincide up to a multiplicative constant, as they mutually dominate each other. Hence, definitions (5.2) and (5.5) are indeed equivalent.

### 5.1.3 Convergence of $\xi^{\text{AI}}$ to $\mu^{\text{AI}}$

In Section 3.9.2 we proved the following entropy inequality:

$$\sum_{i=1}^N (y_i - z_i)^2 \;\leq \sum_{i=1}^N y_i \ln\frac{y_i}{z_i} \qquad \text{with} \qquad \sum_{i=1}^N y_i = 1, \;\; \sum_{i=1}^N z_i \leq 1. \tag{5.8}$$

In Section 5.11 we give a different proof since it contains some ideas that could be interesting for their own sake.[4] If we identify $N{=}|\mathcal{X}|$, $i{=}x_k$, $y_i{=}\mu(yx_{<k}\underline{yx}_k)$ and $z_i{=}\xi(yx_{<k}\underline{yx}_k)$, multiply both sides with $\mu(\underline{yx}_{<k})$, take the sum over $x_{<k}$ and $k$ and use the chain rule $\mu(\underline{yx}_{<k})\cdot\mu(yx_{<k}\underline{yx}_k){=}\mu(\underline{yx}_{1:k})$, we get

$$\sum_{k=1}^n \sum_{x_{1:k}} \mu(\underline{yx}_{<k})\Big(\mu(yx_{<k}\underline{yx}_k) - \xi(yx_{<k}\underline{yx}_k)\Big)^2 \leq \sum_{k=1}^n \sum_{x_{1:k}} \mu(\underline{yx}_{1:k}) \ln\frac{\mu(yx_{<k}\underline{yx}_k)}{\xi(yx_{<k}\underline{yx}_k)} = \ldots \tag{5.9}$$

In the r.h.s. we can replace $\sum_{x_{1:k}}\mu(\underline{yx}_{1:k})$ by $\sum_{x_{1:n}}\mu(\underline{yx}_{1:n})$, as the argument of the logarithm is independent of $x_{k+1:n}$. The $k$ sum can now be brought

[4] Actually, a proof similar to the one in Section 5.11 of an inequality similar to (5.8), found in [Hut00], in a sense initiated the whole book.

into the logarithm and converts to a product. Using the chain rule (4.10) for $\mu$ and $\xi$ we get

$$\ldots = \sum_{x_{1:n}} \mu(yx_{1:n}) \ln \prod_{k=1}^{n} \frac{\mu(yx_{<k} yx_k)}{\xi(yx_{<k} yx_k)} = \sum_{x_{1:n}} \mu(yx_{1:n}) \ln \frac{\mu(yx_{1:n})}{\xi(yx_{1:n})} \stackrel{+}{\leq} \ln 2 \cdot K(\mu) \tag{5.10}$$

where we have used the universality property (5.6) of $\xi$ in the last step. The line of reasoning is identical to that in (3.18); the $y$ are, again, pure spectators. This will change when we analyze loss/value bounds analogous to Theorem 3.48.

Bound (5.9)/(5.10) shows that the $\mu$-expected squared difference of $\mu$ and $\xi$ is finite for computable $\mu$. This, in turn, shows that $\xi(yx_{<k} yx'_k)$ converges to $\mu(yx_{<k} yx'_k)$ for $k \to \infty$ with $\mu$-probability 1. If we take a finite product of $\xi$'s and use the chain rule, we see that also $\xi(yx_{<k} yx'_{k:k+r})$ converges to $\mu(yx_{<k} yx'_{k:k+r})$. More generally, by supplementing the results on multistep predictions in Section 3.7.1 with action $y$ as conditions we have[5]

$$\xi(yx_{<k} yx'_{k:m_k}) \stackrel{k \to \infty}{\longrightarrow} \mu(yx_{<k} yx'_{k:m_k}) \begin{cases} \text{i.m.s. if } & h_k \equiv m_k - k + 1 \leq h_{max} < \infty, \\ \text{i.m. for general } h_k \equiv m_k - k + 1. \end{cases} \tag{5.11}$$

This gives hope that the outputs $\dot{y}_k$ of the AIXI model (5.3) could converge to the outputs $\dot{y}_k$ from the AI$\mu$ model (4.17).

We want to call an AI model *universal*, if it is $\mu$-independent (unbiased, model-free) and is able to solve any solvable problem and learn any learnable task. Further, we call a universal model *universally optimal* if there is no program that can solve or learn significantly faster (in terms of interaction cycles). Indeed, the AIXI model is parameter free, $\xi$ converges to $\mu$ (5.11), the AI$\mu$ model is itself optimal, and we expect no other model to converge faster to AI$\mu$ by analogy to SP (Theorem 3.48):

**Claim 5.12 (We expect AIXI to be universally optimal)**

This is our main claim. In a sense, the intention of the remaining sections is to define this statement more rigorously and to give further support.

### 5.1.4 Intelligence Order Relation

We define the $\xi$-expected reward in cycles $k$ to $m$ of a policy $p$ similar to (4.6) and (5.1). We extend the definition to programs $p \notin \dot{P}_k$ that are not consistent with the current history.

$$V_{km}^{p\xi}(\dot{y}\dot{x}_{<k}) := \frac{1}{\mathcal{N}} \sum_{q:q(\dot{y}_{<k})=\dot{x}_{<k}} 2^{-\ell(q)} \cdot V_{km}^{\tilde{p}q} \tag{5.13}$$

[5] Here and elsewhere we interpret $a_k \to b_k$ as an abbreviation for $a_k - b_k \to 0$; $\lim_{k\to\infty} b_k$ itself may not exist.

The normalization $\mathcal{N}$ is again only necessary for interpreting $V_{km}$ as the expected reward but is otherwise unneeded. For consistent policies $p \in \dot{P}_k$ we define $\tilde{p}:=p$. For $p \notin \dot{P}_k$, $\tilde{p}$ is a modification of $p$ in such a way that its outputs are consistent with the current history $\dot{y}\dot{x}_{<k}$, hence $\tilde{p} \in \dot{P}_k$, but unaltered for the current and future cycles $\geq k$. Using this definition of $V_{km}$ we could take the maximum over all policies $p$ in (5.1), rather than only the consistent ones.

**Definition 5.14 (Intelligence order relation)** We call a policy $p$ *more or equally intelligent* than $p'$ and write

$$p \succeq p' \quad :\Leftrightarrow \quad \forall k \forall \dot{y}\dot{x}_{<k} : V_{km_k}^{p\xi}(\dot{y}\dot{x}_{<k}) \geq V_{km_k}^{p'\xi}(\dot{y}\dot{x}_{<k}),$$

i.e. if $p$ yields in any circumstance higher $\xi$-expected reward than $p'$.

As the algorithm $p^\xi$ behind the AIXI agent maximizes $V_{km_k}^{p\xi}$, we have $p^\xi \succeq p$ for all $p$. The AIXI model is hence the most intelligent agent w.r.t. $\succeq$. Relation $\succeq$ is a universal order relation in the sense that it is free of any parameters (except $m_k$) or specific assumptions about the environment. A proof that $\succeq$ is a reliable intelligence order (which we believe to be true) would prove that AIXI is universally optimal. We could further ask: How useful is $\succeq$ for ordering policies of practical interest with intermediate intelligence, or how can $\succeq$ help to guide toward constructing more intelligent systems with reasonable computation time? An effective intelligence order relation $\succeq^c$ will be defined in Section 7.2, which is more useful from a practical point of view.

## 5.2 On the Optimality of AIXI

In this section we outline ways toward an optimality proof of AIXI, which will be followed thereafter, but not always to the end. Sources of inspiration are the SP loss bounds proven in Chapter 3 and optimality criteria from the adaptive control literature (mainly) for linear systems [KV86]. The value bounds for AIXI are expected to be, in a sense, weaker than the SP loss bounds because the problem class covered by AIXI is much larger than the class of induction problems. Convergence of $\xi$ to $\mu$ has already been proven, but is not sufficient to establish convergence of the behavior of the AIXI model to the behavior of the AI$\mu$ model. We will focus on the following approaches toward a general optimality proof:

**What is meant by universal optimality.** The first step is to investigate what we can expect from AIXI, i.e. what is meant by *universal optimality*. A "learner" (like AIXI) may converge to the optimal informed decision-maker (like AI$\mu$) in several senses. Possibly relevant concepts from statistics are consistency, self-tunability, self-optimization, efficiency, unbiasedness, asymptotic or finite convergence [KV86], Pareto optimality, and some others defined in

Section 5.3. Some concepts are stronger than necessary; others are weaker than desirable but are suitable to start with. Self-optimization is defined as the asymptotic convergence of the average true value $\frac{1}{m}V_{1m}^{p^\xi\mu}$ of AIXI to the optimal value $\frac{1}{m}V_{1m}^{*\mu}$. Apart from convergence speed, self-optimization of AIXI would most closely correspond to the loss bounds proven for SP. We investigate which properties are desirable and under which circumstances the AIXI model satisfies these properties. We will show that no universal model, including AIXI, can in general be self-optimizing. On the other hand, we show that AIXI is Pareto optimal in the sense that there is no other policy which performs better or equal in all environments, and strictly better in at least one.

**Limited environmental classes.** The problem of defining and proving general value bounds becomes more feasible by considering, in a first step, restricted concept classes. We analyze AIXI for known classes (Markovian, factorizable, predictive, game, optimization, and supervised learning environments) in Chapter 6 and for the classes (forgetful, relevant, asymptotically learnable, farsighted, uniform, pseudo-passive, and passive) defined in Section 5.3.

**Generalization of AIXI to general Bayes mixtures.** Another approach is to generalize AIXI to AI$\xi$, where $\xi()=\sum_{\nu\in\mathcal{M}}w_\nu\nu()$ is a general Bayes mixture of distributions $\nu$ in some class $\mathcal{M}$. If $\mathcal{M}$ is the multi-set of enumerable chronological semimeasures, then AI$\xi$ coincides with AIXI. If $\mathcal{M}$ contains only passive environments, then AI$\xi$ reduces to the $\Lambda_\xi$ predictor, which has been shown to perform well in Section 3.4.1. We show that these loss/value bounds generalize to wider classes, at least asymptotically. Promising classes are, again, the ones described in Section 5.3. Especially for ergodic MDPs we show that AI$\xi$ is self-optimizing. Obviously, the least we must demand from $\mathcal{M}$ to have a chance of finding a self-optimizing policy is that there exists some self-optimizing policy at all. This necessary condition will also be sufficient. More generally, the key is not to prove absolute results for specific problem classes, but to prove relative results of the form "if there exists a policy with certain desirable properties, then AI$\xi$ also possesses these desirable properties". If there are tasks that cannot be solved by any policy, AI$\xi$ cannot be blamed for failing. Note that in this approach we have for each environmental class $\mathcal{M}$ a corresponding model AI$\xi$, whereas in the previous approach the same AIXI model is analyzed for all environmental classes.

**Optimality by construction.** A possible further approach toward an optimality "proof" is to regard AIXI as *optimal by construction*. This perspective is common in various (simpler) settings. For instance, in bandit problems, where pulling arm $i$ leads to reward 1 (0) with unknown probability $p_i$ $(1-p_i)$, the traditional Bayesian solution to the uncertainty about $p_i$ is to assume a uniform (or Beta) prior over $p_i$ and to maximize the (subjectively) expected reward sum over multiple trials. The exact solution (in terms of

Gittins indices) is widely regarded as "optimal", although justified alternative approaches exist. Similarly, but simpler, assuming a uniform subjective prior over the Bernoulli parameter $p_{(i)} \in [0,1]$, one arrives at the reasonable, but more controversial, Laplace rule for predicting i.i.d. sequences. AIXI is similar in the sense that the unknown $\mu \in \mathcal{M}$ is the analogue of the unknown $p \in [0,1]$, and the prior beliefs $w_\nu = 2^{-K(\nu)}$ justified by Occam's razor are the analogue of a uniform distribution over $[0,1]$. In the same sense as Gittins' solution to the bandit problem and Laplace' rule for Bernoulli sequences, AIXI may also be regarded as optimal by construction. Theorems relating AIXI to AI$\mu$ would not be regarded as optimality proofs of AIXI, but just as how much harder it becomes to operate when $\mu$ is unknown, i.e. the achievements of the first three approaches are simply reinterpreted.

## 5.3 Value Bounds and Separability Concepts

### 5.3.1 Introduction

The values $V_{km}$ associated with the AI systems correspond roughly to the negative error measure $-E_n^\Theta$ or negative loss $-L_n^\Lambda$ of the SP systems. In SP, we were interested in small bounds for the error excess $E_n^{\Theta_\xi} - E_n^\Theta$. Unfortunately, simple value bounds for AIXI in terms of $V_{km}$ analogous to the loss bound in Theorem 3.48 do not hold. We even have difficulties in specifying what we can expect to hold for AIXI or any AI system that claims to be universally optimal. Consequently, we cannot have a proof if we don't know what to prove. In SP, the only important property of $\mu$ for proving error bounds was its complexity $K(\mu)$. We will see that in the AI case, there are no useful bounds in terms of $K(\mu)$ only. We either have to study restricted problem classes or consider bounds depending on other properties of $\mu$, rather than on its complexity only. In the following, we will exhibit the difficulties by two examples and introduce concepts that may be useful for proving value bounds. Despite the difficulties in even claiming useful value bounds, we nevertheless firmly believe that the order relation in Definition 5.14 correctly formalizes the intuitive meaning of intelligence and, hence, that the AIXI agent is universally optimal.

### 5.3.2 (Pseudo) Passive $\mu$ and the HeavenHell Example

In the following we choose $m_k = m$. We want to compare the true, i.e. $\mu$-expected, value $V_{1m}^\mu$ of a $\mu$-independent universal policy $p^{best}$ with any other policy $p$. Naively, we might expect the existence of a policy $p^{best}$ that maximizes $V_{1m}^\mu$, apart from additive corrections of lower order for $m \to \infty$,

$$V_{1m}^{p^{best}\mu} \geq V_{1m}^{p\mu} - o(...) \quad \forall \mu, p. \tag{5.15}$$

Such policies are sometimes called self-optimizing [KV86]. Note that $V_{1m}^{*\mu} \geq V_{1m}^{p\mu} \forall p$, but $* = p^\mu$ is not a candidate for a universal $p^{best}$, as it depends on $\mu$. On the other hand, the policy $p^\xi$ of the AIXI agent maximizes $V_{1m}^\xi$ by definition ($p^\xi \succeq p$). As $V_{1m}^\xi$ is thought to be a guess of $V_{1m}^\mu$, we might expect $p^{best} = p^\xi$ to approximately maximize $V_{1m}^\mu$, i.e. (5.15) to hold. Let us consider the problem class (set of environments) $\{\mu_0,\mu_1\}$ with $\mathcal{Y}=\mathcal{R}=\{0,1\}$ and $r_k = \delta_{iy_1}$ in environment $\mu_i$. The first action $y_1$ decides whether you go to heaven with all future rewards $r_k$ being 1 (good), or to hell with all future rewards being 0 (bad). Note that $\mu_i$ are (deterministic, non-ergodic) MDPs:

$\mu_i \quad =$

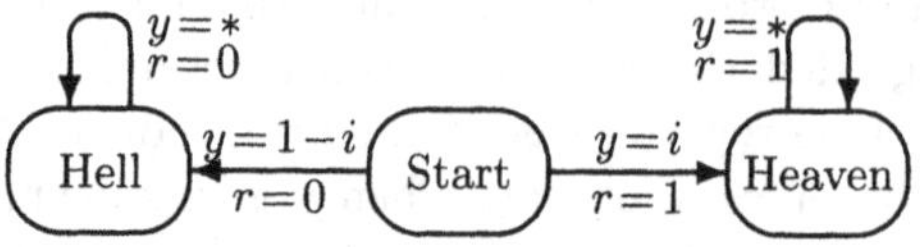


It is clear that if $\mu_i$, i.e. $i$ is known, the optimal policy $p^{\mu_i}$ is to output $y_1 = i$ in the first cycle with $V_{1m}^{p^{\mu_i}\mu} = m$. On the other hand, any unbiased policy $p^{best}$ independent of the actual $\mu$ either outputs $y_1=1$ or $y_1=0$. Independent of the actual choice $y_1$, there is always an environment ($\mu=\mu_{1-y_1}$) for which this choice is catastrophic ($V_{1m}^{p^{best}\mu}=0$). No single agent can perform well in both environments $\mu_0$ *and* $\mu_1$. The r.h.s. of (5.15) equals $m-o(m)$ for $p=p^\mu$. For all $p^{best}$ there is a $\mu$ for which the l.h.s. is zero. (The situation remains the same if we add a third action allowing to stay in the start state. See also Problem 5.2 for a similar two-state environment.) We have shown that no $p^{best}$ can satisfy (5.15) for all $\mu$ and $p$, so we cannot expect $p^\xi$ to do so. Nevertheless, there are problem classes for which (5.15) holds, for instance, SP and CF and ergodic MDPs. For SP, (5.15) is just a reformulation of Theorem 3.48 with an appropriate choice for $p^{best}$ (which differs from $p^\xi$, see next section). We expect (5.15) to hold for all inductive problems in which the environment is not influenced by the output of the agent. (Of course, the reward feedback $r_k$ depends on the agent's output. What we mean is, like in sequence prediction, that the true sequence is not influenced by the agent.) We want to call these $\mu$ *passive* or *inductive* environments. Further, we want to call $\mathcal{M}$ and $\mu\in\mathcal{M}$ satisfying (5.15) with $p^{best}=p^\xi$ *pseudo-passive.* So we expect inductive $\mu$ to be pseudo-passive.

### 5.3.3 The OnlyOne Example

Let us give a further example to demonstrate the difficulties in establishing value bounds. Let $\mathcal{X}=\mathcal{R}=\{0,1\}$ and $|\mathcal{Y}|$ be large. We consider all (deterministic) environments in which a single complex output $y^*$ is correct ($r=1$) and all others are wrong ($r=0$). The problem class $\mathcal{M}$ is defined by

$$\mathcal{M} := \{\mu_{y^*} : y^* \in \mathcal{Y},\ K(y^*) =_\lfloor \log_2|\mathcal{Y}|_\rfloor\}, \quad \text{where} \quad \mu_{y^*}(y\!x_{<k}y_k\underline{1}) := \delta_{y_k y^*}\ \forall k.$$

There are $N \stackrel{\times}{=} |\mathcal{Y}|$ such $y^*$. The only way a $\mu$-independent policy $p$ can find the correct $y^*$, is by trying one $y$ after the other in a certain order. In the first $N-1$ cycles, at most $N-1$ different $y$ are tested. As there are $N$ different possible $y^*$, there is always a $\mu\in\mathcal{M}$ for which $p$ gives erroneous outputs in the first $N-1$ cycles. The number of errors is $E_\infty^p \geq N-1 \stackrel{\times}{=} |\mathcal{Y}| \stackrel{\times}{=} 2^{K(y^*)} \stackrel{\times}{=} 2^{K(\mu)}$ for this $\mu$. As this is true for any $p$, it is also true for the AIXI model, hence $E_k^{p^\xi} \leq 2^{K(\mu)}$ is the best possible error bound we can expect that depends on $K(\mu)$ only. Actually, we will derive such a bound in Section 6.2 for SP. Unfortunately, as we are mainly interested in the cycle region $k \ll |\mathcal{Y}| \stackrel{\times}{=} 2^{K(\mu)}$ (see Section 4.3.2) this bound is vacuous. There are no interesting bounds for deterministic $\mu$ depending on $K(\mu)$ only, unlike the SP case. Bounds must either depend on additional properties of $\mu$ or we have to consider specialized bounds for restricted problem classes. The case of probabilistic $\mu$ is similar. Whereas for SP there are useful bounds in terms of $E_k^{\Theta_\mu}$ and $K(\mu)$, there are no such bounds for AIXI. Again, this is not a drawback of AIXI since for no unbiased AI system could the errors/rewards be bound in terms of $K(\mu)$ and the errors/rewards of AI$\mu$ only.

There is a way to make use of gross (e.g. $2^{K(\mu)}$) bounds. Assume that after a reasonable number of cycles $k$, the information $\dot{x}_{<k}$ perceived by the AIXI agent contains a lot of information about the true environment $\mu$. The information in $\dot{x}_{<k}$ might be coded in any form. Let us assume that the complexity $K(\mu|\dot{x}_{<k})$ of $\mu$, under the condition that $\dot{x}_{<k}$ is known, is of order 1. Consider a theorem, bounding the sum of rewards or of other quantities over cycles $1...\infty$ in terms of $f(K(\mu))$ for a function $f$ with $f(O(1))=O(1)$, like $f(n)=2^n$. Then, there will be a bound for cycles $k...\infty$ in terms of $\approx f(K(\mu|\dot{x}_{<k}))=O(1)$. Hence, a bound like $2^{K(\mu)}$ can be replaced by small bound $\approx 2^{K(\mu|\dot{x}_{<k})}=O(1)$ after $k$ cycles. All one has to show/ensure/assume is that enough information about $\mu$ is presented (in any form) in the first $k$ cycles. In this way, even a gross bound could become useful. In Section 6.5 we use a similar argument to prove that AIXI is able to learn supervised (cf. Problems 3.13 and 6.3).

### 5.3.4 Asymptotic Learnability

In the following, we weaken (5.15) in the hope of getting a bound applicable to wider problem classes than the passive one. Consider the I/O sequence $\dot{y}_1\dot{x}_1...\dot{y}_n\dot{x}_n$ caused by AIXI. On history $\dot{y}\dot{x}_{<k}$, AIXI will output $\dot{y}_k \equiv \dot{y}_k^\xi$ in cycle $k$. Let us compare this to $\dot{y}_k^\mu$ which AI$\mu$ would output, still on the same history $\dot{y}\dot{x}_{<k}$ produced by AIXI. As AI$\mu$ maximizes the $\mu$-expected value, AIXI causes lower (or at best equal) $V_{km_k}^\mu$ if $\dot{y}_k^\xi$ differs from $\dot{y}_k^\mu$. Let $D_{n\mu\xi} := \mathbf{E}[\sum_{k=1}^n 1-\delta_{\dot{y}_k^\mu,\dot{y}_k^\xi}]$ be the $\mu$-expected number of suboptimal choices of AIXI, i.e. outputs different from AI$\mu$ in the first $n$ cycles. One might weigh the deviating cases by their severity. In particular, when the $\mu$-expected rewards $V_{km_k}^{p\mu}$ for $\dot{y}_k^\xi$ and $\dot{y}_k^\mu$ are equal or close to each other, this should be taken into account in a definition of $D_{n\mu\xi}$, e.g. by a weight factor $[V_{km}^{*\mu}(y\!x_{<k})-V_{km}^{p^\xi\mu}(y\!x_{<k})]$. These details do not matter in the following qualitative discussion. The important difference to (5.15) is that here we stick to the history produced by AIXI

and count a wrong decision as, at most, one error. The wrong decision in the HeavenHell example in the first cycle no longer counts as losing $m$ rewards, but counts as one wrong decision. In a sense, this is fairer. One shouldn't blame somebody too much who makes a single wrong decision for which he just has too little information available to make a correct decision. The AIXI model deserves to be called asymptotically optimal if the probability of making a wrong decision tends to zero, i.e. if

$$D_{n\mu\xi}/n \to 0 \quad \text{for} \quad n \to \infty, \quad \text{i.e.} \quad D_{n\mu\xi} \;=\; o(n). \tag{5.16}$$

We say that $\mu$ can be *asymptotically learned* (by AIXI) if (5.16) is satisfied. We claim that AIXI (for $m_k \to \infty$) can asymptotically learn every problem $\mu$ of relevance, i.e. AIXI is asymptotically optimal. We included the qualifier *of relevance*, since there may be strange $\mu$ spoiling (5.16), but we expect those $\mu$ to be irrelevant from the perspective of AI. In the field of machine learning, there are many asymptotic learnability theorems, which are often not too difficult to prove. So a proof of (5.16) might also be feasible. Unfortunately, asymptotic learnability theorems are often too weak to be useful from a practical point of view. Nevertheless, they point in the right direction.

### 5.3.5 Uniform $\mu$

From the convergence (5.11) of $\xi \to \mu$ we might expect $V_{km_k}^{p\xi} \to V_{km_k}^{p\mu}$ for all $p$, and hence we might also expect $\dot{y}_k^\xi$ defined in (5.3) to converge to $\dot{y}_k^\mu$ defined in (4.17) for $k \to \infty$. The first problem is that if the $V_{km_k}$ for the different choices of $y_k$ are nearly equal, then even if $V_{km_k}^{p\xi} \approx V_{km_k}^{p\mu}$, $\dot{y}_k^\xi \neq \dot{y}_k^\mu$ is possible due to the noncontinuity of $\text{argmax}_{y_k}$. This can be cured by a weighted $D_{n\mu\xi}$ as described above. More serious is the second problem we explain for $h_k = 1$ and $\mathcal{X} = \mathcal{R} = \{0,1\}$. For $\dot{y}_k^\xi \equiv \text{argmax}_{y_k} \xi(\dot{y}\dot{r}_{<k} y_k \underline{1})$ to converge to $\dot{y}_k^\mu \equiv \text{argmax}_{y_k} \mu(\dot{y}\dot{r}_{<k} y_k \underline{1})$, it is not sufficient to know that $\xi(\dot{y}\dot{r}_{<k} \dot{y}\dot{\underline{r}}'_k) \to \mu(\dot{y}\dot{r}_{<k} \dot{y}\dot{\underline{r}}'_k)$, as proven in (5.11). We need convergence not only for the true output $\dot{y}_k$ and arbitrary reward $\dot{r}'_k$, but also for alternative outputs $y_k$. $\dot{y}_k^\xi$ converges to $\dot{y}_k^\mu$ if $\xi$ converges uniformly to $\mu$, i.e. if in addition to (5.11)

$$\big|\mu(yx_{<k} y'_k \underline{x}'_k) - \xi(yx_{<k} y'_k \underline{x}'_k)\big| \;<\; c\cdot\big|\mu(yx_{<k} \underline{yx}_k) - \xi(yx_{<k} \underline{yx}_k)\big| \quad \forall y'_k x'_k \tag{5.17}$$

holds for some constant $c$ (at least in a $\mu$-expected sense). We call $\mu$ satisfying (5.17) *uniform*. For uniform $\mu$ one can show (5.16) with appropriately weighted $D_{n\mu\xi}$ and bounded horizon $h_k \le h_{max}$. Unfortunately there are relevant $\mu$ that are not uniform.

### 5.3.6 Other Concepts

In the following, we briefly mention some further concepts. A *Markovian* $\mu$ is defined as depending only on the last cycle, i.e. $\mu(yx_{<k} \underline{yx}_k) =$

$\mu_k(x_{k-1}\underline{yx}_k)$. We say $\mu$ is *generalized ($l^{th}$-order) Markovian*, if $\mu(\underline{yx}_{<k}\underline{yx}_k) = \mu_k(x_{k-l}yx_{k-l+1:k-1}\underline{yx}_k)$ for fixed $l$. This property has some similarities to *factorizable* $\mu$ defined in (4.26). If further $\mu_k \equiv \mu_1 \forall k$, $\mu$ is called *stationary*. *Ergodic* Markov decision processes are defined in Section 5.6. Further, we call $\mu$ ($\xi$) *forgetful* if $\mu(yx_{<k}\underline{yx}_k)$ ($\xi(yx_{<k}\underline{yx}_k)$) becomes independent of $yx_{<l}$ for fixed $l$ and $k \to \infty$ with $\mu$-probability 1 (cf. Problems 2.5 and 5.13). Further, we say $\mu$ is *farsighted* if $\lim_{m_k \to \infty} \dot{y}_k^{(m_k)}$ exists. More details are given in Section 5.7, where we also give an example of a farsighted $\mu$ for which nevertheless the limit $m_k \to \infty$ makes no sense.

### 5.3.7 Summary

We have introduced several concepts that might be useful for proving value bounds, including forgetful, relevant, asymptotically learnable, farsighted, uniform, (generalized) Markovian, factorizable and (pseudo-)passive $\mu$. We have sorted them here, approximately in the order of decreasing generality. We will call them *separability concepts*. The more general (like relevant, asymptotically learnable and farsighted) $\mu$ will be called weakly separable, the more restrictive (like (pseudo-)passive and factorizable) $\mu$ will be called strongly separable, but we will use these qualifiers in a more qualitative, rather than rigid sense. Other (nonseparability) concepts are deterministic $\mu$ and, of course, the class of all chronological $\mu$.

## 5.4 Value-Related Optimality Results

### 5.4.1 The AI$\rho$ Models: Preliminaries

In Chapter 4 we gave verbal definitions of the AI$\mu$ model (Definition 4.4) and of value functions (Definition 4.5) and derived different mathematical expressions for them. The AIXI model involved similar expressions with $\mu$ replaced by $M \stackrel{\times}{=} \xi_U = \xi$. The following definitions summarize all formulas for general environment $\rho$, and general environmental classes $\mathcal{M}$ and general weights $w_\nu$ in iterative form.

**Definition 5.18 ($\rho$-Value function)** We define the *value* of policy $p$ in environment $\rho$ given history $yx_{<k}$, or shorter, the $\rho$-value of $p$ given $yx_{<k}$, as

$$V_{km}^{p\rho}(yx_{<k}) := \sum_{x_{k:m}} (r_k + ... + r_m)\rho(yx_{<k}\underline{yx}_{k:m})_{|y_{1:m}=p(x_{<m})}$$

**Definition 5.19 (Functional AI$\rho$ model)** The AI$\rho$ model is defined as the policy $p^\rho$ that maximizes the (total) value $V_\rho^p := V_{1m}^{p\rho}(\epsilon)$:

$$p^\rho := \arg\max_p V_\rho^p, \qquad V_{km}^{*\rho}(y\!x_{<k}) := V_{km}^{p^\rho\rho}(y\!x_{<k}).$$

**Theorem 5.20 (Iterative AI$\rho$ model)** The $\rho$-optimal policy $p^\rho$ and its value $V_{km}^{*\rho}(y\!x_{<k})$ can be explicitly written as

$$\dot{y}_k = \arg\max_{y_k}\sum_{x_k}\max_{y_{k+1}}\sum_{x_{k+1}} ... \max_{y_m}\sum_{x_m}(r_k+ ... +r_m)\cdot\rho(\dot{y}\!\dot{x}_{<k}\underline{y\!x}_{k:m}),$$

$$V_{km}^{*\rho}(y\!x_{<k}) = \max_{y_k}\sum_{x_k}\max_{y_{k+1}}\sum_{x_{k+1}} ... \max_{y_m}\sum_{x_m}(r_k+ ... +r_m)\cdot\rho(y\!x_{<k}\underline{y\!x}_{k:m}).$$

Furthermore, $V_{km}^{*\rho}(y\!x_{<k}) \geq V_{km}^{p\rho}(y\!x_{<k})\ \forall p$.

The proof is the same as in Chapter 4 with $\mu$ replaced by $\rho$. The following property of $V_\rho$ is crucial.

**Theorem 5.21 (Linearity and convexity of $V_\rho$ in $\rho$)** $V_\rho^p$ is a linear function in $\rho$ and $V_\rho^*$ is a convex function in $\rho$ in the sense that

$$V_\xi^p = \sum_{\nu\in\mathcal{M}} w_\nu V_\nu^p \quad\text{and}\quad V_\xi^* \leq \sum_{\nu\in\mathcal{M}} w_\nu V_\nu^*, \quad\text{where}\quad \xi(\underline{y\!x}_{1:m}) = \sum_{\nu\in\mathcal{M}} w_\nu \nu(\underline{y\!x}_{1:m})$$

**Proof.** Linearity is obvious from the Definition 5.18 of $V_\rho^p$. Convexity follows from $V_\xi^* \equiv V_\xi^{p^\xi} = \sum_\nu w_\nu V_\nu^{p^\xi} \leq \sum_\nu w_\nu V_\nu^*$, where the first equality is just Definition 5.19, the second equality uses linearity of $V_\rho^{p^\xi}$ just proven, and the last inequality follows from the dominance (5.20) and positivity of the weights $w_\nu$. □

One loose interpretation of the convexity is that a mixture can never increase performance.

### 5.4.2 Pareto Optimality of AI$\xi$

This subsection shows Pareto optimality of AI$\xi$ analogous to SP. The total $\mu$-expected reward $V_\mu^{p^\xi}$ of policy $p^\xi$ of the AI$\xi$ model is of central interest in judging the performance of AI$\xi$. We know that there *are* policies (e.g. $p^\mu$ of AI$\mu$) with higher $\mu$-value ($V_\mu^* \geq V_\mu^{p^\xi}$). In general, every policy based on an estimate $\rho$ of $\mu$ that is closer to $\mu$ than is $\xi$ outperforms $p^\xi$ in environment $\mu$, simply because it is more tailored toward $\mu$. On the other hand, such a system probably performs worse than $p^\xi$ in other environments. Since we do not know $\mu$ in advance, we may ask whether there exists a policy $p$ with better or equal performance than $p^\xi$ in *all* environments $\nu\in\mathcal{M}$ and a strictly better performance for one $\nu\in\mathcal{M}$. This would clearly render $p^\xi$ suboptimal. We show that there is no such $p$.

**Definition 5.22 (Pareto optimal policies)** A policy $\tilde{p}$ is called Pareto optimal if there is no other policy $p$ with $V_\nu^p\geq V_\nu^{\tilde{p}}$ for all $\nu\in\mathcal{M}$ and strict inequality for at least one $\nu$.

**Theorem 5.23 (Pareto optimality of $p^\xi$)** AI$\xi$ alias $p^\xi$ is Pareto optimal.

**Proof.** We want to arrive at a contradiction by assuming that $p^\xi$ is not Pareto optimal, i.e. by assuming the existence of a policy $p$ with $V_\nu^p\geq V_\nu^{p^\xi}$ for all $\nu\in\mathcal{M}$ and strict inequality for at least one $\nu$:

$$V_\xi^p \;=\; \sum_\nu w_\nu V_\nu^p \;>\; \sum_\nu w_\nu V_\nu^{p^\xi} \;=\; V_\xi^{p^\xi} \;\equiv\; V_\xi^* \;\geq\; V_\xi^p$$

The two equalities follow from linearity of $V_\rho$ (5.21). The strict inequality follows from the assumption and $w_\nu>0$. The last inequality follows from the fact that $p^\xi$ maximizes by definition the universal value (5.20). The contradiction $V_\xi^p>V_\xi^p$ proves Pareto optimality of AI$\xi$. □

Pareto optimality should be regarded as a necessary condition for an agent aiming to be optimal. From a practical point of view, a significant increase of $V$ for many environments $\nu$ may be desirable, even if this causes a small decrease of $V$ for a few other $\nu$. The impossibility of such a "balanced" improvement is a more demanding condition on $p^\xi$ than pure Pareto optimality. The next theorem shows that $p^\xi$ is also balanced Pareto optimal in the following sense:

**Theorem 5.24 (Balanced Pareto optimality)**

$$\Delta_\nu := V_\nu^{p^\xi} - V_\nu^{\tilde{p}}, \quad \Delta := \sum_{\nu\in\mathcal{M}} w_\nu \Delta_\nu \quad\Rightarrow\quad \Delta\geq 0.$$

This implies the following: Assume $\tilde{p}$ has lower value than $p^\xi$ on environments $\mathcal{L}\subset\mathcal{M}$ by a total weighted amount of $\Delta_\mathcal{L}:=\sum_{\lambda\in\mathcal{L}}w_\lambda\Delta_\lambda$. Then $\tilde{p}$ can have higher value on $\eta\in\mathcal{H}:=\mathcal{M}\setminus\mathcal{L}$, but the improvement is bounded by $\Delta_\mathcal{H}:=|\sum_{\eta\in\mathcal{H}}w_\eta\Delta_\eta|\leq\Delta_\mathcal{L}$. In particular, $|\Delta_\eta|\leq w_\eta^{-1}\max_{\lambda\in\mathcal{L}}\Delta_\lambda$.

This means that a weighted value increase $\Delta_\mathcal{H}$ by using $\tilde{p}$ instead of $p^\xi$ is compensated by an at least as large weighted decrease $\Delta_\mathcal{L}$ on other environments. If the decrease is small, the increase can also only be small. In the special case of only a single environment with decreased value $\Delta_\lambda$, the increase is bound by $\Delta_\eta \le \frac{w_\lambda}{w_\eta}|\Delta_\lambda|$, i.e. a decrease by an amount $\Delta_\lambda$ can only cause an increase by at most the same amount times a factor $\frac{w_\lambda}{w_\eta}$. For the choice of the weights $w_\nu \sim 2^{-K(\nu)}$, a decrease can only cause a smaller increase in simpler environments, but a scaled increase in more complex environments. Finally note that pure Pareto optimality in the sense of Theorem 5.23 follows from balanced Pareto optimality in the special case of no decrease $\Delta_\mathcal{L} \equiv 0$.

**Proof.** $\Delta \ge 0$ follows from $\Delta = \sum_\nu w_\nu [V_\nu^{p^\xi} - V_\nu^{\tilde{p}}] = V_\xi^{p^\xi} - V_\xi^{\tilde{p}} \ge 0$, where we have used linearity of $V_\rho$ (Theorem 5.21) and dominance $V_\xi^{p^\xi} \ge V_\xi^p$ (Theorem 5.20). The remainder of Theorem 5.24 is obvious from $0 \le \Delta = \Delta_\mathcal{L} - \Delta_\mathcal{H}$ and by bounding the weighted average $\Delta_\eta$ by its maximum. □

### 5.4.3 Self-Optimizing Policy $p^\xi$ w.r.t. Average Value

We have argued in Section 5.3, Eq. (5.15) that there is no (universal) policy $\tilde{p}$ (independent of the actual environment $\mu$) for which

$$\frac{1}{m}V_{1m}^{\tilde{p}\mu} \to \frac{1}{m}V_{1m}^{*\mu} \qquad \text{for} \qquad m \to \infty, \tag{5.25}$$

and hence also $p^\xi$ cannot converge to the optimal policy $p^\mu$ in this sense. On the other hand, we know from Section 2.4 that convergence of this type (and even stronger) holds for SP. Section 5.3 suggested investigating restricted environmental classes. In the following we consider the generalized AI$\xi$ model (with $\xi() = \sum_{\nu \in \mathcal{M}} w_\nu \nu()$) and restricted classes $\mathcal{M}$. The least we must demand from $\mathcal{M}$ to have a chance that

$$\frac{1}{m}V_{1m}^{p^\xi\mu} \to \frac{1}{m}V_{1m}^{*\mu} \quad \text{for} \quad m \to \infty \tag{5.26}$$

is that there exists some policy $\tilde{p}$ at all with this property (5.25). Luckily, this necessary condition will also be sufficient. This is another (asymptotic) optimality property of (generalized) AI$\xi$. *If* universal convergence in the sense of (5.25) is possible at all in a class of environments $\mathcal{M}$, then AI$\xi$ converges in the same sense (5.26). We will call policies $\tilde{p}$ with a property like (5.26) *self-optimizing* [KV86].

The following two lemmas pave the way for proving the convergence theorem:

**Lemma 5.27 (Value difference relation)**

$$0 \le V_\nu^* - V_\nu^{\tilde{p}} =: \Delta_\nu \quad \Rightarrow \quad 0 \le V_\nu^* - V_\nu^{p^\xi} \le \frac{1}{w_\nu}\Delta \quad \text{with} \quad \Delta := \sum_{\nu \in \mathcal{M}} w_\nu \Delta_\nu$$

**Proof.** The following sequence of inequalities proves the lemma:

$$0 \le w_\nu[V_\nu^* - V_\nu^{p^\xi}] \le \textstyle\sum_\nu w_\nu [V_\nu^* - V_\nu^{p^\xi}] \le \sum_\nu w_\nu [V_\nu^* - V_\nu^{\tilde{p}}] = \sum_\nu w_\nu \Delta_\nu \equiv \Delta$$

In the first and second inequality we used $w_\nu \geq 0$ and $V_\nu^* - V_\nu^{p^\xi} \geq 0$. The last inequality follows from $\sum_\nu w_\nu V_\nu^{p^\xi} = V_\xi^{p^\xi} \equiv V_\xi^* \geq V_\xi^{\tilde{p}} = \sum_\nu w_\nu V_\nu^{\tilde{p}}$. □

We also need some results for averages of functions $\delta_\nu(m)$ converging to zero.

**Lemma 5.28 (Convergence of averages)** For $\delta(m) := \sum_{\nu\in\mathcal{M}} w_\nu \delta_\nu(m)$ the following holds (we only need $\sum_\nu w_\nu \leq 1$):

(*i*) $\delta_\nu(m) \leq f(m)\ \forall\nu$ implies $\delta(m) \leq f(m)$.

(*ii*) $\delta_\nu(m) \stackrel{m\to\infty}{\longrightarrow} 0\ \ \forall\nu$ implies $\delta(m) \stackrel{m\to\infty}{\longrightarrow} 0$ if $0 \leq \delta_\nu(m) \leq c$.

(*iii*) $\delta(m) \leq \max\limits_\nu \delta_\nu(m)$.

(*iv*) $\delta_\nu(m) = O(f(m))\ \forall\nu$ implies $\delta(m) = O(f(m))$ if $\mathcal{M}$ is finite.

**Proof.** (*i*) immediately follows from $\delta(m) = \sum_\nu w_\nu \delta_\nu(m) \leq \sum_\nu w_\nu f(m) \leq f(m)$.

(*ii*) We choose some order on $\mathcal{M}$ and some $\nu_0 \in \mathcal{M}$ large enough such that $\sum_{\nu\geq\nu_0} w_\nu \leq \frac{\varepsilon}{c}$. Using $\delta_\nu(m) \leq c$ this implies

$$\sum_{\nu\geq\nu_0} w_\nu \delta_\nu(m) \ \leq\ \sum_{\nu\geq\nu_0} w_\nu c \ \leq\ \varepsilon.$$

Furthermore, the assumption $\delta_\nu(m) \to 0$ means that there is an $m_{\nu\varepsilon}$ depending on $\nu$ and $\varepsilon$ such that $\delta_\nu(m) \leq \varepsilon$ for all $m \geq m_{\nu\varepsilon}$. This implies

$$\sum_{\nu\leq\nu_0} w_\nu \delta_\nu(m) \ \leq\ \sum_{\nu\leq\nu_0} w_\nu \varepsilon \ \leq\ \varepsilon \quad \text{for all} \quad m \ \geq\ \max_{\nu\leq\nu_0}\{m_{\nu\varepsilon}\} \ =:\ m_\varepsilon.$$

$m_\varepsilon < \infty$, since the maximum is over a finite set. Together we have

$$\delta(m) \ \equiv\ \sum_{\nu\in\mathcal{M}} w_\nu \delta_\nu(m) \ \leq\ 2\varepsilon \quad \text{for} \quad m \ \geq\ m_\varepsilon \quad \Rightarrow \quad \delta(m) \to 0 \quad \text{for} \quad m \to \infty$$

since $\varepsilon$ was arbitrary and $\delta(m) \geq 0$.

(*iii*) $\delta(m) \equiv \sum\limits_\nu w_\nu \delta_\nu(m) \leq \sum\limits_\nu w_\nu \max\limits_\nu \delta_\nu(m) \leq \max\limits_\nu \delta_\nu(m)$.

(*iv*) From $\delta_\nu(m) \leq c_\nu f(m)$ it follows that

$$\delta(m) \ \leq\ \sum_\nu w_\nu c_\nu f(m) \ \leq\ (\max_\nu c_\nu) f(m) (\sum_\nu w_\nu) \ \leq\ c_{max} f(m)$$

with $c_{max} := \max_{\nu\in\mathcal{M}} c_\nu$ being finite, since $\mathcal{M}$ is finite by assumption. □

Without the boundedness assumption in (*ii*) on $\delta_\nu(m)$, $\delta(m)$ may not only not converge, but may not even exist. The stronger finiteness assumption on $\mathcal{M}$ is necessary to obtain the speed of convergence in (*iv*) (see Problem 5.6).

**Theorem 5.29 (Self-optimizing policy $p^\xi$ w.r.t. average value)** If there exists a sequence of policies $\tilde{p}_m$, $m=1,2,3,...$ with value within $\Delta(m)$ to optimum for all environments $\nu\in\mathcal{M}$, then, save for a constant factor, this also holds for the sequence of universal policies $p_m^\xi$, i.e.

$$(i) \quad \text{If } \exists\tilde{p}_m\forall\nu: V_{1m}^{*\nu}-V_{1m}^{\tilde{p}_m\nu}\leq\Delta(m) \quad\Rightarrow\quad V_{1m}^{*\mu}-V_{1m}^{p_m^\xi\mu}\leq\tfrac{1}{w_\mu}\Delta(m).$$

If there exists a sequence of self-optimizing policies $\tilde{p}_m$ in the sense that their expected average reward $\frac{1}{m}V_{1m}^{\tilde{p}_m\nu}$ converges to the optimal average $\frac{1}{m}V_{1m}^{*\nu}$ for all environments $\nu\in\mathcal{M}$, then this also holds for the sequence of universal policies $p_m^\xi$, i.e.

$$(ii) \quad \text{If } \exists\tilde{p}_m\forall\nu: \tfrac{1}{m}V_{1m}^{\tilde{p}_m\nu}\stackrel{m\to\infty}{\longrightarrow}\tfrac{1}{m}V_{1m}^{*\nu} \quad\Rightarrow\quad \tfrac{1}{m}V_{1m}^{p_m^\xi\mu}\stackrel{m\to\infty}{\longrightarrow}\tfrac{1}{m}V_{1m}^{*\mu}.$$

The beauty of this theorem is that if universal convergence in the sense of (5.25) is possible at all in a class of environments $\mathcal{M}$, then AI$\xi$ converges (in the same sense (5.26)). The necessary condition of convergence is also sufficient. The unattractive point is that this is not an asymptotic convergence statement for $V_{km}^{p^\xi\mu}$ of a single policy $p^\xi$ for $k\to\infty$ for some fixed $m$, and in fact no such theorem could be true, since always $k\leq m$. The theorem merely says that under the stated conditions the average value of AI$\xi(m)$ can be arbitrarily close to optimum for sufficiently large (pre-chosen) horizon $m$. This weakness will be resolved in the next section.

**Proof.** $(i)$ For $\Delta_{(\nu)}$ as defined in Lemma 5.27, assuming $\Delta_\nu(m)\leq f(m)$ implies $\Delta(m)\leq f(m)$ by Lemma 5.28$(i)$. Inserting this into Lemma 5.27 we get $(i)$ (recovering the $m$ dependence and finally renaming $f\rightsquigarrow\Delta$).

$(ii)$ We define $\delta_\nu(m):=\frac{1}{m}\Delta_\nu(m)=\frac{1}{m}[V_\nu^*-V_\nu^{\tilde{p}}]$. Since we generally assumed bounded rewards $0\leq r\leq r_{max}$ (Assumption 4.28) we have

$$V_\nu^*\leq mr_{max} \quad\text{and}\quad V_\nu^{\tilde{p}}\geq 0 \quad\Rightarrow\quad \Delta_\nu\leq mr_{max} \quad\Rightarrow\quad 0\leq\delta_\nu(m)\leq c:=r_{max}.$$

The premise in $(ii)$ is that $\delta_\nu(m)=\frac{1}{m}[V_{1m}^{*\nu}-V_{1m}^{\tilde{p}\nu}]\to 0$, which implies

$$0 \;\leq\; \tfrac{1}{m}[V_{1m}^{*\nu}-V_{1m}^{p^\xi\nu}] \;\leq\; \tfrac{1}{w_\nu}\tfrac{\Delta(m)}{m} \;=\; \tfrac{1}{w_\nu}\delta(m) \;\to\; 0.$$

The inequalities follow from Lemma 5.27 and convergence to zero from Lemma 5.28$(ii)$. This proves Theorem 5.29$(ii)$. □

By using Lemma 5.28$(iv)$ instead of Lemma 5.28$(i)$ we can similarly prove that $V_{1m}^{*\nu}-V_{1m}^{\tilde{p}_m\nu}=O(\Delta(m))\;\forall\nu$ implies $V_{1m}^{*\mu}-V_{1m}^{p_m^\xi\mu}=O(\Delta(m))$ in case of finite $\mathcal{M}$. In Section 5.6 we show that a converging $\tilde{p}$ exists for ergodic MDPs, and hence $p^\xi$ converges in this environmental class too (in the sense of Theorem 5.29).

## 5.5 Discounted Future Value Function

We now shift our focus from the total value $V_{1m}$ and $m \to \infty$ to the future value (value-to-go) $V_{k?}$ and $k \to \infty$. The reasons are at least twofold.

First, we want to compare the future value of the optimal informed policy $p^\mu$ to the universal learner $p^\xi$. We regard the first $k$ cycles as a grace period in which $p^\xi$ learns and after which it performs well. The HeavenHell example of Section 5.3 shows that one cannot avoid that a learner gets trapped. We do not want to exclude trapping environments in our analysis from the very beginning, since there could be interesting structure and behavior in the traps themselves (even hell or heaven may reward intelligent behavior). One possibility is to compare future values $V_{k?}^{p^\xi \mu}$ with $V_{k?}^{*\mu}$ on the same (fictitious) history $y\!x_{<k}$. This addresses questions like: If $p^\xi$ gets trapped in a (structured) hell, does it perform as well as $p^\mu$ when put in hell?

Second, we want to get rid of the horizon parameter $m$. In the last section we showed a convergence theorem for $m \to \infty$, but a specific policy $p^\xi$ is defined for all times relative to a fixed horizon $m$. Current time $k$ is moving, but $m$ is fixed. Actually, to use $k \to \infty$ arguments we *have* to get rid of $m$, since $k \leq m$. This is the reason for the question mark in $V_{k?}$ above. The dynamic horizon $m_k$ introduced earlier was convenient to discuss qualitative properties, but does not lead to a consistent model.

We eliminate the horizon by discounting the rewards $r_k \rightsquigarrow \gamma_k r_k$ with $\gamma_k \geq 0$ and $\sum_{i=1}^\infty \gamma_i < \infty$ and letting $m \to \infty$. The analogue of $m$ is now an effective horizon $h_k^{eff}$, which may be defined by $\sum_{i=k}^{k+h_k^{eff}} \gamma_i \sim \sum_{i=k+h_k^{eff}}^\infty \gamma_i$ (see Section 5.7 for a detailed discussion of the horizon problem). Furthermore, we renormalize $V_{k\infty}$ by $\sum_{i=k}^\infty \gamma_i$ and denote it by $V_{k\gamma}$. It can be interpreted as a future expected weighted-average reward. Furthermore, we extend the definition to probabilistic policies $\pi$ (see Problem 4.2).

**Definition 5.30 (Discounted AI$\rho$ model and value)** We define the $\gamma$ discounted weighted-average future *value* of (probabilistic) policy $\pi$ in environment $\rho$ given history $y\!x_{<k}$, or shorter, the $\rho$-value of $\pi$ given $y\!x_{<k}$, as

$$V_{k\gamma}^{\pi\rho}(y\!x_{<k}) := \frac{1}{\Gamma_k} \lim_{m\to\infty} \sum_{y\!x_{k:m}} (\gamma_k r_k + ... + \gamma_m r_m)\rho(y\!x_{<k}\underline{y\!x}_{k:m})\pi(y\!x_{<k}\underline{y}x_{k:m})$$

with $\Gamma_k := \sum_{i=k}^\infty \gamma_i$. The discounted AI$\rho$ model is defined as the policy $p^\rho$ that maximizes the future value $V_{k\gamma}^{\pi\rho}$:

$$p^\rho := \arg\max_\pi V_{k\gamma}^{\pi\rho}, \quad V_{k\gamma}^{*\rho} := V_{k\gamma}^{p^\rho \rho} = \max_\pi V_{k\gamma}^{\pi\rho} \geq V_{k\gamma}^{\pi\rho} \, \forall\pi.$$

**Remarks.**

- $\pi(\dot{y}\dot{x}_{<k}\underline{y}\underline{x}_{k:m})$ is actually independent of $x_m$.
- Normalization of $V_{k\gamma}$ by $\Gamma_k$ does not affect the policy $p^\rho$.
- The definition of $p^\rho$ is independent of $k$ (in the sense of Problem 5.7).
- Without normalization by $\Gamma_k$ the future values would converge to zero for $k\to\infty$ in every environment for every policy.
- For an MDP environment, a stationary policy, and geometric discounting, the future value is independent of $k$ and reduces to the well-known MDP value function.
- There is always a deterministic optimizing policy $p^\rho$ (which we use).
- For a deterministic policy there is exactly one $y_{k:m}$ for each $x_{k:m}$ with $\pi\neq 0$. The sum over $y_{k:m}$ drops in this case.
- An iterative representation as in Theorem 5.20 is possible.
- Setting $\gamma_k=1$ for $k\leq m$ and $\gamma_k=0$ for $k>m$ gives back the undiscounted AI$\rho$ model (5.19) with $V_{1\gamma}^{p\rho}=\frac{1}{m}V_{1m}^{p\rho}$.
- $V_{k\gamma}$ and $w_k^\nu$ (see below) depend on the realized history $\dot{y}\dot{x}_{<k}$.

Similarly to the previous section, one can prove the following properties:

**Theorem 5.31 (Linearity and convexity of $V_\rho$ in $\rho$)** $V_{k\gamma}^{\pi\rho}$ is a linear function in $\rho$ and $V_{k\gamma}^{*\rho}$ is a convex function in $\rho$ in the sense that

$$V_{k\gamma}^{\pi\xi} = \sum_{\nu\in\mathcal{M}} w_k^\nu V_{k\gamma}^{\pi\nu} \quad\text{and}\quad V_{k\gamma}^{*\xi} \leq \sum_{\nu\in\mathcal{M}} w_k^\nu V_{k\gamma}^{*\nu},$$

$$\text{where}\quad \xi(\dot{y}\dot{x}_{<k}\underline{y}\underline{x}_{k:m}) = \sum_{\nu\in\mathcal{M}} w_k^\nu\, \nu(\dot{y}\dot{x}_{<k}\underline{y}\underline{x}_{k:m}) \quad\text{with}\quad w_k^\nu := w_\nu \frac{\nu(\dot{y}\dot{\underline{x}}_{<k})}{\xi(\dot{y}\dot{\underline{x}}_{<k})}$$

The conditional representation of $\xi$ can easily be proven by dividing the definition of $\xi(\underline{y}\underline{x}_{1:m})$ (5.21) by $\xi(\dot{y}\dot{\underline{x}}_{<k})$ and by using the chain rule. The posterior weight $w_k^\nu$ may be interpreted as the posterior belief in $\nu$ and is related to learning aspects of policy $p^\xi$ (see Section 3.2.3).

**Theorem 5.32 (Pareto optimality w.r.t. discounted value)** For every $k$ and history $\dot{y}\dot{x}_{<k}$ the following holds: $p^\xi$ is Pareto optimal in the sense that there is no other policy $\pi$ with $V_{k\gamma}^{\pi\nu}\geq V_{k\gamma}^{p^\xi\nu}$ for all $\nu\in\mathcal{M}$ and strict inequality for at least one $\nu$.

**Lemma 5.33 (Value difference relation)**

$$0 \leq V_{k\gamma}^{*\nu} - V_{k\gamma}^{\tilde{\pi}_k\nu} =: \Delta_k^\nu \quad\Rightarrow\quad 0 \leq V_{k\gamma}^{*\nu} - V_{k\gamma}^{p^\xi\nu} \leq \frac{1}{w_k^\nu}\Delta_k$$

with $\Delta_k := \sum_{\nu\in\mathcal{M}} w_k^\nu\Delta_k^\nu$, where all quantities depend on history $\dot{y}\dot{x}_{<k}$.

The proof of Theorem 5.32 and Lemma 5.33 follows the same steps as for Theorem 5.23 and Lemma 5.27 with appropriate replacements. The proof of the analogue of the convergence Theorem 5.29 involves one additional step.

**Theorem 5.34 (Self-optimizing policy $p^\xi$ w.r.t. discounted value)** For any $\mathcal{M}$, if there exists a sequence of self-optimizing policies $\tilde{\pi}_k$ $k=1,2,3,...$ in the sense that their expected weighted-average reward $V_{k\gamma}^{\tilde{\pi}_k\nu}$ converges for $k\to\infty$ with $\mu$-probability one to the optimal value $V_{k\gamma}^{*\nu}$ for all environments $\nu\in\mathcal{M}$, then this also holds for the universal policy $p^\xi$ in the $\mu$-environment, i.e.

$$\text{If } \exists\tilde{\pi}_k\forall\nu : V_{k\gamma}^{\tilde{\pi}_k\nu} \stackrel{k\to\infty}{\longrightarrow} V_{k\gamma}^{*\nu} \quad \text{w.}\nu\text{.p.1} \quad \Rightarrow \quad V_{k\gamma}^{p^\xi\mu} \stackrel{k\to\infty}{\longrightarrow} V_{k\gamma}^{*\mu} \quad \text{w.}\mu\text{.p.1.}$$

The probability qualifier refers to the historic perceptions $x_{<k}$. The historic actions $y_{<k}$ are arbitrary.

The conclusion is valid for action histories $y_{<k}$ if the condition is satisfied for this action history. Since we need the conclusion for the $p^\xi$-action history, which is hard to characterize, we usually need to prove the condition for *all* action histories. Theorem 5.34 is a powerful result: An inconsistent sequence of probabilistic policies $\tilde{\pi}_k$ suffices to prove the existence of a consistent deterministic policy $p^\xi$. A result similar to Theorem 5.29($i$) also holds for the discounted case, roughly saying that $V^{\tilde{\pi}}-V^*=O(\Delta(k))$ implies $V^{p^\xi}-V^*=\frac{1}{\varepsilon}O(\Delta(k))$ with $\mu$-probability $1-\varepsilon$ for finite $\mathcal{M}$.

**Proof.** We define $\delta_\nu(k) := \Delta_k^\nu = V_{k\gamma}^{*\nu} - V_{k\gamma}^{\tilde{\pi}\nu}$. Since we generally assumed bounded rewards $0\leq r\leq r_{max}$ (4.28) and $V_{k\gamma}^{*\nu}$ is a weighted average of rewards we have

$$V_{k\gamma}^{*\nu} \leq r_{max} \quad \text{and} \quad V_{k\gamma}^{\tilde{\pi}\nu} \geq 0 \quad \Rightarrow \quad 0 \leq \delta_\nu(k) = \Delta_k^\nu \leq c := r_{max}$$

The premise in Theorem 5.34 is that $\delta_\nu(k)=V_{k\gamma}^{*\nu}-V_{k\gamma}^{\tilde{\pi}\nu}\to 0$, for $k\to\infty$ which implies

$$0 \leq V_{k\gamma}^{*\nu} - V_{k\gamma}^{p^\xi\nu} \leq \frac{1}{w_k^\mu}\Delta_k = \frac{1}{w_k^\mu}\delta(k).$$

The inequalities follow from Lemma 5.33, and $\delta(k)$ converges to zero (w.$\mu$.p.1) by Lemma 5.28($ii$). What is new and what remains to be shown is that $w_k^\mu$ is bounded from below. We show that $z_{k-1} := \frac{w_\mu}{w_k^\mu} = \frac{\xi(y\!x_{<k})}{\mu(y\!x_{<k})} \geq 0$ converges to a finite value, which completes the proof. Let $\mathbf{E}$ denote the $\mu$ expectation. Then

$$\mathbf{E}[z_k|x_{<k}] = \sum_{x_k}{}' \mu(y\!x_{<k}y\!x_k)\frac{\xi(y\!x_{1:k})}{\mu(y\!x_{1:k})} = \frac{\sum'_{x_k}\xi(y\!x_{<k}y\!x_k)\xi(y\!x_{<k})}{\mu(y\!x_{<k})} \leq \frac{\xi(y\!x_{<k})}{\mu(y\!x_{<k})} = z_{k-1}.$$

$\sum'_{x_k}$ runs over all $x_k$ with $\mu(y\!x_{1:k})\neq 0$. The first equality holds w.$\mu$.p.1. In the second equality we used the chain rule twice. $\mathbf{E}[z_k|x_{<k}]\leq z_{k-1}$ shows that $-z_k$

is a semi-martingale. Since $-z_k$ is non-positive, [Doo53, Thm.4.1$s(i)$,p324] implies that $-z_k$ converges for $k\to\infty$ to a finite value w.$\mu$.p.1, which completes the proof. (If $\mu$ and $\xi$ are lower semicomputable, then boundedness of $z_{k-1}$ follows without the use of martingales from $z_{k-1}=\frac{\xi(y\!x_{<k})}{\mu(y\!x_{<k})}\stackrel{\times}{\leq}\frac{\xi_U(y\!x_{<k})}{\mu(y\!x_{<k})}\leq c<\infty$, where the first inequality follows from the universality of $\xi_U$ and the second inequality holds for all $\mu$.M.L.-random sequences.) □

We want to give an intuitive reason for the necessity of the probability qualifier. Assume that the true environment is $\mu$, but choose a history $x_{<k}$ sampled from $\nu\neq\mu$. This implies that $\xi$ converges to $\nu$ for $k\to\infty$. This means that at a fixed but large time $k$, $\xi$ is very close to $\nu$. It is very hard (takes large $h_k^{eff}$) to get rid of this wrong bias and to become close to $\mu$ later. In the limit this is not possible at all.

The following continuity properties for the discounted values hold:

**Theorem 5.35 (Continuity of discounted value)** The values $V_{k\gamma}^{\pi\mu}$ and $V_{k\gamma}^{*\mu}$ are continuous in $\mu$, and $V_{k\gamma}^{p^{\hat\mu}\mu}$ is continuous in $\hat\mu$ at $\hat\mu=\mu$ w.r.t. a conditional 1-norm in the following sense:

If $\sum_{x_k}|\mu(y\!x_{<k}y\!\underline{x}_k)-\hat\mu(y\!x_{<k}y\!\underline{x}_k)|\leq\varepsilon\ \forall y\!x_{<k}y_k\ \forall k\geq k_0,$ then

$$(i)\quad |V_{k\gamma}^{\pi\mu}-V_{k\gamma}^{\pi\hat\mu}|\leq\delta(\varepsilon),$$

$$(ii)\quad |V_{k\gamma}^{*\mu}-V_{k\gamma}^{*\hat\mu}|\leq\delta(\varepsilon),$$

$$(iii)\quad |V_{k\gamma}^{*\mu}-V_{k\gamma}^{p^{\hat\mu}\mu}|\leq 2\delta(\varepsilon)$$

for all $k\geq k_0$ and $y\!x_{<k}$, where $\delta(\varepsilon)=r_{max}\cdot\min_{n\geq k}\{(n-k)\varepsilon+\frac{\Gamma_n}{\Gamma_k}\}\stackrel{\varepsilon\to 0}{\longrightarrow}0.$

Care has to be taken in the interpretation and use of this theorem: It cannot be used to conclude $V_{k\gamma}^{p^\xi\mu}\to V_{k\gamma}^{*\mu}$, since $\xi\to\mu$ does not hold for all $y\!x_{1:\infty}$, but only for $\mu$-random ones (more precisely w.$\mu$.p.1). The condition in Theorem 5.35 cannot be weakened, since $p^\xi$ is not self-optimizing if $\mathcal{M}$ does not admit self-optimizing policies. Furthermore, continuity is not uniform in $k$, which prevents us from using this theorem in the proof of Theorem 5.38. Finally, note that $V_{k\gamma}^{p^{\hat\mu}\mu}$ can be discontinuous in $\hat\mu$ at $\hat\mu\neq\mu$. On the positive side, continuity holds for any $\mu$ and $\gamma$; no structural assumptions have to be made. By setting $\gamma_k=1$ for $k\leq m$ and $\gamma_k=0$ for $k>m$ we also get continuity of $V_{km}^{p\hat\mu}$, $V_{km}^{*\hat\mu}$, and $V_{km}^{p^{\hat\mu}\mu}$ with $\delta(\varepsilon)\leq r_{max}\varepsilon(m-k+1)^2$ (set $n=m+1$). For geometric discount $\gamma_k=\gamma^k$ the theorem holds with $\delta(\varepsilon)=\frac{r_{max}\varepsilon}{1-\gamma}$ (which follows from the second last bound on $\Delta_k$ in the proof below for $n=\infty$).

**Proof.** $(i)$ $V_{k\gamma}$ can be represented recursively like in the undiscounted case as

$$\Gamma_k V_{k\gamma}^{\pi\rho}(y\!x_{<k}) = \sum_{y\!x_k}\pi(y\!x_{<k}\underline{y}_k)\rho(y\!x_{<k}y\!\underline{x}_k)[\gamma_k r_k+\Gamma_{k+1}V_{k+1,\gamma}^{\pi\rho}(y\!x_{1:k})],$$

which can easily be verified by induction. The absolute difference of two values can be written as

$$\begin{aligned}
\Delta_k &:= \Gamma_k|V_{k\gamma}^{\pi\mu} - V_{k\gamma}^{\pi\hat\mu}| = \Big|\sum_{yx}\pi\cdot\mu\cdot(\gamma r+\Gamma V) - \sum_{yx}\pi\cdot\hat\mu\cdot(\gamma r+\Gamma\hat V)\Big| \\
&\leq \sum_y \pi\Big|\sum_x(\mu-\hat\mu)\gamma r + \Gamma\sum_x(\mu V-\hat\mu\hat V)\Big| \\
&= \sum_y \pi\Big|\sum_x(\mu-\hat\mu)\gamma r + \tfrac{1}{2}\Gamma\sum_x(\mu-\hat\mu)(V+\hat V) + \tfrac{1}{2}\Gamma\sum_x(\mu+\hat\mu)(V-\hat V)\Big| \leq ...
\end{aligned}$$

where we have suppressed all indices and arguments of all variables and functions. We upper-bound the last expression by pulling in the absolute bars. Using (in this order) $0\leq r\leq r_{max}$, $\sum_x|\mu-\hat\mu|\leq\varepsilon$ (by assumption), $0\leq V+\hat V\leq 2r_{max}$, $|V-\hat V|\leq\max_{yx}|V-\hat V|$, $\sum_x(\mu+\hat\mu)=2$, $\sum_y\pi=1$, we get

$$\begin{aligned}
... &\leq \varepsilon\gamma r_{max} + \Gamma\varepsilon r_{max} + \Gamma\max_{yx}|V-\hat V| \\
&= \varepsilon\Gamma_k r_{max} + \max_{yx_k}\Delta_{k+1} \leq ... \\
&\leq \varepsilon r_{max}\sum_{i=k}^{n-1}\Gamma_i + \max_{yx_{k:n-1}}\Delta_n \quad \leq \quad \varepsilon r_{max}(n-k)\Gamma_k + \Gamma_n r_{max}.
\end{aligned}$$

In the second line we used $\gamma_k+\Gamma_{k+1}=\Gamma_k$ and the definition of $\Delta_k$. In the third line we recursively inserted the bound for $\Delta_i$, $i=k+1,...,n-1$ we just derived. In the final expression we used $\Gamma_i\leq\Gamma_k$ for $i\geq k$ and $|V-\hat V|\leq r_{max}$. This bound on $\Delta_k$ is valid for all $n\geq k$, so we may take the minimum over $n\geq k$. This leads to $\Delta_k\leq\Gamma_k\delta(\varepsilon)$ where $\delta(\varepsilon)$ was defined in Theorem 5.35. This proves $(i)$.

$(ii)$ For any two real-valued functions $f,\hat f$ over some domain $\mathcal{D}$ with $|f(x)-\hat f(x)|\leq\delta\,\forall x\in\mathcal{D}$ we also have $|f_{max}-\hat f_{max}|\leq\delta$, where $f_{max}:=\max_{x\in\mathcal{D}}f(x)$ and $\hat f_{max}:=\max_{x\in\mathcal{D}}\hat f(x)$, since $f(x)\leq\hat f(x)+\delta\leq\hat f_{max}+\delta\,\forall x$. Hence $f_{max}\leq\hat f_{max}+\delta$, and similarly $\hat f_{max}\leq f_{max}+\delta$. For $x=\pi$, $\mathcal{D}=\{\pi\}$, $f(x)=V_{k\gamma}^{\pi\mu}$, $\hat f(x)=V_{k\gamma}^{\pi\hat\mu}$ we get $(ii)$ from $(i)$.

$(iii)$ Noting that $V_{k\gamma}^{*\hat\mu}\equiv V_{k\gamma}^{p^{\hat\mu}\hat\mu}$ we get $|V_{k\gamma}^{*\mu}-V_{k\gamma}^{p^{\hat\mu}\mu}|=|V_{k\gamma}^{*\mu}-V_{k\gamma}^{*\hat\mu}+V_{k\gamma}^{p^{\hat\mu}\hat\mu}-V_{k\gamma}^{p^{\hat\mu}\mu}|\leq|V_{k\gamma}^{*\mu}-V_{k\gamma}^{*\hat\mu}|+|V_{k\gamma}^{p^{\hat\mu}\hat\mu}-V_{k\gamma}^{p^{\hat\mu}\mu}|\leq 2\delta(\varepsilon)$ by $(ii)$ and $(i)$, which proves $(iii)$.

To prove $\delta(\varepsilon)\to 0$ we replace $\min_n$ by $n\sim\varepsilon^{-1/2}$ and get $0\leq\delta(\varepsilon)\leq r_{max}\cdot((n-k)\varepsilon+\frac{\Gamma_n}{\Gamma_k})\stackrel{\varepsilon\to 0}{\longrightarrow}0$, since $n\to\infty$, $\Gamma_n\stackrel{n\to\infty}{\longrightarrow}0$, and $n\varepsilon\to 0$. □

The next theorem shows that, for a given policy $p$ and history generated by $p$ and $\mu$, i.e. on-policy, the future universal value $V_{...}^{p\xi}$ converges to the true value $V_{...}^{p\mu}$.

**Theorem 5.36 (Convergence of universal to true value)** If the history $y\!x_{<k}$ is generated by policy $p$ (and environment $\mu$), and $V_{k..}^{p..} = V_{k..}^{p..}(y\!x_{<k})$, then the universal undiscounted future value $V_{km_k}^{p\xi}$ with bounded dynamic horizon $h_k = m_k - k + 1 \le h_{max}$ converges i.m.s. to the true value $V_{km_k}^{p\mu}$, and the discounted future value $V_{k\gamma}^{p\xi}$ converges i.m. to $V_{k\gamma}^{p\mu}$ for *any* summable discount sequence $\gamma_k$. In detail:

$$(i) \quad \begin{aligned} |V_{km}^{p\xi} - V_{km}^{p\mu}| &\le (m-k+1) r_{max} a_{k:m}, \\ |V_{k\gamma}^{p\xi} - V_{k\gamma}^{p\mu}| &\le r_{max}\sqrt{2d_{k:\infty}}. \end{aligned}$$

$$(ii) \quad \begin{aligned} \textstyle\sum_{k=1}^{\infty} \mathbf{E}(V_{km_k}^{p\xi} - V_{km_k}^{p\mu})^2 &\le 2h_{max}^3 r_{max}^2 D_\infty, \\ \mathbf{E}(V_{k\gamma}^{p\xi} - V_{k\gamma}^{p\mu})^2 &\le 2r_{max}^2 (D_\infty - D_{k-1}) \stackrel{k\to\infty}{\longrightarrow} 0. \end{aligned}$$

$$(iii) \quad \begin{aligned} V_{km_k}^{p\xi} &\stackrel{k\to\infty}{\longrightarrow} V_{km_k}^{p\mu} \quad \text{i.m.s.} \quad \text{if } h_{max} < \infty, \\ V_{k\gamma}^{p\xi} &\stackrel{k\to\infty}{\longrightarrow} V_{k\gamma}^{p\mu} \quad \text{i.m.} \quad \text{for any } \gamma. \end{aligned}$$

$$\begin{aligned} a_{k:m} &:= \textstyle\sum_{x_{k:m}} |\mu(y\!x_{<k} \underline{y\!x}_{k:m}) - \xi(y\!x_{<k} \underline{y\!x}_{k:m})|, \\ d_{k:m} &:= \textstyle\sum_{x_{k:m}} \mu(y\!x_{<k} \underline{y\!x}_{k:m}) \ln \frac{\mu(y\!x_{<k} \underline{y\!x}_{k:m})}{\xi(y\!x_{<k} \underline{y\!x}_{k:m})}, \end{aligned}$$

and $D_k := d_{1:k} \le \ln w_\mu^{-1} < \infty$ are defined as in Section 3.7.1 with actions $y_{1:m} = p(x_{<m})$ as additional conditions.

**Proof.** $(i)_{top}$ follows from $|V_{km}^{p\xi} - V_{km}^{p\mu}| = \Big|\sum_{x_{k:m}} (r_k + ... + r_m)[\xi() - \mu()]\Big| \le$

$$\sum_{x_{k:m}} (r_k + ... + r_m)|\xi() - \mu()| \le (m-k+1) r_{max} \sum_{x_{k:m}} |\xi() - \mu()| = (m-k+1) r_{max} a_{k:m}$$

where $\rho() = \rho(y\!x_{<k} \underline{y\!x}_{k:m})_{|y_{1:m} = p(x_{<m})}$. $(i)_{bottom}$ is shown similarly: Let $V_{km\gamma}$ be the discounted future value $V_{k\gamma}$ but cut after cycle $m$. We have

$$\begin{aligned} |V_{km\gamma}^{p\xi} - V_{km\gamma}^{p\mu}| &= \frac{1}{\Gamma_k}\Big|\sum_{x_{k:m}} (\gamma_k r_k + ... + \gamma_m r_m)[\xi() - \mu()]\Big| \le ... \\ &... \le r_{max} a_{k:m} \le r_{max}\sqrt{2d_{k:m}}. \end{aligned}$$

In the last step we used $a_{k:m} \le \sqrt{2d_{k:m}}$ (see Lemma 3.11 or Section 3.7.1). $(i)_{bottom}$ follows by taking the limit $m \to \infty$, which exists since $V_{km\gamma}$ and $d_{k:m}$ are monotone increasing in $m$ and bounded. $(ii)_{top}$ follows from $(i)_{top}$, and $m_k - k + 1 \le h_{max}$, and bound (3.71) with $t \rightsquigarrow k$, $n_t \rightsquigarrow m_k$, $h \rightsquigarrow h_{max}$, and $y$ as additional conditions. $(ii)_{bottom}$ follows from $(i)_{bottom}$ and $\mathbf{E}[d_{k:n}] = D_n - D_{k-1}$ (see Section 3.7.1). $(iii)$ follows from $(ii)$ by Definition 3.8 of convergence i.m.(s.) □

Convergence of the average values $\frac{1}{h_k}V_{km_k}^{p\xi} \to \frac{1}{h_k}V_{km_k}^{p\mu}$ also holds, i.m.s. for bounded horizon, and i.m. for arbitrary horizon. Note also that if the history is generated by $p=p^\xi$, then $(iii)$ implies $V_{k\gamma}^{*\xi} \to V_{k\gamma}^{p^\xi\mu}$. Hence the universal value $V_{k\gamma}^{*\xi}$ can be used to estimate the true value $V_{k\gamma}^{p^\xi\mu}$, without any assumptions on $\mathcal{M}$ and $\gamma$. Nevertheless, maximization of $V_{k\gamma}^{p\xi}$ may asymptotically differ from maximization of $V_{k\gamma}^{p\mu}$, since $V_{k\gamma}^{p\xi} \not\to V_{k\gamma}^{p\mu}$ for $p \neq p^\xi$ is possible (and also $V_{k\gamma}^{*\xi} \not\to V_{k\gamma}^{*\mu}$, see Section 5.3 and Problem 5.2).

## 5.6 Markov Decision Processes (MDP)

From all possible environments, Markov (Decision) Processes are probably the most intensively studied ones. To give an example, we apply Theorems 5.29 and 5.34 to ergodic Markov decision processes (MDPs), but we will be quite brief.

**Definition 5.37 (Ergodic Markov decision processes)** We call $\mu$ a (stationary) *Markov decision process* (MDP) if the probability of observing $o_k \in \mathcal{O}$ and reward $r_k \in \mathcal{R}$, given history $y\!x_{<k}y_k$ only depends on the last action $y_k \in \mathcal{Y}$ and the last observation $o_{k-1}$, i.e. if $\mu(y\!x_{<k}y_k\underline{x}_k) = \mu(o_{k-1}y_k\underline{x}_k)$, where $x_k \equiv o_k r_k$. In this case $o_k \in \mathcal{O}$ is called a *state*, $\mathcal{O}$ the *state space*, and $\mu(o_{k-1}y_k\underline{x}_k)$ is a product of the *transition matrix* $\mu(o_{k-1}y_k\underline{o}_k)$ and the (probabilistic) *reward function* $\mu(o_{k-1}y_k o_k \underline{r}_k)$.[6] An MDP $\mu$ is called *ergodic* if there exists a policy under which every state is visited infinitely often with probability 1. Let $\mathcal{M}_{\text{MDP}}$ be the set of MDPs and $\mathcal{M}_{\text{MDP1}}$ be the set of ergodic MDPs. If the transition matrix $\mu(o_{k-1}y_k\underline{o}_k)$ is independent of the action $y_k$, the MDP is a *Markov process*; if $\mu(o_{k-1}y_k\underline{x}_k)$ is independent of $o_{k-1}$ we have an *i.i.d.* process.

Stationary MDPs $\mu$ have stationary optimal policies $p^\mu$ in case of geometric discount, mapping the same state/observation $o_k$ always to the same action $y_k$. On the other hand, a mixture $\xi$ of MDPs is itself not an MDP, i.e. $\xi \notin \mathcal{M}_{\text{MDP}}$, which implies that $p^\xi$ is, in general, not a stationary policy. The definition of ergodicity given here is least demanding, since it only demands the existence of a single policy under which the Markov process is ergodic. Often, stronger assumptions, e.g. that every policy is ergodic or that a stationary distribution exists, are made. We now show that there are self-optimizing policies for the class of ergodic MDPs in the following sense.

[6] Alternatively, one could assume $\mathcal{X}$ to be the state space, i.e. include the rewards in the states. The transition matrix then has the form $\mu(x_{k-1}y_k\underline{x}_k)$ with action $y_k \in \mathcal{Y}$ leading to state $x_k \in \mathcal{X}$ from state $x_{k-1} \in \mathcal{X}$ [Hut02b]. The advantage is that the reward $r_k = r(x_k)$ is now a known deterministic function only depending on the current state, but for the prize of a (possibly much) larger state space.

**Theorem 5.38 (Self-optimizing policies for ergodic MDPs)** There exist self-optimizing policies $\tilde{p}_m$ for the class of ergodic MDPs in the sense that

$$(i)\quad \exists\tilde{p}_m\forall\nu\in\mathcal{M}_{\text{MDP1}}:\tfrac{1}{m}V_{1m}^{*\nu}-\tfrac{1}{m}V_{1m}^{\tilde{p}_m\nu}\;=\;O(m^{-1/3}).$$

In the discounted case, if the discount sequence $\gamma_k$ has unbounded effective horizon $h_k^{\gamma eff}\stackrel{k\to\infty}{\longrightarrow}\infty$, then there exist self-optimizing policies $\tilde{\pi}_k$ for the class of ergodic MDPs in the sense that

$$(ii)\quad \exists\tilde{\pi}_k\forall\nu\in\mathcal{M}_{\text{MDP1}}:V_{k\gamma}^{\tilde{\pi}_k\nu}\stackrel{k\to\infty}{\longrightarrow}V_{k\gamma}^{*\nu}\quad\text{for any history } \dot{y}\dot{x}_{<k}\quad\text{if } \tfrac{\gamma_{k+1}}{\gamma_k}\to 1.$$

There is much literature on constructing and analyzing self-optimizing learning algorithms in MDP environments. The assumptions on the structure of the MDPs vary, but all include some form of ergodicity, often stronger than Definition 5.37, demanding that the Markov process is ergodic under *every* policy. See, for instance, [KV86, BT96]. Note also that in (*ii*) the history need not be generated by $\tilde{\pi}_k$ and/or $\nu$. Indeed, the application we are most interested in here, is to action histories generated by $p^\xi$. We will only briefly outline one algorithm satisfying Theorem 5.38 without trying to optimize performance.

**Proof.** Let $T_{ss'}^a\equiv\nu(sa\underline{s}')$ be the probability of reaching state $s'\in\mathcal{S}:=\mathcal{O}$ from state $s$ under action $a\in\mathcal{A}:=\mathcal{Y}$, and let $R_{ss'}^a:=\sum_{r\in\mathcal{R}}r\cdot\nu(sas'\underline{r})$ be the expected reward of transition $s\stackrel{a}{\to}s'$. With this notation the value function can be written as $V_{1m}^{p\nu}=\sum_{s_{1:m}}(R_{s_0s_1}^{a_1}+...+R_{s_{m-1}s_m}^{a_m})\cdot T_{s_0s_1}^{a_1}\cdot...\cdot T_{s_{m-1}s_m}^{a_m}$, where $a_k=p(s_{<k})$ is the action of policy $p$ at time $k$ ($s_0\notin\mathcal{S}$ is some special initial "state").

For (*i*) one can choose a policy $\tilde{p}_m$ that performs (uniformly) random actions in cycles $1...k_0-1$ with $1\ll k_0\ll m$ and that follows thereafter the optimal policy based on an estimate of the transition matrix $T_{ss'}^a$ from the initial $k_0-1$ cycles. The existence of an ergodic policy implies that for every pair of states $s_{start},s\in\mathcal{S}$ there is a sequence of actions and transitions of length at most $|\mathcal{S}|-1$ such that state $s$ is reached from state $s_{start}$. The probability that the "right" transition occurs is at least $T_{min}$ with $T_{min}$ being the smallest nonzero transition probability in $T$. The probability that a random action is the "right" action is at least $|\mathcal{A}|^{-1}$. So the probability of reaching a state $s$ in $|\mathcal{S}|-1$ cycles via a random policy is at least $(T_{min}/|\mathcal{A}|)^{|\mathcal{S}|-1}$. In state $s$ action $a$ is taken with probability $|\mathcal{A}|^{-1}$ and leads to state $s'$ with probability $T_{ss'}^a\geq T_{min}$. Hence, the expected number of transitions $s\stackrel{a}{\to}s'$ to occur in the first $k_0$ cycles is $\geq\frac{k_0}{|\mathcal{S}|}(T_{min}/|\mathcal{A}|)^{|\mathcal{S}|}\sim k_0$. Hence, for $T_{ss'}^a\neq 0$, the accuracy of the frequency estimate $\hat{T}_{ss'}^a$ of $T_{ss'}^a$ and $\hat{R}_{ss'}^a$ of $R_{ss'}^a$ is $\sim k_0^{-1/2}$, while for $T_{ss'}^a=0$ the estimate $\hat{T}_{ss'}^a=0$ is exact and $\hat{R}_{ss'}^a$ is not needed. In summary: similar MDPs lead to "similar" optimal policies, which lead to similar values. More precisely, one can show (see Problem 5.12) that $\hat{T}-T\sim k_0^{-1/2}$ and $\hat{R}-R\sim k_0^{-1/2}$

implies the same accuracy in the average value, i.e. $|\frac{1}{m}V_{k_0m}^{\tilde{p}_m\nu}-\frac{1}{m}V_{k_0m}^{*\nu}|\sim k_0^{-1/2}$, where $\tilde{p}_m$ is the optimal policy based on $\hat{T}$ and $*$ is the optimal policy based on $T(=\nu)$. Since $\frac{1}{m}V_{1k_0}\sim\frac{k_0}{m}$, $(i)$ follows (with probability 1) by setting $k_0\sim m^{2/3}$. The policy $\tilde{p}_m$ can be derandomized, showing $(i)$ for sure.

The discounted case $(ii)$ can be proven similarly. The history $y\!x_{<k}$ is simply ignored and the analogue to $m\to\infty$ is $h_k^{eff}\to\infty$ for $k\to\infty$, which is ensured by $\frac{\gamma_{k+1}}{\gamma_k}\to 1$. Let $\tilde{\pi}_k$ be the policy that performs (uniformly) random actions in cycles $k...k_0-1$ with $k\ll k_0\ll h_k^{eff}$ and that follows thereafter the optimal policy[7] based on an estimate $\hat{T}$ of the transition matrix $T$ from cycles $k...k_0-1$. The existence of an ergodic policy, again, ensures that the expected (after derandomization for sure) number of transitions $s\xrightarrow{a}s'$ occurring in cycles $k...k_0-1$ is proportional to $\Delta:=k_0-k$. The accuracy of the frequency estimate $\hat{T}$ of $T$ and $\hat{R}$ of $R$ is $\sim\Delta^{-1/2}$, which implies by a strengthening of Theorem 5.35$(iii)$ for ergodic MDPs similar to Problem 5.12 that

$$V_{k_0\gamma}^{\tilde{\pi}_k\nu}\to V_{k_0\gamma}^{*\nu} \quad \text{for} \quad \Delta=k_0-k\to\infty, \tag{5.39}$$

where $\tilde{\pi}_k$ is the optimal policy based on $\hat{T}$, and $*$ is the optimal policy based on $T(=\nu)$. It remains to be shown that the achieved reward in the random phase $k...k_0-1$ gives a negligible contribution to $V_{k\gamma}$. The following implications for $k\to\infty$ are easy to show:

$$\frac{\gamma_{k+1}}{\gamma_k}\to 1 \Rightarrow \frac{\gamma_{k+\Delta}}{\gamma_k}\to 1 \Rightarrow \frac{\Gamma_{k+\Delta}}{\Gamma_k}\to 1 \Rightarrow \frac{1}{\Gamma_k}\sum_{i=k}^{k_0-1}\gamma_i r_i\leq\frac{r_{max}}{\Gamma_k}[\Gamma_{k+\Delta}-\Gamma_k]\to 0$$

Since convergence to zero is true for all fixed finite $\Delta$ it is also true for sufficiently slowly increasing $\Delta(k)\to\infty$. This shows that the contribution of the first $\Delta$ rewards $r_k,...,r_{k_0-1}$ to $V_{k\gamma}$ is negligible. Together with (5.39) this shows $V_{k\gamma}^{\tilde{\pi}_k\nu}\to V_{k\gamma}^{*\nu}$ for $k_0:=k+\Delta(k)$. □

The rate of convergence $m^{-1/3}$ rather than $m^{-1/2}$ in the undiscounted case may be a bit surprising. If we would explore for $k_0=m$ steps and ask for the accuracy of the value function estimate afterwards, i.e. for $k>m$, we would get $\sim k_0^{-1/2}=m^{-1/2}$, but since we are considering $\frac{1}{m}V_{1m}$ *including* the history $1...k_0$ we must get a worse result, namely $m^{-1/3}$. Actually, the result can be improved $O(\frac{\log m}{m})$ (Problem 5.12). Although in the discounted case, we consider the future value $V_{k\gamma}$, the situation is nevertheless similar. Exploration takes place in cycles $k...k_0-1$; history $y\!x_{<k}$ is not exploited.

The conditions $\Gamma_k<\infty$ and $\frac{\gamma_{k+1}}{\gamma_k}\to 1$ on the discount sequence are, for instance, satisfied for $\gamma_k=1/k^2$, so the theorem is not vacuous. The popular geometric discount $\gamma_k=\gamma^k$ fails the latter condition; it has finite effective horizon. Section 5.7 gives a detailed account of the discount and horizon issues.

[7] Note that for non-geometric discounts as here, optimal policies are, in general, *not* stationary.

Together with Theorems 5.29 and 5.34, Theorem 5.38 immediately implies that AI$\xi$ is self-optimizing for the class of ergodic MDPs.

**Corollary 5.40 (AI$\xi$ is self-optimizing for ergodic MDPs)** If $\mathcal{M}$ is a countable class of ergodic MDPs, and $\xi := \sum_{\nu\in\mathcal{M}} w_\nu \nu$, then AI$\xi$ alias $p_m^\xi$ maximizing $V_{1m}^{p\xi}$ and $p^\xi$ maximizing $V_{k\gamma}^{\pi\xi}$ are self-optimizing in the sense that

$$\forall\nu\in\mathcal{M} : \frac{1}{m}V_{1m}^{p_m^\xi\nu} \stackrel{m\to\infty}{\longrightarrow} \frac{1}{m}V_{1m}^{*\nu} \quad \text{and} \quad V_{k\gamma}^{p^\xi\nu} \stackrel{k\to\infty}{\longrightarrow} V_{k\gamma}^{*\nu} \quad \text{if} \quad \frac{\gamma_{k+1}}{\gamma_k}\to 1.$$

If $\mathcal{M}$ is finite, then the speed of the first convergence is at least $O(m^{-1/3})$.

**Continuous classes $\mathcal{M}$.** There are uncountably many ergodic MDPs. Since we have restricted our development to countable classes $\mathcal{M}$ we had to give the corollary for a countable subset of $\mathcal{M}_{\text{MDP1}}$. We may choose $\mathcal{M}$ as the set of all ergodic MDPs with rational (or computable) transition probabilities. In this case $\mathcal{M}$ is a dense subset of $\mathcal{M}_{\text{MDP1}}$ that is, from a practical point of view, sufficiently rich. On the other hand, it is possible to extend the theory to continuously parameterized families of environments $\mu_\theta$ and $\xi = \int d\theta\, w_\theta \mu_\theta$. Under some mild (differentiability and existence) conditions, most results of this book remain valid in some form, especially Corollary 5.40 for *all* ergodic MDPs $\mathcal{M}_{\text{MDP1}}$.

**Bayesian self-optimizing policy.** AI$\xi_{\text{MDP1}}$ with unbounded horizon is a purely Bayesian self-optimizing consistent policy for ergodic MDPs. The policies of all other known approaches are either hand-crafted, like those in the proof of Theorem 5.38, or are Bayesian with a pre-chosen horizon $m$ or with geometric discounting $\gamma$ with finite effective horizon [KV86, BT96]. The combined conditions $\Gamma_k<\infty$ and $\frac{\gamma_{k+1}}{\gamma_k}\to 1$ allowed a consistent self-optimizing Bayesian policy based on mixtures.

**Bandits.** For instance, consider the popular class of bandits $B$. In a two-armed bandit problem you pull repeatedly one out of two levers, resulting in a gain of 1\$ with probability $p_i$ for arm number $i$. The game can be described as an MDP with parameters $p_i$. If the $p_i$ are unknown, Corollary 5.40 shows that AI$\xi_B$ yields asymptotically optimal payoff for discounted unbounded-horizon bandits.

**Other environmental classes.** Bandits, i.i.d. processes, classification tasks, and many more are all special (degenerate) cases of ergodic MDPs, for which Corollary 5.40 shows that $p^\xi$ is self-optimizing. But the existence of self-optimizing policies is not limited to (subclasses of ergodic) MDPs. Certain classes of POMDPs, $k^{th}$-order ergodic MDPs, factorizable environments, repeated games, and prediction problems are not MDPs, but nevertheless admit self-optimizing policies (to be shown elsewhere), and hence the corresponding Bayes optimal mixture policy $p^\xi$ is self-optimizing by Theorems 5.29 and 5.34.

**Restricted policy classes.** The development in this and the last paragraphs can be scaled down to restricted classes of policies $\mathcal{P}$. If one defines $V^* = \text{argmax}_{p\in\mathcal{P}} V^p$ all theorems remain valid, more or less unchanged. For instance, consider a finite class of quickly computable policies. For MDPs, $\xi$ is quickly computable, and $V_\xi^p$ can be (efficiently) computed by Monte Carlo sampling. Maximizing over the finitely many policies $p\in\mathcal{P}$ selects the asymptotically best policy $p^\xi$ from $\mathcal{P}$ for all ergodic MDPs.

**Outlook.** Future research could be the derivation of non-asymptotic bounds, possibly along the lines of Chapter 3. To get good bounds one may have to exploit extra properties of the environments, like the mixing rate of MDPs [KS98]. Finally, instead of convergence of the expected reward sum, convergence with high probability of the actual reward sum would be interesting to study (cf. Problem 3.4).

## 5.7 The Choice of the Horizon

The only significant arbitrariness in the AI$\xi$ model lies in the choice of the horizon function $h_k \equiv m_k - k + 1$. We discuss some choices that seem to be natural and give preliminary conclusions at the end. We will not discuss ad hoc choices of $h_k$ for specific problems (like the discussion in Section 6.3 in the context of finite strategic games). We are interested in universal choices of $m_k$.

**Fixed horizon.** If the lifetime of the agent is known to be $m$, which is in practice always large but finite, then the choice $m_k = m$ maximizes correctly the expected future reward. Lifetime $m$ is usually not known in advance, as in many cases the time we are willing to run an agent depends on the quality of its outputs. For this reason, it is often desirable that good outputs are not delayed too much, if this results in a marginal reward increase only. This can be incorporated by damping the future rewards. If, for instance, the probability of survival in a cycle is $\gamma < 1$, an exponential damping (geometric discount) $r_k := r'_k \cdot \gamma^k$ is appropriate, where $r'_k$ are bounded, e.g. $r'_k \in [0,1]$. Expression (5.3) converges for $m_k \to \infty$ in this case ($\dot{y}_k = \text{argmax}_{y_k} \lim_{m_k\to\infty} V^{*\xi}_{km_k}(\dot{y}\dot{x}_{<k}y_k)$ exists). But this does not solve the problem, as we introduced a new arbitrary time scale $(1-\gamma)^{-1}$. Every damping introduces a time scale. Taking $\gamma \to 1$ is prone to the same problems as $m_k \to \infty$ in the undiscounted case.

**General discounting.** Geometric discounting does not solve the horizon problem, but the idea of discounting is fruitful. Let $r_k := \gamma_k r'_k$ with $\gamma_k > 0$ and $r'_k \in [0,1]$. If $\Gamma_k := \sum_{i=k}^\infty \gamma_i < \infty$, then $V^{p\rho}_{k\gamma} := \frac{1}{\Gamma_k} \lim_{m\to\infty} V^{p\rho}_{km}$ exists. Rewards $r_{k+h}$ give only a small contribution to $V^{p\rho}_{k\gamma}$ for large $h$, since $\gamma_{k+h} \stackrel{h\to\infty}{\longrightarrow} 0$. The instantaneous effective horizon may be defined as the $\hat{h}$ for which $\gamma_{k+\hat{h}}$ is only half (or more generally a fraction $\beta < 1$) of $\gamma_k$. Formally, we may define

$\hat{h}_k^\beta := \min\{h \geq 0 : \gamma_{k+h} \leq \beta\gamma_k\}$. For any summable convex discount sequence $\gamma_k$ we have $\hat{h}_k^\beta \leq c \cdot k$ for some constant $c$ independent of $k$. A better definition for the $\beta$-effective horizon is the $h$ for which the cumulative discount $\Gamma_{k+h} \approx \beta\Gamma_k$, or more formally, $h_k^\beta := \min\{h \geq 0 : \Gamma_{k+h} \leq \beta\Gamma_k\}$. Approximating the infinite reward sum in $V_{k\gamma}$ by the first $h_k^\beta$ terms introduces an error of at most $\beta r_{max}$. We define *the effective horizon* by $h_k^{eff} := h_k^{\beta=1/2}$. Table 5.41 shows effective horizons for various types of discounts $\gamma_k$.

**Table 5.41 (Effective horizons)** The table shows the effective horizons $\hat{h}_k^\beta := \min\{h \geq 0 : \gamma_{k+h} \leq \beta\gamma_k\}$ and $h_k^\beta := \min\{h \geq 0 : \Gamma_{k+h} \leq \beta\Gamma_k\}$ for various types of discounts $\gamma_k$.

| Horizons | $\gamma_k$ | $\hat{h}_k^\beta$ | $\Gamma_k = \sum_{i=k}^\infty \gamma_i$ | $h_k^\beta$ | $h_k^{eff} = h_k^{\beta=1/2}$ |
|---|---|---|---|---|---|
| finite | 1 for $k \leq m$, 0 for $k > m$ | $m-k+1$ | $m-k+1$ | $\lceil(1-\beta)(m-k+1)\rceil$ | $\lceil\frac{1}{2}(m-k+1)\rceil$ |
| geometric | $\gamma^k,\ 0 \leq \gamma < 1$ | $\lceil\frac{\ln\beta}{\ln\gamma}\rceil$ | $\frac{\gamma^k}{1-\gamma}$ | $\lceil\frac{\ln\beta}{\ln\gamma}\rceil$ | $\approx \frac{\ln 2}{1-\gamma}$ for $\gamma \approx 1$ |
| power | $k^{-1-\varepsilon},\ \varepsilon > 0$ | $\sim(\beta^{-\frac{1}{1+\varepsilon}}-1)k$ | $\sim\frac{1}{\varepsilon}k^{-\varepsilon}$ | $\sim(\beta^{-1/\varepsilon}-1)k$ | $\propto k$ |
| harmonic≈ | $\frac{1}{k\ln^{1+\varepsilon}k},\ \varepsilon > 0$ | $\sim(\beta^{-1}-1)k$ | $\sim\frac{1}{\varepsilon}(\ln k)^{-\varepsilon}$ | $\sim k^{\beta^{-1/\varepsilon}}$ | $\sim k^{2^{1/\varepsilon}}$ |
| universal | $2^{-K(k)}$ | $\approx k$ on average | decreases slower than any computable function | increases faster than any computable function | |

**Dynamic horizon (universal & harmonic discounting).** The largest horizon with guaranteed finite and enumerable reward sum can be obtained by the universal discount $\gamma_k = 2^{-K(k)}$ (or the monotone variant $\gamma_k = \min_{i \leq k} 2^{-K(i)}$). This discount results in a truly farsighted agent with effective horizon that grows faster than any computable function. It is somewhat similar to a near-harmonic discount $\gamma_k = [k\log^2 k]^{-1}$, since $2^{-K(k)} \leq 1/k$ for most $k$ and $2^{-K(k)} \geq c/(k\log^2 k)$ (see Theorem 2.10(*ii*)), but the latter leads to $h_k^{eff} \sim k^2$. Similarly, the time scale invariant power damping $\gamma_k = k^{-1-\varepsilon}$ introduces a dynamic time scale. In cycle $k$ the contribution of cycle $2^{\frac{1}{1+\varepsilon}} \cdot k$ is damped by a factor $\frac{1}{2}$. The instantaneous effective horizon $\hat{h}_k$ in this case is $\sim k$, the maximum possible. The choice $h_k = \alpha \cdot k$ with $\alpha \sim 2^{\frac{1}{1+\varepsilon}}$ qualitatively models the same behavior. We have not introduced an arbitrary time scale $m$, but limited the farsightedness to some multiple (or fraction) of the length of the current history. This avoids the preselection of a global time scale $m$ or $\frac{1}{1-\gamma}$. This choice has some appeal, as it seems that humans of age $k$ years usually do not plan their lives for more than, perhaps, the next $k$ years ($\alpha_{human} \approx 1$). From a practical point of view this model might serve all needs, but from a theoretical point we feel uncomfortable with such a limitation in the horizon from the very beginning. Note that we have to choose $\alpha = O(1)$ because otherwise we would again introduce a number $\alpha$ that has to be justified. We favor the universal discount $\gamma_k = 2^{-K(k)}$, since it allows us, if desired, to "mimic" all other more greedy behaviors based on other discounts $\gamma_k$ by choosing $r_k \in [0, c\cdot\gamma_k] \subseteq [0, 2^{-K(k)}]$.

**Infinite horizon.** The naive limit $m_k\to\infty$ in (5.3) may turn out to be well defined and the previous discussion superfluous. In the following, we suggest a limit that is always well defined (for finite $\mathcal{Y}$). Let $\dot{y}_k^{(m_k)}$ be defined as in (5.3) with dependence on $m_k$ made explicit. Further, let $\dot{\mathcal{Y}}_k^{(m)}:=\{\,\dot{y}_k^{(m_k)}:m_k\geq m\}$ be the set of outputs in cycle $k$ for the choices $m_k=m,m+1,m+2,...$. Because $\dot{\mathcal{Y}}_k^{(m)}\supseteq\dot{\mathcal{Y}}_k^{(m+1)}\neq\{\}$, we have $\dot{\mathcal{Y}}_k^{(\infty)}:=\bigcap_{m=k}^{\infty}\dot{\mathcal{Y}}_k^{(m)}\neq\{\}$. We define the $m_k=\infty$ model to output any $\dot{y}_k^{(\infty)}\in\dot{\mathcal{Y}}_k^{(\infty)}$. This is the best output consistent with some arbitrary large choice of $m_k$. Choosing the lexicographically smallest $\dot{y}_k^{(\infty)}\in\dot{\mathcal{Y}}_k^{(\infty)}$ would correspond to the lower limit $\underline{\lim}_{m\to\infty}\dot{y}_k^{(m)}$, which always exists (for finite $\mathcal{Y}$). Generally $\dot{y}_k^{(\infty)}\in\dot{\mathcal{Y}}_k^{(\infty)}$ is unique, i.e. $|\dot{\mathcal{Y}}_k^{(\infty)}|=1$ iff the naive limit $\lim_{m\to\infty}\dot{y}_k^{(m)}$ exists. Note that the limit $\lim_{m\to\infty}V_{km}^*(y\!x_{<k})$ need not exist for this construction.

**Average reward and differential gain.** Taking the raw average reward $(r_k+...+r_m)/(m-k+1)$ and $m\to\infty$ also does not help: consider an arbitrary policy for the first $k$ cycles and the/an optimal policy for the remaining cycles $k+1...\infty$. In e.g. i.i.d. environments the limit exists, but all these policies give the same average value, since changing a finite number of terms does not affect an infinite average. In MDP environments with a single recurrent class one can define the relative or differential gain [Ber95b]. In more general environments (we are interested in) the differential gain can be infinite, which is acceptable, since differential gains can still be totally ordered. The major problem is the *existence* of the differential gain, i.e. whether it converges for $m\to\infty$ in $\mathbb{R}\cup\{\infty\}$ at all (and does not oscillate). This is just the old convergence problem in slightly different form.

**Immortal agents are lazy.** The construction above leads to a mathematically elegant, no-parameter AI$\xi$ model. Unfortunately this is not the end of the story. The limit $m_k\to\infty$ can cause undesirable results in the AI$\mu$ model for special $\mu$, which might also happen in the AI$\xi$ model whatever we define $m_k\to\infty$. Consider an agent who for every $\sqrt{l}$ consecutive days of work, can thereafter take $l$ days of holiday. Formally, consider $\mathcal{Y}=\mathcal{X}=\mathcal{R}=\{0,1\}$. Output $y_k=0$ shall give reward $r_k=0$ and output $y_k=1$ shall give $r_k=1$ iff $\dot{y}_{k-l-\sqrt{l}}...\dot{y}_{k-l}=0...0$ for some $l$, i.e. the agent can achieve $l$ consecutive positive rewards if there was a preceding sequence of length at least $\sqrt{l}$ with $y_k=r_k=0$. If the lifetime of the AI$\mu$ agent is $m$, it outputs $\dot{y}_k=0$ in the first $s$ cycles and then $\dot{y}_k=1$ for the remaining $s^2$ cycles with $s$ such that $s+s^2=m$. This will lead to the highest possible total reward $V_{1m}=s^2=m+\frac{1}{2}-\sqrt{m+\frac{1}{4}}$. Any fragmentation of the 0 and 1 sequences would reduce $V_{1m}$, e.g. alternatingly working for 2 days and taking 4 days off would give $V_{1m}=\frac{2}{3}m$. For $m\to\infty$ the AI$\mu$ agent can and will delay the point $s$ of switching to $\dot{y}_k=1$ indefinitely and always output 0 leading to total reward 0, obviously the worst possible behavior for the agent. The AI$\xi$ agent will explore the above rule after a while of trying $y_k=0/1$ and then applies the same behavior as the AI$\mu$

agent, since the simplest rules covering past data dominate $\xi$. For finite $m$ this is exactly what we want, but for infinite $m$ the AI$\xi$ model (probably) fails, just as the AI$\mu$ model does. The good point is that this is not a weakness of the AI$\xi$ model in particular, as AI$\mu$ fails too. The bad point is that $m_k \to \infty$ has far-reaching consequences, even when starting from an already very large $m_k = m$. This is because the $\mu$ of this example is highly nonlocal in time, i.e. it may violate one of our weak separability conditions.

**Conclusions.** We are not sure whether the choice of $m_k$ is of marginal importance, as long as $m_k$ is chosen sufficiently large and of low complexity, $m_k = 2^{2^{16}}$ for instance, or whether the choice of $m_k$ will turn out to be a central topic for the AI$\xi$ model or for the planning aspect of any AI system in general. We suppose that the limit $m_k \to \infty$ for the AI$\xi$ model results in correct behavior for weakly separable $\mu$. A proof of this conjecture, if true, would probably give interesting insights.

## 5.8 Outlook

**Expert advice approach.** We considered expected performance bounds for predictions based on Solomonoff's prior. The other, dual, and currently very popular approach, is 'prediction with expert advice' (PEA) invented by Littlestone and Warmuth [LW89] and Vovk [Vov92]. Whereas PEA performs well in any environment, but only relative to a given set of experts, our $\Lambda_\xi$ predictor competes with *any* other predictor, but only in expectation for environments with computable distribution (see Section 3.7.4). It seems philosophically less compromising to make assumptions on prediction strategies than on the environment, however weak. There are also a few results on expert-based *active* learning [ACBFS02, BEYL04]. One could investigate whether PEA can be generalized to the case of a general active agent, which would result in a model dual to AIXI. We believe the answer to be negative, which on the positive side would show the necessity of Occam's razor assumption, and the distinguishedness of AIXI.

**Actions as random variables.** The uniqueness for the choice of the generalized $\xi$ (2.26) in the AIXI model could be explored. From the originally many alternatives, which could all be ruled out, there is one alternative that still seems possible. Instead of defining $\xi$ as in (2.26) one could treat the agent's actions $y$ also as universally distributed random variables and then conditionalize $\xi$ on $y$ by the chain rule (see Problem 5.1).

**Structure of AIXI.** The algebraic properties and the structure of AIXI could be investigated in more depth (we already saw that the value $V_\mu^p$ is a linear function in $\mu$ and $V_\mu^*$ is a convex function in $\mu$). This would extract the essentials from AIXI, which finally could lead to an axiomatic characterization of AIXI. The benefit is as in any axiomatic approach. It would clearly exhibit the assumptions, separate the essentials from technicalities, simplify understanding and, most important, guide in finding proofs.

**Posterization.** Many properties of Kolmogorov complexity, Solomonoff's prior, and (policies based on) Bayes mixtures remain valid after "posterization". By posterization we mean replacing $V_{1m}$, $w_\nu$, $K(\nu)$, $\nu(y\!x_{1:m})$, etc. by the posteriors $V_{km}$, $w_k^\nu$, $K(\nu|y\!x_{<k})$, $\nu(y\!x_{<k}y\!x_{k:m})$, etc. Strangely enough, for $w_\nu$ chosen as $2^{-K(\nu)}$ it is not true that $w_k^\nu \sim 2^{-K(\nu|y\!x_{<k})}$. If this property were true, weak bounds as the one proven in Section 6.2 (which is too weak to be of practical importance) could be boosted to practical bounds of order one. Hence, it is of high impact to rescue the posterization property in some way. It may be valid when grouping together essentially equal distributions $\nu$ (cf. Problems 3.13 and 6.3).

## 5.9 Conclusions

All tasks that require intelligence to be solved can naturally be formulated as a maximization of some expected utility in the framework of agents. We gave an explicit expression (4.17) of such a decision-theoretic agent. The main remaining problem is the unknown prior probability distribution $\mu^{\text{AI}}$ of the environment(s). Conventional learning algorithms are unsuitable, because they can neither handle large (unstructured) state spaces nor do they converge in the theoretically minimal number of cycles nor can they handle non-stationary environments appropriately. On the other hand, the universal semimeasure $\xi$ (2.26), based on ideas from algorithmic information theory, solves the problem of the unknown prior distribution for induction problems. No explicit learning procedure is necessary, as $\xi$ automatically converges to $\mu$. We unified the theory of universal sequence prediction with the decision-theoretic agent by replacing the unknown true prior $\mu^{\text{AI}}$ by an appropriately generalized universal semimeasure $\xi^{\text{AI}}$. We gave strong arguments that the resulting AI$\xi$ model is universally optimal. Furthermore, possible solutions to the horizon problem were discussed. In Chapter 6 we present a number of problem classes, and outline how the AI$\xi$ model can solve them. They include sequence prediction, strategic games, function minimization and, especially, how AI$\xi$ learns to learn supervised. In Chapter 7 we develop a modified time-bounded (computable) AI$\xi^{tl}$ version.

## 5.10 Converting Functions into Chronological Semimeasures

To complete the proof of the universality (5.6) of $M$ we need to convert enumerable functions $\psi : \mathbb{B}^* \to \mathbb{R}^+$ into enumerable chronological semimeasures $\rho : (\mathcal{Y} \times \mathcal{X})^* \to \mathbb{R}^+$ with certain additional properties, where $\mathcal{X}$ and $\mathcal{Y}$ are countable (finite or infinite). The proof given here follows [LV97, p273], but is slightly more formal and compact. Every enumerable function like $\psi$ and $\rho$ can be approximated from below by definition[8] by primitive recursive functions $\varphi : \mathbb{B}^* \times \mathbb{N} \to \mathbb{Q}^+$ and $\phi : (\mathcal{Y}\times\mathcal{X})^* \times \mathbb{N} \to \mathbb{Q}^+$ with $\psi(s) = \sup_t \varphi(s,t)$ and $\rho(s) = \sup_t \phi(s,t)$ and recursion parameter $t$. For arguments of the form $s = y\!x_{1:n}$ we recursively (in $n$) construct $\phi$ from $\varphi$ as follows:

$$\varphi'(y\!x_{1:n},t) := \begin{cases} \varphi(y\!x_{1:n},t) & \text{for } x_n < t, \\ 0 & \text{for } x_n \geq t, \end{cases} \qquad \varphi'(\epsilon,t) := \varphi(\epsilon,t), \tag{5.42}$$

$$\phi(\epsilon,t) := \max_{0\leq i\leq t}\left\{\varphi'(\epsilon,i) : \varphi'(\epsilon,i) \leq 1\right\}, \tag{5.43}$$

$$\phi(y\!\underline{x}_{1:n},t) := \max_{0\leq i\leq t}\left\{\varphi'(y\!x_{1:n},i) : \textstyle\sum_{x_n<t}\varphi'(y\!x_{1:n},i) \leq \phi(y\!\underline{x}_{<n},t)\right\}. \tag{5.44}$$

By $x_n < t$ we mean that the natural number associated with string $x_n$ is smaller than $t$. According to (5.42) with $\varphi$ also $\varphi'$ as well as $\sum_{x_n}\varphi'$ are primitive recursive functions. Further, if we allow $t=0$ we have $\varphi'(s,0)=0$. This ensures that $\phi$ is a total function.

In the following we prove by induction over $n$ that $\phi$ is a primitive recursive chronological semimeasure monotone increasing in $t$. All necessary properties hold for $n=0$ ($y\!x_{1:0}=\epsilon$) according to (5.43). For general $n$ assume that the induction hypothesis is true for $\phi(y\!\underline{x}_{<n},t)$. We can see from (5.44) that $\phi(y\!\underline{x}_{1:n},t)$ is monotone increasing in $t$. $\phi$ is total as $\varphi'(y\!x_{1:n},i=0)=0$ satisfies the inequality. By assumption $\phi(y\!x_{<n},t)$ is primitive recursive, hence with $\sum_{x_n}\varphi'$ also the order relation $\sum\varphi' \leq \phi$ is primitive recursive. This ensures that the nonempty finite set $\{\varphi' : \sum\varphi' \leq \phi\}_i$ and its maximum $\phi(y\!\underline{x}_{1:n},t)$ are primitive recursive. Further, $\phi(y\!\underline{x}_{1:n},t) = \varphi'(y\!x_{1:n},i)$ for some $i$ with $i \leq t$ independent of $x_n$. Thus, $\sum_{x_n}\phi(y\!\underline{x}_{1:n},t) = \sum_{x_n}\varphi'(y\!x_{1:n},i) \leq \phi(y\!\underline{x}_{<n},t)$, which is the condition for $\phi$ being a chronological semimeasure. Inductively we have proved that $\phi$ is indeed a primitive recursive chronological semimeasure monotone increasing in $t$.

In the following we show that every (by definition total) enumerable chronological semimeasure $\rho$ can be enumerated by some $\phi$. By definition of enumerability there exist primitive recursive functions $\tilde\varphi$ with $\rho(s) = \sup_t\tilde\varphi(s,t)$. The function $\varphi(s,t) := (1-{}^1\!/_t)\cdot\max_{i<t}\tilde\varphi(s,i)$ also enumerates $\rho$ but has the additional advantage of being strictly monotone increasing in $t$.

$\varphi'(y\!x_{1:n},\infty) = \varphi(y\!x_{1:n},\infty) = \rho(y\!x_{1:n})$ by definition (5.42). $\phi(\epsilon,t) = \varphi'(\epsilon,t)$ by (5.43) and the fact that $\varphi'(\epsilon,i-1) < \varphi'(\epsilon,i) \leq \varphi(\epsilon,i) \leq \rho(\epsilon) \leq 1$, hence $\phi(\epsilon,\infty) = \rho(\epsilon)$. $\phi(y\!\underline{x}_{1:n},t) \leq \varphi'(y\!x_{1:n},t)$ by (5.44), hence $\phi(y\!\underline{x}_{1:n},\infty) \leq \rho(y\!\underline{x}_{1:n})$. We prove the opposite direction $\phi(y\!\underline{x}_{1:n},\infty) \geq \rho(y\!x_{1:n})$ by induction over $n$. We have

[8] Defining enumerability as the supremum of total primitive recursive functions is more suitable for our purpose than the equivalent definition as a limit of monotone increasing partial recursive functions. In terms of Turing machines, the recursion parameter is the time after which a computation is terminated.

$$\sum_{x_n}\varphi'(y\!x_{1:n},i) \leq \sum_{x_n}\varphi(y\!x_{1:n},i) < \sum_{x_n}\varphi(y\!x_{1:n},\infty) = \sum_{x_n}\rho(y\!x_{1:n}) \leq \rho(y\!x_{<n}). \tag{5.45}$$

The strict monotony of $\varphi$ and the semimeasure property of $\rho$ have been used. By induction hypothesis $\lim_{t\to\infty}\phi(y\!x_{<n},t)\geq\rho(y\!x_{<n})$ and (5.45) for sufficiently large $t$ we have $\phi(y\!x_{<n},t)>\sum_{x_n}\varphi'(y\!x_{1:n},i)$. The condition in (5.44) is, hence, satisfied and therefore $\phi(y\!x_{1:n},t)\geq\varphi'(y\!x_{1:n},i)$ for sufficiently large $t$, especially $\phi(y\!x_{1:n},\infty)\geq\varphi'(y\!x_{1:n},i)$ for all $i$. Taking the limit $i\to\infty$ we get $\phi(y\!x_{1:n},\infty)\geq\varphi'(y\!x_{1:n},\infty)=\rho(y\!x_{1:n})$.

Combining all results, we have shown that the constructed $\phi(\cdot,t)$ are primitive recursive chronological semimeasures which are monotone increasing in $t$, and which converge to the enumerable chronological semimeasure $\rho$. This finally proves the enumerability of the set of enumerable chronological semimeasures.

## 5.11 Proof of the Entropy Inequality

We show[9] that

$$(1-\alpha)^2+\sum_{i=1}^n(y_i-x_i)^2 \;\leq\; \sum_{i=1}^n y_i\ln\frac{y_i}{x_i}$$

$$\text{with}\quad y_i\geq 0,\quad x_i>0,\quad \sum_{i=1}^n y_i=1,\quad \sum_{i=1}^n x_i=\alpha\leq 1$$

and with $0\ln 0:=0$, or equivalently that

$$\sum_{i=1}^n f(x_i,y_i)\geq(1-\alpha)^2 \quad\text{with}\quad f(x,y):=y\ln\frac{y}{x}-(y-x)^2,\quad f:(0,1]\times[0,1]\to I\!R. \tag{5.46}$$

The proof of the case $n=2$ will not be repeated here, as it is elementary and well known. We will reduce the general case $n>2$ to the case $n=2$. It is enough to show that $\sum f\geq 0$ at all extremal points and "at" the boundary.

The boundary is the set of all $(\boldsymbol{x},\boldsymbol{y})$ where one $x_i$ or one $y_i$ is or tends to zero. If one $y_i=0$ we can reduce (5.46) to $n-1$ (with a different $\alpha'=\alpha-x_i$) since $f(x_i,0)\geq(1-\alpha)^2-(1-\alpha')^2$. If one $x_i\to 0$ then $f(x_i,y_i)\to\infty$. As $f$ is bounded from below ($f>-2$), $\sum f$ tends to infinity and (5.46) is satisfied. Hence (5.46) is satisfied "at" the boundary.

The extrema in the interior are found by differentiation. To include the boundary conditions we add Lagrange multipliers $\lambda$ and $\mu$

$$L(\boldsymbol{x},\boldsymbol{y}) \;:=\; \sum_{i=1}^n f(x_i,y_i)+\lambda\cdot\Big(\alpha-\sum_{i=1}^n x_i\Big)+\mu\cdot\Big(1-\sum_{i=1}^n y_i\Big). \tag{5.47}$$

[9] We will not explicate every subtlety and only sketch the proof (cf. Section 3.9.1).

The extrema are at $\partial L/\partial x_i = \partial L/\partial y_i = 0$, i.e. at

$$\lambda = \partial_{x_i} f(x_i, y_i), \qquad \mu = \partial_{y_i} f(x_i, y_i). \tag{5.48}$$

Assume that $(\boldsymbol{x}^*,\boldsymbol{y}^*)$ is a solution of (5.48). We can determine $(\lambda^*,\mu^*)$ for this solution by inserting e.g. the first component $(x_1^*,y_1^*)$ into (5.48). But all other components of $(\boldsymbol{x}^*,\boldsymbol{y}^*)$ must be consistent with (5.48) too. Let us assume that for given $(\lambda^*,\mu^*)$ there are $m<\infty$ different solutions of (5.48), i.e. $(\boldsymbol{x}^*,\boldsymbol{y}^*)$ consists only of $m$ different components $(\tilde{x}_k,\tilde{y}_k)$ with $1\leq k\leq m$ where each component has multiplicity $n_k\geq 1$. Define $\bar{x}_k := n_k\tilde{x}_k$ and $\bar{y}_k := n_k\tilde{y}_k$. We have

$$\sum_{k=1}^m n_k = n, \qquad \sum_{k=1}^m \bar{y}_k = \sum_{k=1}^m n_k\tilde{y}_k = \sum_{i=1}^n y_i = 1, \qquad \sum_{k=1}^m \bar{x}_k = \alpha.$$

Equal components in (5.46) can be grouped together

$$\begin{aligned}
\sum_{i=1}^n f(x_i^*, y_i^*) &= \sum_{k=1}^m n_k \Big[\tilde{y}_k \ln\frac{\tilde{y}_k}{\tilde{x}_k} - (\tilde{y}_k - \tilde{x}_k)^2\Big] \\
&\geq \sum_{k=1}^m \Big[n_k\tilde{y}_k \ln\frac{n_k\tilde{y}_k}{n_k\tilde{x}_k} - n_k^2(\tilde{y}_k - \tilde{x}_k)^2\Big] \\
&= \sum_{k=1}^m \Big[\bar{y}_k \ln\frac{\bar{y}_k}{\bar{x}_k} - (\bar{y}_k - \bar{x}_k)^2\Big] \\
&= \sum_{k=1}^m f(\bar{x}_k, \bar{y}_k) \overset{?}{\geq} (1-\alpha)^2.
\end{aligned}$$

So we have reduced the case $\sum_{i=1}^n$ in a somewhat unconventional way to the case $\sum_{k=1}^m$ with $m$ being the number of solutions of (5.48). One can show that there are at most two solutions for $y_i/x_i$ by considering the equation for $\lambda+\mu$, and from this that $m\leq 2$. In the following we present a more abstract and general proof method.

We will show that $R(x,y) := f(x,y) - \lambda x - \mu y$ has at most two (non-degenerate) extrema, which will in turn proof that $\partial_x f = \lambda$, $\partial_y f = \mu$ has at most two solutions. Let us consider $R$ on a curve connecting two extrema $g(t) := R(x(t),y(t))$, $x(0)=\tilde{x}_k$, $x(1)=\tilde{x}_l$, $y(0)=\tilde{y}_k$, $y(1)=\tilde{y}_l$, $0\leq t\leq 1$. From $g'(0)=g'(1)=0$ we know that there is a $t_0\in[0,1]$ with $g''(t_0)=0$. That is, every connecting curve between two extrema contains a point in which $R$ has curvature zero in one direction and hence zero Gauss curvature $G$. The support of $R$ is divided by the curve(s) $G(x,y)=0$ into zones. Each zone can contain at most one extremum.

$$\begin{aligned}
G(x,y) &:= \det\begin{pmatrix} \partial_x^2 R & \partial_x\partial_y R \\ \partial_y\partial_x R & \partial_y^2 R \end{pmatrix} = (\partial_x^2 f)(\partial_y^2 f) - (\partial_x\partial_y f)^2 \\
&= \Big(\frac{y}{x^2} - 2\Big)\Big(\frac{1}{y} - 2\Big) - \Big(-\frac{1}{x} + 2\Big)^2 = -\frac{2}{x^2 y}(x-y)^2.
\end{aligned}$$

This is zero for $x=y$ only. The support of $f$ is divided into two zones, $x<y$ and $x>y$. The (infinitely many) degenerate extrema for $\tilde{x}_k=\tilde{y}_k$ give no contribution to (5.46) ($f(x,x)=0$) and are hence irrelevant. So there are at most two nontrivial solutions of $\partial_x f=\lambda$, $\partial_y f=\mu$, hence $m\leq 2$.

In summary we have reduced (5.46) for general $n$ to $m=2$, which can be shown by elementary means. □

## 5.12 History & References

Most references relevant to this chapter have already been given in previous chapters or in the main text of this chapter. Below we only remark on and give references to two further, in the context of AI$\xi$ interesting, topics: protocols in probability theory and bandit problems.

**Paradoxes, sample spaces, protocols, and incompleteness.** There are many paradoxes in probability theory, like the Petersburg, Monty Hall, Simpson, Newcomb, rich uncle, and three prisoners [Szé86, EF98, Res01, Mos65]. Some of them are not particularly related to probability theory, but just to the improper use of math in general. Probably the most interesting paradoxes directly related to probability theory concern the awareness and choice of sample spaces and protocols (see [GH02] and references therein, especially [Sha85]). If one phrases these paradoxes within the AI$\mu$ model, one automatically has to be aware of and choose a suitable sample space and protocol. If this procedure uniquely determines $\mu$, the paradox is solved. If not, the problem description was not complete, i.e. the description is consistent with a whole set $\mathcal{M}$ of possible environments. This incompleteness can sometimes be overcome by symmetry or maximum entropy arguments (see [GH02] and Section 2.3.4). In general, a universal prior $\xi_\mathcal{M}$ and the predictions/actions of the SP$\xi_\mathcal{M}$/AI$\xi_\mathcal{M}$ model represent a satisfactory solution to the paradoxes, solving the sample space, protocol, *and* incompleteness problem.

**Bandit problems.** Bandit problems arose historically from the desire to optimally assign treatments to patients. They are prototypical problems for the so-called exploration versus exploitation dilemma. They were originally introduced by Robbins [Rob52]. One out of several arms (treatments) can be chosen, leading to a possible reward (success). The goal is to maximize one's reward in repeated trials. The simplest model is to assume that arm $i$ leads to reward 1 (0) with probability $p_i$ $(1-p_i)$, where the probabilities are unknown. The traditional Bayesian solution to the uncertainty about $p_i$ is to assume a (second-order Beta) prior over $p_i$. The goal is to maximize the (exponentially) discounted reward sum. A closed solution can be given in terms of Gittins indices [GJ74, Git89]. For regular discount sequences these strategies are not self-optimizing [BF85, KV86]. Many efficient heuristic self-optimizing approaches exist. In a minimax approach one tries to find strategies that lead to the highest expected reward in the worst case over unknown chances $p_i$ [Vog60]. A complete worst-case approach without any probabilistic assumption on the environment can be found in [ACBFS02]. The default textbook on bandits is [BF85] and on Gittins indices is [Git89].

## 5.13 Problems

**5.1 (Actions as random variables)** [C35oi] Instead of defining $\xi^{\text{AI}}(\underline{y}x_{1:n})$ as a universal distribution over perceptions $x_{1:n}$ conditioned under actions $y_{1:n}$ as in (5.2), one may think of the following alternative definition: We use a universal distribution over perceptions *and* actions and then conditionalize to the actions, i.e. $\xi_{alt}^{\text{AI}}(\underline{y}x_{1:n}) := M(\underline{yx}_{1:n})/\sum_{x_{1:n}} M(\underline{yx}_{1:n})$, where $M$ is Solomonoff's prior (2.21) (we could use $\xi_U$ as well). One motivation for doing so is to regard $M$ as a prior belief in the whole arrangement of agent+environment. The major problem with this approach is that $\xi_{alt}^{\text{AI}}$ is not enumerable. More precisely, the presented definition does not lead to an enumeration procedure for $\xi_{alt}^{\text{AI}}$. This does not necessarily imply non-enumerability of $\xi_{alt}^{\text{AI}}$. Whether $\xi_{alt}^{\text{AI}} \stackrel{\times}{=} \xi^{\text{AI}}$ is also an open problem (cf. Problem 2.6). If true it would imply universality of $\xi_{alt}^{\text{AI}}$ and convergence to computable $\mu^{\text{AI}}$. This alternative approach also allows conditionalization w.r.t. the perceptions and to determine $M(\underline{yx}_{<k}\underline{y}_k)$, which may be interpreted as the agent's own belief in selecting action $y_k$. But an actual action selection based on this probability would lead to a poorly performing agent, which differs from the "optimal" action $y_k^\xi$ and $y_k^{\xi_{alt}}$ via the expectimax expression. Could $M(\underline{yx}_{<k}\underline{y}_k)$ nevertheless be close to the action of $p^\xi$ and/or $p^{\xi_{alt}}$ for large $k$, justifying the above interpretation of $M$? (cf. Section 8.5.2 on multi-agent systems.) Prove or disprove the stated open questions, conjectures, and assertions.

**5.2 (Absorbing two-state environment)** [C15ui] The HeavenHell example of Section 5.3.2 was a three-state MDP that did not allow for self-optimizing policies. Here, a similar two-state MDP shall be analyzed in more detail. Let $\mathcal{M}=\{\mu_0,\mu_1,...\}$ with $w_0=w_1=\frac{1}{2}(1-\beta)>0$, $\sum_{i\geq 2} w_i=\beta\geq 0$, $r_k\in\mathcal{R}=\{0,{}^2/_3,1\}$, and $\mathcal{Y}=\{a,b\}$. Environments $\mu_0$ and $\mu_1$ are deterministic MDPs defined as $\mu_i =$ [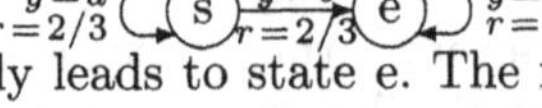

], i.e. initially being in state s, action $b$ irrevocably leads to state e. The reward in state s is 2/3. Environments $\mu_0$ and $\mu_1$ differ only in the reward in state e, which is $r=0$ in environment $\mu_0$, and $r=1$ in environment $\mu_1$. Show that there is no policy that is self-optimizing in $\mu_0$ *and* $\mu_1$. Now, consider the case $\beta=0$: Determine $V_{\mu_i}^p$ for $i\in\{0,1\}$ and $V_\xi^p$ for all policies $p$. The results only depend on the first time action $b$ is taken (if at all) and on the farsightedness $m$. Now determine $V_{\mu_i}^*$, $p^\xi$, and $V_{\mu_i}^{p^\xi}$. The results show that $p^\xi$ is not self-optimizing ($V_{\mu_1}^{p^\xi} \not\to V_{\mu_1}^*$). Generalize the latter result to $0<\beta<\frac{1}{6}(1-\frac{1}{m})$ with environments $\mu_i$, $i\geq 2$ defined arbitrarily, to $V_{km}^{\cdots}$, and to $V_{k\gamma}^{\cdots}$.

**5.3 (Computing $\rho$)** [C30u] Show that for every enumerable chronological semimeasure $\rho$ there exists a Turing machine $T$ of length $K(\rho)$ that computes it, i.e. $\rho(y\!x_{1:n})=\sum_{q:T(q,y_{1:n})=x_{1:n}}2^{-\ell(q)}$ and $\ell(T)\stackrel{\pm}{=}K(\rho)$ (see (5.7) for context). Hint: Adapt and improve Lemma 4.3.4 of [LV97, p255].

**5.4 (Pareto optimality)** [C30u] We have shown Pareto optimality of AI$\xi$ with $\xi$ given in the form $\xi=\sum_\nu w_\nu\nu$. Show that AIXI with $M$ defined as in (5.2) is also Pareto optimal. Compare with or use the results of Problem 2.2.

**5.5 (Pareto optimality)** [C30oi] We define policy $p$ to be equivalent to policy $p'$ if both policies lead to the same value $V_\nu$ in all environments $\nu$, i.e. if $V_\nu^p=V_\nu^{p'}\;\forall\nu\in\mathcal{M}$. Are all Pareto optimal policies equivalent to some (Pareto optimal) mixture policy $p^\xi$ for certain weights $w_\nu$? A positive answer to this question implies that one can restrict the search of optimal policies to mixture policies. Try to find necessary and/or sufficient conditions that make the above question true.

**5.6 (Convergence of averages)** [C20u] Let $\delta(m):=\sum_{\nu\in\mathcal{M}}w_\nu\delta_\nu(m)$ and $\sum_{\nu\in\mathcal{M}}w_\nu\leq 1$ as in Lemma 5.28. Show that the boundedness assumption $0\leq\delta_\nu(m)\leq c$ in Lemma 5.28(*ii*) is necessary for $\delta_\nu(m)\stackrel{m\to\infty}{\longrightarrow}0$ to imply existence and/or convergence of $\delta(m)\to 0$. Show that $\delta_\nu(m)=O(f(m))\;\forall\nu\in\mathcal{M}$ does not necessarily imply $\delta(m)=O(f(m))$ if $\mathcal{M}$ is infinite, even for bounded $\delta_\nu$ (cf. Lemma 5.28(*iv*)).

Hints: For instance, for $\mathcal{M}\cong I\!N$ and $\delta_\nu(m):=\{{(mw_\nu)^{-1}\;for\;m\geq\nu \atop 0\;\;else}$, we have $\delta_\nu(m)\stackrel{m\to\infty}{\longrightarrow}0$, but $\delta(m)\equiv 1\not\to 0$. For $\delta_\nu(m):=\frac{1}{mw_\nu}\stackrel{m\to\infty}{\longrightarrow}0$, $\delta(m)$ does not even exist. For $1\geq\delta_\nu(m):=\frac{1}{1+mw_\nu}=O(\frac{1}{m})$ and $w_\nu=\frac{1}{\nu(\nu+1)}$, we have $\delta(m)\geq\frac{1}{\sqrt{2m}}$. Another interesting example is $\delta_\nu(m):=\mathrm{e}^{-m/\nu}\leq 1$, which decays exponentially in $m$ for every $\nu$, but for $w_\nu:=\frac{1}{\nu(\nu+1)}$, $\delta(m)$ decays only harmonically. Show that $\delta(m)\geq\frac{1}{2m}(1-\mathrm{e}^{-m})\geq\frac{1}{4m}$ for $m\geq 1$ (easy) and even $\delta(m)\geq\frac{1}{2m}$ (harder).

**5.7 (Domain of definitions)** [C20u] Several subtleties concerning the domain of definition and existence have been ignored. First, Definition 5.19 defined $p^\rho$ only on histories produced by $p^\rho$ itself. Given a history $y_{<k}\neq\dot y_{<k}$ one has to generalize the definition similarly to Definition 5.30. Even in this generalized form $p^\rho$ is only defined for histories that occur with nonzero $\rho$-probability. Show that $p^\mu$ and $p^\xi$ are defined for all histories that have nonzero $\mu$-probability. Use this to verify the soundness of the definitions and theorems in this chapter.

**5.8 (Self-optimizing policies for geometric discounting)** [C30ui] Show that there are self-optimizing policies in ergodic MDPs even for geometric discounting ($\frac{\gamma_{k+1}}{\gamma_k}\not\to 1$). On the other hand, the Bayes mixture policy $p^\xi$ is not self-optimizing for bandit problems with geometric discounting. Since bandits are special ergodic MDPs, these results seem to contradict Theorem 5.34. Clarify this paradox and discuss the implications. Hint: Histories $y\!x_{<k}$ are policy dependent.

**5.9 (Relevant and non-computable environments $\mu$)** [C30oi] Assume feedback $x$ consists of three parts $x = x'r'x''$, future I/O is completely independent of $x''$, $x'r'$ is sampled from a computable distribution, $x''$ from a (possibly) non-computable distribution. Show that $M$ multiplicatively dominates $\mu'$, where $\mu'$ is the true distribution $\mu$ modified in a way such that $V_\mu = V_{\mu'}$ but $\mu'$ is computable. This suggests that the computability assumption on $\mu$ may be weakened to the (for AIXI) *relevant* parts of the environment. Formulate and prove all this rigorously and generalize it to less trivial cases, where the relevant computable and the irrelevant non-computable information in $x$ cannot be factored so easily.

**5.10 (Self-optimizing environments)** [C35u] Ergodic MDPs admit self-optimizing policies, which implies that $p^\xi_{\text{MDP1}}$ is self-optimizing (see Section 5.6). Show that bandits, i.i.d. processes, and classification tasks are special (degenerate) cases of ergodic MDPs. The existence of self-optimizing policies is not limited to (subclasses of ergodic) MDPs. Suitably define ergodic partially observable MDPs (ergodic POMDPs) and $k^{th}$-order ergodic MDPs and show that these classes also admit self-optimizing policies. Furthermore, show that factorizable environments, defined in Section 4.3.1, admit self-optimizing policies.

**5.11 (Belief contamination)** [C30ui] Consider an environmental class $\mathcal{M}$ that admits self-optimizing policies. Theorem 5.34 shows that $p^\xi$ is self-optimizing in the sense of $\lim_{k\to\infty}[V^{*\nu}_{k\gamma} - V^{p^\xi\nu}_{k\gamma}] = 0$. The Bayes mixture $\xi := \sum_{\nu\in\mathcal{M}} w_\nu \nu$ expresses the degree of belief $w_\nu$ in environment $\nu \in \mathcal{M}$. We want to study the effect of additionally believing in some $\rho \notin \mathcal{M}$ with some small probability $\alpha$. The new belief prior is $\xi' := (1-\alpha)\xi + \alpha\rho$. Show that a belief $\alpha$ in $\rho$ much smaller than the belief $w_\mu$ in the true environment $\mu \in \mathcal{M}$ causes only a small corruption of the self-optimizing property. More precisely, $\limsup_{k\to\infty}[V^{*\mu}_{k\gamma} - V^{p^{\xi'}\mu}_{k\gamma}] \leq \frac{\alpha\cdot r_{max}}{(1-\alpha)w_\mu}$. Construct examples for which $V^{p^\xi\mu}_{k\gamma} - V^{p^{\xi'}\mu}_{k\gamma} = \frac{\alpha\cdot r_{max}}{(1-\alpha)w_\mu}$. This shows that the upper bound cannot be improved in general and that a belief contamination $\alpha$ of magnitude comparable to $w_\mu$ can completely degrade performance.

**5.12 (Continuity of value for ergodic MDPs)** [C40usm] Let $\mu$ and $\hat\mu$ be MDPs that are "close" to each other in the sense that the transition matrix $\hat T^a_{ss'} = \hat\mu(sa\underline{s}')$ is close to $T^a_{ss'} = \mu(sa\underline{s}')$, and the reward function $\hat R^a_{ss'} = \sum_{r\in\mathcal{R}} r\cdot\hat\mu(sas'\underline{r})$ is close to $R^a_{ss'} = \sum_{r\in\mathcal{R}} r\cdot\mu(sas'\underline{r})$ in the sense that $\varepsilon := \max_{ss'a}\{|T^a_{ss'} - \hat T^a_{ss'}|, |R^a_{ss'} - \hat R^a_{ss'}|\}$ is "small". Furthermore, let $p$ be a stationary policy, i.e. $a_k := p(s_{<k}) = p(s_{k-1})$. Show properties $(i)-(vii)$ of the value function(s)

- $(o)$ Condition: $T$ is ergodic –or– $T^a_{ss'} = 0$ implies $\hat T^a_{ss'} = 0$.
- $(i)$ $V^{pT}_{1m} = \sum_{s_{1:m}} (R^{a_1}_{s_0 s_1} + ... + R^{a_m}_{s_{m-1}s_m}) \cdot T^{a_1}_{s_0 s_1} \cdot ... \cdot T^{a_m}_{s_{m-1}s_m}$ (where $s_0$ is a special initial "state").

(*ii*) $\mathrm{v}_T^p := \lim_{m\to\infty}\frac{1}{m}V_{1m}^{pT}$ exists.
(*iii*) $|\mathrm{v}_T^p - \mathrm{v}_{\hat{T}}^p| = O(\varepsilon)$ if (*o*).
(*iv*) $p^T := \mathrm{argmax}_p \mathrm{v}_T^p$ can be chosen stationary. $\mathrm{v}_T^* := \max_p \mathrm{v}_T^p = \mathrm{v}_T^{p^T}$.
(*v*) $|\mathrm{v}_T^* - \mathrm{v}_{\hat{T}}^*| = O(\varepsilon)$ if (*o*).
(*vi*) $|\mathrm{v}_T^p - \frac{1}{m}V_{1m}^{pT}| = O(\frac{1}{m})$.
(*vii*) $|\frac{1}{m}V_{1m}^{p^T T} - \frac{1}{m}V_{1m}^{*T}| = O(\frac{1}{m})$ if (*o*).
(*viii*) $|\frac{1}{m}V_{1m}^{p^{\hat{T}} T} - \frac{1}{m}V_{1m}^{*T}| = 3\cdot O(\frac{1}{m}) + 2\cdot O(\varepsilon)$ if (*o*).

The "constant" factor hidden in $O()$ depends on $T$, but is independent of $m$, $\varepsilon$, and $\hat{T}$. Note that $\mathrm{argmax}_p V_{1m}^{pT}$ may be non-stationary. (*viii*) with $k_0 m$ instead of $1m$ and $\varepsilon \sim k_0^{-1/2} \sim m^{-1/3}$ was used in the proof of Theorem 5.38(*i*).

Finally, show that $\mathrm{v}_T^{p^{\hat{T}}} = \mathrm{v}_T^*$ (exactly) for sufficiently small $\varepsilon > 0$. If $\hat{T},\hat{R}$ are frequency estimates of $T,R$, then the probability that $\varepsilon$ is not small decreases exponentially with the sample size $k_0$. Together with the choice $k_0 \propto \log m$, improve Theorem 5.38(*i*) to $O(\frac{\log m}{m})$.

Hints: (*i*) immediate from definitions. (*ii*) follows from the existence of $\bar{T} := \lim_{m\to\infty}\frac{1}{m}\sum_{k=0}^{m-1}(T)^k$ [Ber95b, p187]. (*iii*) For ergodic $T$ all rows of $\bar{T}$ coincide with the stationary distribution, which is proportional to some column of the adjoint matrix of $T$, which itself is a polynomial in (the components) of $T$. (*iv*) similar to [Ber95b, p191]. (*v*) from (*iii*) and (*iv*). (*vi*) similar (*ii*). (*vii*) similar (*iv*). For (*viii*) chain (*vi*)+(*iii*)+(*v*)+(*vi*)+(*vii*) and use the triangle inequality.

**5.13 (Ergodic versus forgetful environments)** [C20s/C10u] Forgetful environments were defined in Section 5.3.6 as being asymptotically independent of the history. Ergodic MDPs were defined in Section 5.6 as visiting every state infinitely often. An environment is called acyclic if the probability of infinitely repeating cycles is zero. Show that every acyclic ergodic MDP is forgetful, but not every forgetful MDP is ergodic. Note also that forgetfulness is a broader concept than ($k^{th}$-order) MDPs.

**5.14 (Uniform mixture of MDPs)** [C30usi] In the following you are asked to derive explicit expressions for $\xi^{\text{MDP}}$ for uniform prior belief $w$. Let $\mu_T \in \mathcal{M}_{\text{MDP}}$ be a (completely observable) MDP with transition matrix $T$. $T_{ss'}^a$ is the probability of going to state $s' \in \mathcal{X}$ under action $a \in \mathcal{Y}$ if currently in state $s \in \mathcal{X}$. Given a policy that determines the actions $a_t$, the probability of action-observation history $a_1 s_1 ... a_n s_n$ after cycle $n$ is $\mu_T(a_1\underline{s}_1...a_n\underline{s}_n) = T_{s_0 s_1}^{a_1} \cdot ... \cdot T_{s_{n-1}s_n}^{a_n}$, where $s_0$ is some initial state (randomization over the initial state may be performed). Reward is a given function of state ($r_k = r(s_k)$). Optimal policies/actions follow from the recursive Bellman equations (4.29) or the explicit expectimax expression of the AI$\mu$ model (4.17). Assume now that we only know that the true environment is an MDP, but nothing more, i.e. $T \in \mathcal{T} := \{T : T_{ss'}^a \geq 0, \sum_{s'} T_{ss'}^a = 1\}$ is unknown. Since $\mathcal{T}$ is continuous, the Bayes mixture $\xi$ has the form $\xi(a\underline{s}_{1:n}) := \int_{\mathcal{T}} w_T \mu_T(a\underline{s}_{1:n}) dT$.

(*i*) Assume a uniform prior belief over $T$, i.e. $w_T \propto 1$ and the measure $dT$ is the uniform measure on the polytope $\mathcal{T}$. Compute the integral and show that the ratio $\xi(a\!s_{<n} a\underline{s}_n) = \xi(a\underline{s}_{1:n})/\xi(a\underline{s}_{<n}) = N^{a_n}_{s_{n-1}s_n}/(\sum_{s'} N^{a_n}_{s_{n-1}s'} + S - 1)$, where $S = |\mathcal{X}|$ is the number of states and $N^a_{ss'}$ is the historical number of transitions from $s$ to $s'$ under action $a$, (including the transition from $s_0$ if $s = s_0$ and to $s_n$ if $s' = s_n$). This is just Laplace' law of succession [Lap12] (see Problem 2.11), one for each $(ass')$-tuple. For instance, initially all transitions are equally plausible $\xi(a_1\underline{s}_1) = \frac{1}{S}$.

(*ii*) Show that, although the class $\mathcal{T}$ is continuous and contains non-ergodic environments, the Bayes optimal policy $p^\xi$ is self-optimizing for ergodic environments $\mu_T \in \mathcal{M}_{\text{MDP1}}$. The intuitive reason is that $\mathcal{T}$ is compact and the non-ergodic environments have measure zero.

(*iii*) Model-based reinforcement learning algorithms try to estimate $T$ from past experience. Give an expression for the posterior believe $w_T(a\!s_{1:n}) \propto \mu_T(a\underline{s}_{1:n})$ in transition $T$. Note that this is a (complex) distribution over $T$, while most reinforcement learning algorithms only estimate a single (e.g. a most likely) $T$. Show that the expected transition probability $\mathbf{E}[T^a_{ss'}|a\!s_{1:n}] := \int_{\mathcal{T}} T^a_{ss'} w_T(a\!s_{1:n}) dT = (N^a_{ss'}+1)/(\sum_{s'} N^a_{ss'} + S)$ coincides with the relative historic occurrence of $(ass')$. Show that policy $p^\xi$ based on (4.17) appropriately explores the environment, while the popular policy based on $\mathbf{E}[T]$ or other point estimates like Maximum Likelihood lack exploration.

(*iv*) Assume we know that the environment is a *deterministic* MDP, i.e. $\mathcal{T} = \{T : T^a_{ss'} \in \{0,1\}, \sum_{s'} T^a_{ss'} = 1\}$. Repeat (*i*)–(*iii*) with this (now discrete) $\mathcal{T}$. Is the corresponding $p^\xi$ self-optimizing?

(*v*) Assume now that for every action $a$ there exists a mirror "undo" action $\bar{a}$ in the sense that $T^a_{ss'} = T^{\bar{a}}_{s's}$. Repeat (*i*)–(*iii*) for the set of all deterministic "symmetric" MDPs. Is the corresponding $p^\xi$ self-optimizing? An example is a robot moving in a (noiseless) environment, like a maze. Non-symmetric MDPs contain one-way streets or doors, which are missing in symmetric MDPs.

(*vi*) Incorporate further knowledge of the form $T^a_{ss'} = 0/1$ for some $(ass')$ and repeat (*i*)–(*iii*). For example, if we know that the environment is an $l \times l$ grid maze, and transitions are a priori only possible between neighboring cells, we know that $T^a_{ss'} = 0$ if $s$ and $s'$ are not neighboring grid cells.

(*vii*) Explore the difficulties when extending the considerations in (*i*)–(*iii*) to POMDPs, potentially with variable state space size $S$, e.g. with prior $w_S \propto S^{-2}$.

**5.15 (Effective horizons)** [C25u] Derive the expressions for effective horizons presented in Table 5.41.

**5.16 (Effect of discounting)** [C20u] Consider the MDP $\mu$ = [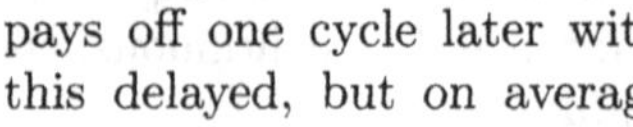

], where $s$ is the initial state. Taking action $a$ forever gives reward 1 per cycle. Taking action $b$ first gives no reward but pays off one cycle later with reward $2+\delta > 2$. Depending on the discount this delayed, but on average higher reward may be favorable. Show that for $\gamma$-geometric discount, $V_{1\gamma}^{a^\infty\mu}=1$ and $V_{1\gamma}^{b^\infty\mu}=\frac{(2+\delta)\gamma}{1+\gamma}$, hence action $b$ is favorable iff $\gamma>\frac{1}{1+\delta}$. Show that for the power discount $\gamma_k=\frac{1}{k^2}$ and small $\delta>0$, the optimal policy performs action $a$ for the first $k\approx\frac{2}{\delta}$ cycles and action $b$ thereafter. In both cases the critical effective horizon is $h_k^{eff}\sim\frac{1}{\delta}$.

*Ideas matter.*

*Approximate the solution, not the problem.*

— Richard Sutton

Richard Sutton

# 6 Important Environmental Classes



In this and the following chapter we define $\xi = \xi_U \overset{\times}{=} M$ to be Solomonoff's prior, i.e. AI$\xi$=AIXI. In order to give further support for the universality and optimality of the AI$\xi$ theory, we apply AI$\xi$ in this chapter to a number of problem classes. They include sequence prediction, strategic games, function minimization and, especially, how AI$\xi$ learns to learn supervised. For some classes we give concrete examples to illuminate the scope of the problem class.

We first formulate each problem class in its natural way (when $\mu^{\text{problem}}$ is known) and then construct a formulation within the AI$\mu$ model and prove its equivalence. We then consider the consequences of replacing $\mu$ by $\xi$. The main goal is to understand why and how the problems are solved by AI$\xi$. We only highlight special aspects of each problem class. Sections 6.2–6.6 together should give a better picture of the AI$\xi$ model. We do not study every aspect for every problem class. The sections may be read selectively, and they are not essential to understand the remaining chapters.

## 6.1 Repetition of the AI$\mu/\xi$ Models

In the last chapter we unified sequential decision theory with the theory of universal induction to a model of artificial intelligence, which we claimed to be universal and superior to any other model in various senses. All tasks that require intelligence to be solved can naturally be formulated as a maximization of some expected utility in the framework of agents. The main remaining problem is the unknown prior probability distribution $\mu^{\text{AI}}$ of the environment(s). Conventional learning algorithms are restricted in the sense that they can neither handle large (unstructured) state spaces, nor do they converge in the theoretically minimal number of interaction cycles, nor can they handle non-stationary environments appropriately. On the other hand, the universal semimeasure $\xi$ (2.26), based on ideas from algorithmic information theory, solves the problem of the unknown prior distribution for induction problems. No explicit learning procedure is necessary, as $\xi$ automatically converges to $\mu$. We unified the theory of universal sequence prediction with the decision-theoretic agent by replacing the unknown true prior $\mu^{\text{AI}}$ by an appropriately generalized universal semimeasure $\xi^{\text{AI}}$. For convenience we repeat some definitions and results from previous chapters that we need in this chapter.

Let $\mu(\underline{yx}_{1:k})$ be the true chronological prior probability that the environment reacts with $x_{1:k}$ if provided with actions $y_{1:k}$ from the agent. We define $V^{*\mu}_{k+1,m}(yx_{1:k})$ to be the $\mu$-expected reward sum in cycles $k+1$ to $m$ with outputs $y_i$ generated by agent $p^*$ and responses $x_i$ from the environment. Adding reward $r_k \equiv r(x_k)$ to $V^{*\mu}_{k+1,m}$, we get the value including cycle $k$. probability of $x_k = r_k o_k$, given history $yx_{<k}y_k$, is given by the conditional probability $\mu(yx_{<k}\underline{yx}_k)$. Policy $p^*$ chooses $y_k$ as to maximize the future reward. So the expected reward sum in cycles $k$ to $m$ given $yx_{<k}$ and $y_k$ chosen by $p^*$ is

$$V^{*\mu}_{km}(yx_{<k}) \;=\; \max_{y_k}\sum_{x_k}[r_k + V^{*\mu}_{k+1,m}(yx_{1:k})]\cdot\mu(yx_{<k}\underline{yx}_k). \tag{6.1}$$

Together with the induction start

$$V^{*\mu}_{m+1,m}(yx_{1:m}) \;:=\; 0, \tag{6.2}$$

$V^{*\mu}_{km}$ is completely defined. If $m_k$ is our horizon function of $p^*$ and $\dot{y}\dot{x}_{<k}$ is the actual history in cycle $k$, the output $\dot{y}_k$ of the agent is given by

$$\dot{y}_k \;=\; \arg\max_{y_k} \sum_{x_k} [r_k + V^{*\mu}_{k+1,m_k}(\dot{y}\dot{x}_{<k}yx_k)]\cdot\mu(\dot{y}\dot{x}_{<k}y\underline{x}_k), \tag{6.3}$$

which in turn defines the policy $p^*$. Then the environment responds $\dot{x}_k$ with probability $\mu(\dot{y}\dot{x}_{<k}\dot{y}\underline{\dot{x}}_k)$, and cycle $k+1$ starts. We may unfold the recursion (6.1) further and give $\dot{y}_k$ explicitly as

$$\dot{y}_k \;=\; \arg\max_{y_k}\sum_{x_k}\max_{y_{k+1}}\sum_{x_{k+1}} \;...\; \max_{y_{m_k}}\sum_{x_{m_k}}(r_k+\,...\,+r_{m_k})\cdot\mu(\dot{y}\dot{x}_{<k}y\underline{x}_{k:m_k}). \tag{6.4}$$

This expression has a direct interpretation: The probability of inputs $x_{k:m_k}$ in cycle $k$ when the agent outputs $y_{k:m_k}$ with actual history $\dot{y}\dot{x}_{<k}$ is $\mu(\dot{y}\dot{x}_{<k}y\underline{x}_{k:m_k})$. The future reward in this case is $r_k+...+r_{m_k}$. The best expected reward is obtained by averaging over the $x_i$ ($\sum_{x_i}$) and maximizing over the $y_i$. This has to be done in chronological order to correctly incorporate the dependencies of $x_i$ and $y_i$ on the history. This is essentially the expectimax algorithm/sequence/tree (see Figure 4.13). The AI$\mu$ model is *optimal* in the sense that no other policy leads to higher expected reward. Unfortunately, in AI, the environment $\mu$ is often unknown. The AI$\xi$ model is defined similarly to the AI$\mu$ model, but with the unknown $\mu$ replaced by the (known) universal prior $\xi$:

$$\dot{y}_k \;=\; \arg\max_{y_k}\sum_{x_k}\max_{y_{k+1}}\sum_{x_{k+1}} \;...\; \max_{y_{m_k}}\sum_{x_{m_k}}(r_k+\,...\,+r_{m_k})\cdot\xi(\dot{y}\dot{x}_{<k}y\underline{x}_{k:m_k}) \tag{6.5}$$

$$\text{with}\quad \xi(y\underline{x}_{1:k}) \;:\stackrel{\times}{=}\; \sum_{q:q(y_{1:k})=x_{1:k}} 2^{-\ell(q)}, \tag{6.6}$$

where the sum runs over all chronological environments $q$ satisfying $q(y_{1:k})=x_{1:k}$. Motivations for AI$\xi$ being a good substitute for AI$\mu$ are the convergence of $\xi$ to $\mu$

$$\xi(\dot{y}\dot{x}_{<k}y\underline{x}_{k:m_k}) \longrightarrow \mu(\dot{y}\dot{x}_{<k}y\underline{x}_{k:m_k}) \quad\text{for}\quad k\to\infty, \tag{6.7}$$

Pareto optimality of AI$\xi$, self-optimization of AI$\xi$ for restricted $\xi$, and tight error and loss bounds in the case of passive sequence prediction.

## 6.2 Sequence Prediction (SP)

We introduced the AI$\xi$ model as a unification of ideas of sequential decision theory and universal probability distribution. We might expect AI$\xi$ to behave identically to SP$\xi:=\Theta_\xi$ when faced with a sequence prediction (SP) problem, but things are not that simple, as we will see. Let us repeat the definition in Section 3.3 of the total number of expected erroneous predictions the SP$\rho$ agent $\Theta_\rho$ makes for the first $n$ observations

$$E_n^{\Theta_\rho} := \sum_{k=1}^{n}\sum_{x_{<k}} \mu(\underline{x}_{<k})[1-\mu(x_{<k}\underline{x}_k^{\Theta_\rho})] \quad\text{with}\quad x_k^{\Theta_\rho} := \arg\max_{x_k} \rho(x_{<k}\underline{x}_k). \tag{6.8}$$

The SP$\mu$ agent is best in the sense that $E_n^{\Theta_\mu} \leq E_n^{\Theta_\rho}$ for *any* $\rho$. We showed that the universal predictor SP$\xi$ is not much worse

$$E_n^{\Theta_\xi} - E_n^{\Theta_\rho} \;\leq\; 2D + 2\sqrt{E_n^{\Theta_\rho} D} \;=\; O(\sqrt{E_n^{\Theta_\rho}}), \qquad D \stackrel{+}{\leq} \ln 2 \cdot K(\mu). \tag{6.9}$$

### 6.2.1 Using the AI$\mu$ Model for Sequence Prediction

We saw in Chapter 3 how to predict sequences for known and unknown prior distribution $\mu^{\text{SP}}$. Here we consider binary sequences $z_1 z_2 z_3 ... \in \mathbb{B}^\infty$ with known prior probability $\mu^{\text{SP}}(\underline{z_1 z_2 z_3 ...})$. (We use $z_k$ to avoid notational conflicts with the agent's inputs $x_k$.)

We want to show how the AI$\mu$ model can be used for sequence prediction. We will see that it makes the same prediction as the SP$\mu$ agent. For simplicity we only discuss the special error loss $\ell_{xy} = 1 - \delta_{xy}$, where $\delta$ is the Kronecker symbol, defined as $\delta_{ab} = 1$ for $a = b$ and 0 otherwise. First, we have to specify *how* the AI$\mu$ model should be used for sequence prediction. The following choice is natural:

The system's output $y_k$ is interpreted as a prediction for the $k^{th}$ bit $z_k$ of the string under consideration. This means that $y_k$ is binary ($y_k \in \mathbb{B} =: \mathcal{Y}$). As a reaction of the environment, the agent receives reward $r_k = 1$ if the prediction was correct ($y_k = z_k$), or $r_k = 0$ if the prediction was erroneous ($y_k \neq z_k$). The question is what the observation $o_k$ in the next cycle should be. One choice would be to inform the agent about the correct $k^{th}$ bit of the string and set $o_k = z_k$. But as from the reward $r_k$ in conjunction with the prediction $y_k$, the true bit $z_k = \delta_{y_k r_k}$ can be inferred, this information is redundant. There is no need for this additional feedback. So we set $o_k = \epsilon \in \mathcal{O} = \{\epsilon\}$, thus having $x_k \equiv r_k \in \mathcal{R} \equiv \mathcal{X} = \{0,1\}$. The agent's performance does not change when we include this redundant information; it merely complicates the notation. The prior probability $\mu^{\text{AI}}$ of the AI$\mu$ model is

$$\mu^{\text{AI}}(y_1\underline{x}_1 ... y_k\underline{x}_k) \;=\; \mu^{\text{AI}}(y_1\underline{r}_1 ... y_k\underline{r}_k) \;=\; \mu^{\text{SP}}(\underline{\delta_{y_1r_1} ... \delta_{y_kr_k}}) \;=\; \mu^{\text{SP}}(\underline{z_1 ... z_k}). \tag{6.10}$$

In the following, we will drop the superscripts of $\mu$ because they are clear from the arguments of $\mu$, and the $\mu$ equal in any case. Equation (6.1) for the expected reward reduces to

$$V_{km}^{*\mu}(y\!x_{<k}) \;=\; \max_{y_k} \sum_{r_k} [r_k + V_{k+1,m}^{*\mu}(y\!x_{1:k})] \cdot \mu(\delta_{y_1r_1} ... \delta_{y_{k-1}r_{k-1}} \underline{\delta_{y_kr_k}}). \tag{6.11}$$

The first observation we can make is that for this special $\mu$, $V_{km}^{*\mu}$ only depends on $\delta_{y_ir_i}$, i.e. replacing $y_i$ and $r_i$ simultaneously with their complements does not change the value of $V_{km}^{*\mu}$. We have a symmetry in $y_ir_i$. For $k = m+1$ this is definitely true as $V^{*\mu}_{m+1,m}=0$ in this case (see (6.2)). For $k\leq m$ we prove it by induction. The r.h.s. of (6.11) is symmetric in $y_i r_i$ for $i<k$ because $\mu$ possesses this symmetry and $V^{*\mu}_{k+1,m}$ possesses it by induction hypothesis, so the symmetry holds for the l.h.s., which completes the proof. The prediction $\dot{y}_k$ is

$$\begin{aligned}\dot{y}_k &\stackrel{(a)}{=} \arg\max_{y_k}\sum_{r_k}[r_k + V^{*\mu}_{k+1,m_k}(\dot{y}\dot{x}_{<k}yx_k)]\cdot\mu(\delta_{\dot{y}_1\dot{r}_1}...\delta_{\dot{y}_{k-1}\dot{r}_{k-1}}\underline{\delta_{y_k r_k}}) \\ &\stackrel{(b)}{=} \arg\max_{y_k}\sum_{r_k} r_k\cdot\mu(\delta_{\dot{y}_1\dot{r}_1}...\underline{\delta_{y_k r_k}}) \stackrel{(c)}{=} \arg\max_{y_k}\mu(\dot{z}_1...\dot{z}_{k-1}\underline{y}_k) \qquad (6.12) \\ &\stackrel{(d)}{=} \arg\max_{z_k}\mu(\dot{z}_1...\dot{z}_{k-1}\underline{z}_k).\end{aligned}$$

Equation $(a)$ is the definition of the agent's action (6.3), where we have used (6.10), which gives the r.h.s. of (6.11) with $\max_{y_k}$ replaced by $\text{argmax}_{y_k}$. $\sum_r f(...\delta_{yr}...)$ is independent of $y$ for any function $f$, depending on the combination $\delta_{yr}$ only. Therefore, the $\sum_r V^*\mu$ term is independent of $y_k$ because $V^{*\mu}_{k+1,m}$ as well as $\mu$ depend on $\delta_{y_k r_k}$ only. In $(b)$ we can therefore drop this term, as adding a constant to the argument of $\text{argmax}_{y_k}$ does not change the location of the maximum. In $(c)$ we evaluated the $\sum_{r_k}$. Further, if the true reward to $\dot{y}_i$ is $\dot{r}_i$, the true $i^{th}$ bit of the string must be $\dot{z}_i=\delta_{\dot{y}_i\dot{r}_i}$. Equation $(d)$ is just a renaming.

So, the AI$\mu$ model predicts that $z_k$ that has maximal $\mu$-probability, given $\dot{z}_1...\dot{z}_{k-1}$. This prediction is independent of the choice of $m_k$. It is exactly the prediction scheme of the sequence predictor SP$\mu$ with known prior described in Section 3.3. As this model was optimal, AI$\mu$ is optimal too, i.e. it has the minimal number of expected errors (maximal $\mu$-expected reward) as compared to any other sequence prediction scheme.

From this, it is already clear that the value $V^{*\mu}_{km}$ must be closely related to the expected sequence prediction error $E^{\Theta_\mu}_m$ (6.8). In the following we prove that $V^{*\mu}_{1m}=m-E^{\Theta_\mu}_m$. We rewrite $V^{*\mu}_{km}$ in (6.11) as a function of $z_i$ instead of $y_i r_i$, as it is symmetric in $y_i r_i$. Further, we can pull $V^{*\mu}_{k+1,m}$ out of the maximization, as it is independent of $y_k$, similarly as in (6.12). Renaming the bounded variables $y_k$ and $r_k$, we get

$$V^{*\mu}_{km}(z_{<k}) \;=\; \max_{z_k}\mu(z_{<k}\underline{z}_k)+\sum_{z_k}V^{*\mu}_{k+1,m}(z_{1:k})\cdot\mu(z_{<k}\underline{z}_k). \qquad (6.13)$$

Recursively inserting the l.h.s. into the r.h.s. we get

$$V^{*\mu}_{km}(z_{<k}) \;=\; \sum_{i=k}^{m}\sum_{z_{k:i-1}}\max_{z_i}\mu(z_{<k}\underline{z}_{k:i}). \qquad (6.14)$$

This is most easily proven by induction. For $k=m$ we have $V^{*\mu}_{mm}(z_{<m})=\max_{z_m}\mu(z_{<m}\underline{z}_m)$ from (6.13) and (6.2), which equals (6.14). By induction

hypothesis, we assume that (6.14) is true for $k+1$. Inserting this into (6.13) we get

$$V_{km}^{*\mu}(z_{<k}) = \max_{z_k}\mu(z_{<k}\underline{z}_k) + \sum_{z_k}\left[\sum_{i=k+1}^{m}\sum_{z_{k+1:i-1}}\max_{z_i}\mu(z_{1:k}\underline{z}_{k+1:i})\right]\mu(z_{<k}\underline{z}_k)$$

$$= \max_{z_k}\mu(z_{<k}\underline{z}_k) + \sum_{i=k+1}^{m}\sum_{z_{k:i-1}}\max_{z_i}\mu(z_{<k}\underline{z}_{k:i}),$$

which equals (6.14). This was the induction step, and hence (6.14) is proven. By setting $k=1$ and slightly reformulating (6.14), we get the total expected reward in the first $m$ cycles

$$V_{1m}^{*\mu}(\epsilon) = \sum_{i=1}^{m}\sum_{z_{<i}}\mu(\underline{z}_{<i})\max\{\mu(z_{<i}\underline{0}),\mu(z_{<i}\underline{1})\} = m - E_m^{\Theta_\mu}$$

with $E_m^{\Theta_\mu}$ defined in (6.8).

### 6.2.2 Using the AI$\xi$ Model for Sequence Prediction

Now we want to use the universal AI$\xi$ model instead of AI$\mu$ for sequence prediction and try to derive error bounds analogous to (6.9). Like in the AI$\mu$ case, the agent's output $y_k$ in cycle $k$ is interpreted as a prediction for the $k^{th}$ bit $z_k$ of the string under consideration. The reward is $r_k=\delta_{y_k z_k}$, and there are no other observations $o_k=\epsilon$. What makes the analysis more difficult is that $\xi$ is not symmetric in $y_i r_i \leftrightarrow (1-y_i)(1-r_i)$ and (6.10) does not hold for $\xi$. On the other hand, $\xi^{\text{AI}}$ converges to $\mu^{\text{AI}}$ in the limit (6.7), and (6.10) should hold asymptotically for $\xi$ in some sense. So we expect that everything proven for AI$\mu$ holds approximately for AI$\xi$. The AI$\xi$ model should behave similarly to Solomonoff prediction SP$\xi$. In particular, we expect error bounds similar to (6.9). Making this rigorous seems difficult. Some general remarks were made in the last chapter. Note that bounds like (5.15) cannot hold in general, but could be valid for AI$\xi$ in (pseudo)passive environments.

Here we concentrate on the special case of a deterministic computable environment, i.e. the environment is a sequence $\dot{z}=\dot{z}_1\dot{z}_2...$ with $Km(\dot{z}_1...\dot{z}_n)\leq Km(\dot{z})<\infty$, where $Km(z_{1:n})$ is the length of the shortest (possibly nonhalting) program printing a string starting with $z_{1:n}$. Furthermore, we only consider the simplest horizon model $m_k=k$, i.e. greedily maximize only the next reward. This is sufficient for sequence prediction, as the reward of cycle $k$ only depends on output $y_k$ and not on earlier decisions. This choice is in no way sufficient and satisfactory for the full AI$\xi$ model, as *one* single choice of $m_k$ should serve for *all* AI problem classes. So AI$\xi$ should allow good sequence prediction for some universal choice of $m_k$ and not only for $m_k=k$, which definitely does not suffice for more complicated AI problems. The analysis of this general case is a challenge for the future. For $m_k = k$ the AI$\xi$ model (6.5) with $o_i = \epsilon$ and $r_k \in \{0,1\}$ reduces to

$$\dot{y}_k \;=\; \arg\max_{y_k}\sum_{r_k} r_k\cdot\xi(\dot{y}\dot{r}_{<k}\underline{yr}_k) \;=\; \arg\max_{y_k}\xi(\dot{y}\dot{r}_{<k}y_k\underline{1}) \;=\; \arg\max_{y_k}\xi(\dot{\underline{y}}\dot{\underline{r}}_{<k}y_k\underline{1}). \tag{6.15}$$

The environmental response $\dot{r}_k$ is given by $\delta_{\dot{y}_k\dot{z}_k}$; it is 1 for a correct prediction ($\dot{y}_k = \dot{z}_k$) and 0 otherwise. In the following, we want to bound the number of errors this prediction scheme makes. We need the following inequality:

$$\xi(\underline{yr}_1...\underline{yr}_k) \;>\; 2^{-Km(\delta_{y_1r_1}...\delta_{y_kr_k})-O(1)} \tag{6.16}$$

We have to find a short program in the sum (6.6) calculating $r_1...r_k$ from $y_1...y_k$. If we knew $z_i := \delta_{y_ir_i}$ for $1 \leq i \leq k$, a program of size $O(1)$ could calculate $r_1...r_k = \delta_{y_1z_1}...\delta_{y_kz_k}$. So, combining this program with a shortest coding of $z_1...z_k$ leads to a program $q$ of size $\ell(q) = Km(z_1...z_k)+O(1)$ with $q(y_{1:k}) = r_{1:k}$, which proves (6.16).

Let us now assume that we make a wrong prediction in cycle $k$, i.e. $\dot{r}_k = 0$, $\dot{y}_k \neq \dot{z}_k$. The goal is to show that $\dot{\xi}$ defined by

$$\dot{\xi}_k \;:=\; \xi(\dot{\underline{y}}\dot{\underline{r}}_{1:k}) \;=\; \xi(\dot{y}\dot{\underline{r}}_{<k}\dot{y}_k\underline{0}) \;\leq\; \xi(\dot{y}\dot{\underline{r}}_{<k}) - \xi(\dot{y}\dot{\underline{r}}_{<k}\dot{y}_k\underline{1}) \;<\; \dot{\xi}_{k-1} - \alpha.$$

decreases for every wrong prediction, at least by some $\alpha$. The $\leq$ arose from the fact that $\xi$ is only a semimeasure.

$$\begin{aligned}\xi(\dot{\underline{y}}\dot{\underline{r}}_{<k}\dot{y}_k\underline{1}) \;&\geq\; \xi(\dot{\underline{y}}\dot{\underline{r}}_{<k}(1-\dot{y}_k)\underline{1}) \;\stackrel{\times}{\geq}\; 2^{-Km(\delta_{\dot{y}_1\dot{r}_1}...\delta_{(1-\dot{y}_k)1})} \\ &=\; 2^{-Km(\dot{z}_1...\dot{z}_k)} \;>\; 2^{-Km(\dot{z})-O(1)} \;=:\; \alpha\end{aligned}$$

In the first inequality we used the fact that $\dot{y}_k$ maximizes by definition (6.15) the argument, i.e. $1-\dot{y}_k$ has lower $\xi$ probability than $\dot{y}_k$. Bound (6.16) was applied in the second inequality. The equality holds because $\dot{z}_i = \delta_{\dot{y}_i\dot{r}_i}$ and $\delta_{(1-\dot{y}_k)1} = \delta_{\dot{y}_k0} = \delta_{\dot{y}_k\dot{r}_k} = \dot{z}_k$. The last inequality follows from the definition of $\dot{z}$.

We showed that each erroneous prediction reduces $\dot{\xi}$ by at least the $\alpha$ defined above. Together with $\dot{\xi}_0 = 1$ and $\dot{\xi}_k > 0$ for all $k$ this shows that the agent can make at most $1/\alpha$ errors, since otherwise $\dot{\xi}_k$ would become negative. So the number of wrong predictions $E_\infty^{\text{AI}\xi}$ of agent (6.15) is bounded by

$$E_\infty^{\text{AI}\xi} \;<\; \tfrac{1}{\alpha} \;=\; 2^{Km(\dot{z})+O(1)} \;<\; \infty \tag{6.17}$$

for a computable deterministic environment string $\dot{z}_1\dot{z}_2...$. The intuitive interpretation is that each wrong prediction eliminates at least one program $p$ of size $\ell(p) \stackrel{+}{\lesssim} Km(\dot{z})$. The size is smaller than $Km(\dot{z})$, as larger policies could not mislead the agent to a wrong prediction, since there is a program of size $Km(\dot{z})$ making a correct prediction. There are at most $2^{Km(\dot{z})+O(1)}$ such policies, which bounds the total number of errors.

We derived a finite bound for $E_\infty^{\text{AI}\xi}$, but unfortunately, a rather weak one as compared to (6.9). The reason for the strong bound in the SP case was that every error at least halves $\dot{\xi}$ because the sum of the $\text{argmax}_{x_k}$ arguments was bounded by 1. Here we have

$$\xi(\dot{y}_1\dot{r}_1...\dot{y}_{k-1}\dot{r}_{k-1}0\underline{0}) + \xi(\dot{y}_1\dot{r}_1...\dot{y}_{k-1}\dot{r}_{k-1}0\underline{1}) \le 1,$$
$$\xi(\dot{y}_1\dot{r}_1...\dot{y}_{k-1}\dot{r}_{k-1}1\underline{0}) + \xi(\dot{y}_1\dot{r}_1...\dot{y}_{k-1}\dot{r}_{k-1}1\underline{1}) \le 1,$$

but $\text{argmax}_{y_k}$ runs over the right top and right bottom $\xi$, for which no sum criterion holds.

The AI$\xi$ model would not be sufficient for realistic applications if the bound (6.17) were sharp, but we have the strong feeling (but only weak arguments) that better bounds proportional to $Km(\dot{z})$ analogous to (6.9) exist. The technique used above may not be appropriate for achieving this. One argument for a better bound is the formal similarity between $\text{argmax}_{z_k}\xi(\dot{z}_{<k}\underline{z}_k)$ and (6.15), the other is that no example sequence for which (6.15) makes more than $O(Km(\dot{z}))$ errors is known (see Problem 6.2).

## 6.3 Strategic Games (SG)

### 6.3.1 Introduction

A very important class of problems are strategic games (SG). Game theory considers simple games of chance like roulette, combined with strategy like backgammon, up to purely strategic games like chess or checkers or go. In fact, what is subsumed under game theory is so general that it includes not only a huge variety of game types, but can also describe political and economic competitions and coalitions, Darwinism and many more topics. It seems that nearly every AI problem could be brought into the form of a game. Nevertheless, the intention of a game is that several players perform actions with (partial) observable consequences. The goal of each player is to maximize some utility function (e.g. to win the game). The players are assumed to be rational, taking into account all information they posses. The different goals of the players are usually in conflict. For an introduction into game theory, see [FT91, OR94, RN95, NM44].

If we interpret the AI$\mu$ model as one player and the environment models the other rational player *and* the environment provides the reinforcement feedback $r_k$, we see that the agent-environment configuration satisfies all criteria of a game. On the other hand, the AI models can handle more general situations, since they interact optimally with an environment, even if the environment is not a rational player with conflicting goals.

### 6.3.2 Strictly Competitive Strategic Games

In the following, we restrict ourselves to deterministic, strictly competitive strategic[1] games with alternating moves. Player 1 makes move $y_k$ in round $k$, followed by the move $o_k$ of player 2.[2] So a game with $n$ rounds consists of a sequence of alternating moves $y_1o_1y_2o_2...y_no_n$. At the end of the game in cycle $n$ the game or final board situation is evaluated with $V(y_1o_1...y_no_n)$. Player 1 tries to maximize $V$, whereas player 2 tries to minimize $V$. In the simplest case, $V$ is 1 if player 1 won the game, $V=-1$ if player 2 won and $V=0$ for a draw. We assume a fixed game length $n$ independent of the actual move sequence. For games with variable length but maximal possible number of moves $n$, we could add dummy moves and pad the length to $n$. The optimal strategy (Nash equilibrium) of both players is a minimax strategy

$$\dot{o}_k = \arg\min_{o_k}\max_{y_{k+1}}\min_{o_{k+1}}...\max_{y_n}\min_{o_n} V(\dot{y}_1\dot{o}_1...\dot{y}_k o_k...y_n o_n), \tag{6.18}$$

$$\dot{y}_k = \arg\max_{y_k}\min_{o_k}...\max_{y_n}\min_{o_n} V(\dot{y}_1\dot{o}_1...\dot{y}_{k-1}\dot{o}_{k-1}y_k o_k...y_n o_n). \tag{6.19}$$

But note that the minimax strategy is only optimal if both players behave rationally. If, for instance, player 2 has limited capabilities or makes errors and player 1 is able to discover these (through past moves), he could exploit these weaknesses and improve his performance by deviating from the minimax strategy. At least the classical game theory of Nash equilibria does not take into account limited rationality, whereas the AI$\xi$ agent should.

### 6.3.3 Using the AI$\mu$ Model for Game Playing

In the following, we demonstrate the applicability of the AI$\mu$ model to games. The AI$\mu$ model takes the position of player 1. The environment provides the evaluation $V$. For a symmetric situation we could take a second AI$\mu$ model as player 2, but for simplicity we take the environment as the second player and assume that this environmental player behaves according to the minimax strategy (6.18). The environment serves as a perfect player *and* as a teacher, albeit a very crude one, as it tells the agent at the end of the game only whether it won or lost.

The minimax behavior of player 2 can be expressed by a (deterministic) probability distribution $\mu^{\text{SG}}$ as the following:

$$\mu^{\text{SG}}(y_1\underline{o}_1...y_n\underline{o}_n) := \begin{cases} 1 & \text{if } o_k = \arg\min_{o'_k}...\max_{y'_n}\min_{o'_n} V(y_1o_1...y_ko'_k...y'_no'_n)\ \forall k, \\ 0 & \text{otherwise.} \end{cases} \tag{6.20}$$

[1] In game theory, games like chess are often called 'extensive', whereas 'strategic' is reserved for a different kind of game.

[2] We anticipate notationally the later identification of the moves of player 1/2 with the actions/observations in the AI models.

The probability that player 2 makes move $o_k$ is $\mu^{\text{SG}}(\dot{y}_1\dot{o}_1...\dot{y}_k\underline{o}_k)$, which is 1 for $o_k=\dot{o}_k$ as defined in (6.18) and 0 otherwise.

Clearly, the AI$\mu$ system receives no feedback, i.e. $r_1=...=r_{n-1}=0$, until the end of the game, where it should receive positive/negative/neutral feedback on a win/loss/draw, i.e. $r_n=V(...)$. The environmental prior probability is therefore

$$\mu^{\text{AI}}(y\!x_{1:n}) \;=\; \begin{cases} \mu^{\text{SG}}(y_1\underline{o}_1...y_n\underline{o}_n) & \text{if } r_1...r_{n-1}=0 \text{ and } r_n=V(y_1o_1...y_no_n), \\ 0 & \text{otherwise,} \end{cases} \tag{6.21}$$

where $x_i=r_io_i$. If the environment is a minimax player (6.18) plus a crude teacher $V$, i.e. if $\mu^{\text{AI}}$ is the true prior probability, the question now is, what is the behavior $\dot{y}_k^{\text{AI}}$ of the AI$\mu$ agent. It turns out that if we set $m_k=n$ the AI$\mu$ agent is also a minimax player (6.19) and hence optimal:

$$\begin{aligned} \dot{y}_k^{\text{AI}} &= \arg\max_{y_k}\sum_{o_k}...\max_{y_n}\sum_{o_n} V(\dot{y}\!\dot{o}_{<k}y\!o_{k:n})\cdot\mu^{\text{SG}}(\dot{y}\!\dot{o}_{<k}y\!\underline{o}_{k:n}) \\ &= \arg\max_{y_k}\sum_{o_k}...\max_{y_{n-1}}\sum_{o_{n-1}}\max_{y_n}\min_{o_n} V(\dot{y}\!\dot{o}_{<k}y\!o_{k:n})\cdot\mu^{\text{SG}}(\dot{y}\!\dot{o}_{<k}y\!\underline{o}_{k:n-1}) \\ &= \;...\; = \arg\max_{y_k}\min_{o_{k+1}}...\max_{y_n}\min_{o_n} V(\dot{y}\!\dot{o}_{<k}y\!o_{k:n}) \;=\; \dot{y}_k^{\text{SG}} \end{aligned} \tag{6.22}$$

In the first line we inserted $m_k=n$ and (6.21) into the definition (6.4) of $\dot{y}_k^{\text{AI}}$. This removes all sums over the $r_i$. Further, the sum over $o_n$ gives only a contribution for $o_n=\text{argmin}_{o'_n}V(\dot{y}_1\dot{o}_1...y_no'_n)$ by definition (6.20) of $\mu^{\text{SG}}$. Inserting this $o_n$ gives the second line. Effectively, $\mu^{\text{SG}}$ is reduced to a lower number of arguments and the sum over $o_n$ replaced by $\min_{o_n}$. Repeating this procedure for $o_{n-1},...,o_{k+1}$ leads to the last line, which is just the minimax strategy of player 1 defined in (6.19).

Let us now assume that the game under consideration is played $s$ times. The prior probability then is

$$\mu^{\text{AI}}(y\!x_1...y\!x_{sn}) \;=\; \prod_{r=0}^{s-1}\mu_1^{\text{AI}}(y\!x_{rn+1}...y\!x_{(r+1)n}), \tag{6.23}$$

where we have renamed the prior probability (6.21) for one game to $\mu_1^{\text{AI}}$. Equation (6.23) is a special case of a factorizable $\mu$ (defined in Section 4.3.1) with identical factors $\mu_r=\mu_1^{\text{AI}}$ for all $r$ and equal episode lengths $n_{r+1}-n_r=n$. The AI$\mu$ agent (6.23) for repeated game playing (SGR) also implements the minimax strategy

$$\dot{y}_k^{\text{AI}} \;=\; \arg\max_{y_k}\min_{o_k}...\max_{y_{(r+1)n}}\min_{o_{(r+1)n}} V(\dot{y}\!\dot{o}_{rn+1:k-1}...y\!o_{k:(r+1)n}) \tag{6.24}$$

with $r$ such that $rn<k\leq(r+1)n$ and for any choice of $m_k$ as long as the horizon $h_k\geq n$. This can be proven by using (4.27) and (6.22).

### 6.3.4 Games of Variable Length

We have argued that a single game of variable but bounded length can be padded to a fixed length without effect. We now analyze in a sequence of games the effect of replacing the games with fixed length by games of variable length. The sequence $y_1o_1...y_no_n$ can still be grouped into episodes corresponding to the moves of separated consecutive games, but now the length and total number of games that fit into the $n$ moves depend on the actual moves taken. (If the sum of game lengths does not fit exactly into $n$ moves, we pad the last game appropriately.) $V(y_1o_1...y_no_n)$ equals the number of games where the agent wins minus the number of games where the environment wins. Whenever a loss, win or draw is achieved by the agent or the environment, a new game starts. The player whose turn it would next be, begins the next game. The games are still separated in the sense that the behavior and reward of the current game does not influence the next game. On the other hand, they are slightly entangled, because the length of the current game determines the time of start of the next. As the rules of the game are time invariant, this does not influence the next game directly. If we play a fixed number of games, the games are completely independent, but if we play a fixed number of total moves $n$, the number of games depends on their lengths. This has the following consequences: the better player tries to keep the games short, to win more games in the given time $n$. The poorer player tries to draw the games out, in order to lose fewer games. The better player might further prefer a quick draw, rather than to win a long game. Formally, this entanglement is represented by the fact that the prior probability $\mu$ no longer factorizes. The reduced form (6.24) of $\dot{y}_k^{\text{AI}}$ to one episode is no longer valid. Also, the behavior $\dot{y}_k^{\text{AI}}$ of the agent depends on $m_k$, even if the horizon $h_k$ is chosen larger than the longest possible game. The important point is that the agent realizes that keeping games short/long can lead to increased reward. In practice, a horizon much larger than the average game length should be sufficient to incorporate this effect. The details of games in the distant future do not affect the current game and can, therefore, be ignored. A more quantitative analysis could be interesting, but would lead us too far astray.

### 6.3.5 Using the AI$\xi$ Model for Game Playing

When going from the specific AI$\mu$ model, where the rules of the game are explicitly modeled into the prior probability $\mu^{\text{AI}}$, to the universal model AI$\xi$, we have to ask whether these rules can be learned from the assigned rewards $r_k$. Here, the main reason for studying the case of repeated games (SGR) rather than just one game arises. For a single game there is only one cycle of nontrivial feedback, namely the end of the game, which is too late to be useful except when further games follow.

Even in the case of repeated games, there is only very limited feedback, at most $\log_2 3$ bits of information per game if the 3 outcomes win/loss/draw have the same frequency. So there are at least $O(K(game))$ number of games necessary to learn a game of complexity $K(game)$. Apart from extremely simple games, even this estimate is far too optimistic. As the AI$\xi$ agent has no information about the game to begin with, its moves will be more or less random, and it can win the first few games merely by pure luck. So the probability that the agent loses is near 1, and hence the information content $I$ in the feedback $r_k$ at the end of the game is much less than $\log_2 3$. This situation remains for a very large number of games. But, in principle, every game should be learnable after a very long sequence of games even with only this minimal feedback, as long as $I \not\equiv 0$.

The important point is that no other learning scheme with no extra information can learn the game more quickly than AI$\xi$. We expect this to be true as $\mu^{\text{AI}}$ factorizes in the case of games of fixed length, i.e. $\mu^{\text{AI}}$ satisfies a strong separability condition. In the case of variable game length the entanglement is also low. $\mu^{\text{AI}}$ should still be sufficiently separable, allowing us to formulate and prove good reward bounds for AI$\xi$. Indeed, the situation is significantly *better* for games of variable length. Since initially, AI$\xi$ loses all games, it tries to draw out a loss as long as possible, without having ever experienced or even knowing what it means to win. Initially, AI$\xi$ will make a lot of illegal moves. Since each illegal move will immediately abort the game resulting in (non-delayed) negative reward (loss), AI$\xi$ can quickly learn the typically simple rules concerning legal moves, which usually constitute most of the rules; just the goal rule is missing. After having learned the move-rules, AI$\xi$ learns the (negatively rewarded) losing positions, the positions leading to losing positions, etc., so it can try to draw out losing games. For instance, in chess, avoiding being check mated for 20, 30, 40 moves against a master is already quite an achievement. At this ability stage, AI$\xi$ should be able to win some games by luck, or speculate about a symmetry in the game that check mating the opponent will be positively rewarded. Once having found out the complete rules (moves and goal), AI$\xi$ will right away reason that playing minimax is best, and henceforth beat all grandmasters.

If a (complex) game cannot be learned in this way in a realistic number of cycles, one has to provide more feedback. This could be achieved by intermediate help during the game. The environment could give positive (negative) feedback for every good (bad) move the agent makes. The demand on whether a move is to be valuated as good should be adapted to the gained experience of the agent in such a way that approximately the better half of the moves are valuated as good and the other half as bad, in order to maximize the information content of the feedback.

For more complicated games like chess, even more feedback may be necessary from a practical point of view. One way to increase the feedback far beyond a few bits per cycle is to train the agent by teaching it good moves. This is called supervised learning. Despite the fact that the AI$\mu$ model has only a reward feedback $r_k$, it is able to learn supervised, as will be shown in Section 6.5. Another way would be to start with more simple games containing certain aspects of the true game and to switch to the true game when the agent has learned the simple game.

No other difficulties are expected when going from $\mu$ to $\xi$. Eventually $\xi^{\text{AI}}$ will converge to the minimax strategy $\mu^{\text{AI}}$. In the more realistic case, where the environment is not a perfect minimax player, AI$\xi$ can detect and exploit the weakness of the opponent.

Finally, we want to comment on the input/output space $\mathcal{X}/\mathcal{Y}$ of the AI models. In practical applications, $\mathcal{Y}$ will possibly include also illegal moves. If $\mathcal{Y}$ is the set of moves of, e.g. a robotic arm, the agent could move a wrong figure or even knock over the figures. A simple way to handle illegal moves $y_k$ is by interpreting them as losing moves, which terminate the game. Further, if, e.g. the input $x_k$ is the image of a video camera which makes one shot per move, $\mathcal{X}$ is not the set of moves by the environment but includes the set of states of the game board. The discussion in this section handles this case as well. There is no need to explicitly design the systems I/O space $\mathcal{X}/\mathcal{Y}$ for a specific game.

The discussion above on the AI$\xi$ agent was rather informal for the following reason: game playing (the SG$\xi$ agent) has (nearly) the same complexity as fully general AI, and quantitative results for the AI$\xi$ agent are difficult (but not impossible) to obtain.

## 6.4 Function Minimization (FM)

### 6.4.1 Applications/Examples

There are many problems that can be reduced to function minimization (FM) problems. The minimum of a (real-valued) function $f:\mathcal{Y}\to I\!\!R$ over some domain $\mathcal{Y}$ or a good approximate to the minimum has to be found, usually with some limited resources.

One popular example is the traveling salesman problem (TSP). $\mathcal{Y}$ is the set of different routes between towns, and $f(y)$ the length of route $y\in\mathcal{Y}$. The task is to find a route of minimal length visiting all cities. This problem is NP hard. Getting good approximations in limited time is of great importance in various applications. Another example is the minimization of production costs (MPC), e.g. of a car, under several constraints. $\mathcal{Y}$ is the set of all alternative car designs and production methods compatible with the specifications, and $f(y)$ is the overall cost of alternative $y\in\mathcal{Y}$. A related example is finding materials or (bio)molecules with certain properties (MAT), e.g. solids with minimal electrical resistance or maximally efficient chlorophyll modifications, or aromatic molecules that taste as close as possible to strawberry. We can also ask for nice paintings (NPT). $\mathcal{Y}$ is the set of all existing or imaginable paintings, and $f(y)$ characterizes how much person $A$ likes painting $y$. The agent should present paintings which $A$ likes.

For now, these are enough examples. The TSP is very rigorous from a mathematical point of view, as $f$, i.e. an algorithm of $f$, is usually known. In principle, the minimum could be found by exhaustive search, were it not for computational resource limitations. For MPC, $f$ can often be modeled in a reliable and sufficiently accurate way. For MAT you need very accurate physical models, which might be unavailable or too difficult to solve or implement. For NPT all we have is the judgement of person $A$ on every presented painting. The evaluation function $f$ cannot be implemented without scanning $A$'s brain, which is not possible with today's technology.

So there are different limitations, some depending on the application we have in mind. An implementation of $f$ might not be available, $f$ can only be tested at some arguments $y$ and $f(y)$ is determined by the environment. We want to (approximately) minimize $f$ with as few function calls as possible or, conversely, find an as close as possible approximation for the minimum within a fixed number of function evaluations. If $f$ is available or can quickly be inferred by the agent and evaluation is quick, it is more important to minimize the total time needed to imagine new trial minimum candidates plus the evaluation time for $f$. As we do not consider computational aspects of AI$\xi$ till Section 7.2, we concentrate on the first case, where $f$ is not available or dominates the computational requirements.

### 6.4.2 The Greedy Model FMG$\mu$

The FM model consists of a sequence $\dot{y}_1\dot{z}_1\dot{y}_2\dot{z}_2...$ where $\dot{y}_k$ is a trial of the FM agent for a minimum of $f$, and $\dot{z}_k=f(\dot{y}_k)$ is the true function value returned by the environment. We randomize the model by assuming a probability distribution $\mu(f)$ over the functions. There are several reasons for doing this. We might really not know the exact function $f$, as in the NPT example, and model our uncertainty by the probability distribution $\mu$. What is more important, we want to parallel the other AI classes, like in the SP$\mu$ model, where we always started with a probability distribution $\mu$ that was finally replaced by $\xi$ to get the universal Solomonoff prediction SP$\xi$. We want to do the same thing here. Further, the probabilistic case includes the deterministic case by choosing $\mu(f)=\delta_{ff_0}$, where $f_0$ is the true function. A final reason is that the deterministic case is trivial when $\mu$ and hence $f_0$ are known, as the agent can internally (virtually) check all function arguments and output the correct minimum from the very beginning.

We will assume that $\mathcal{Y}$ is countable or finite and that $\mu$ is a discrete measure, e.g. by taking only computable functions. The probability that the function values of $y_1,...,y_n$ are $z_1,...,z_n$ is then given by

$$\mu^{\text{FM}}(y_1\underline{z}_1...y_n\underline{z}_n) \;:=\; \sum_{f:f(y_i)=z_i\;\forall 1\leq i\leq n} \mu(f). \tag{6.25}$$

We start with a model that minimizes the expectation $z_k$ of the function value $f$ for the next output $y_k$, taking into account previous information:

$$\dot{y}_k := \arg\min_{y_k} \sum_{z_k} z_k \cdot \mu(\dot{y}_1\dot{z}_1...\dot{y}_{k-1}\dot{z}_{k-1}y_k\underline{z}_k).$$

This type of greedy algorithm, just minimizing the next feedback, was sufficient for sequence prediction (SP) and is also sufficient for classification (CF). It is, however, not sufficient for function minimization, as the following example demonstrates.

Take $f:\{0,1\}\to\{1,2,3,4\}$. There are 16 different functions which shall be equiprobable, $\mu(f)=\frac{1}{16}$. The function expectation in the first cycle

$$\langle z_1\rangle := \sum_{z_1} z_1 \cdot \mu(y_1\underline{z}_1) = \tfrac{1}{4}\sum_{z_1} z_1 = \tfrac{1}{4}(1+2+3+4) = 2.5$$

is just the arithmetic average of the possible function values and is independent of $y_1$. Therefore, $\dot{y}_1=0$, if we defined argmin to take the lexicographically first minimum in an ambiguous case like here. Let us assume that $f_0(0)=2$, where $f_0$ is the true environment function, i.e. $\dot{z}_1=2$. The expectation of $z_2$ is then

$$\langle z_2\rangle := \sum_{z_2} z_2 \cdot \mu(02y_2\underline{z}_2) = \begin{cases} 2 & \text{for } y_2=0, \\ 2.5 & \text{for } y_2=1. \end{cases}$$

For $y_2=0$ the agent already knows $f(0)=2$; for $y_2=1$ the expectation is, again, the arithmetic average. The agent will again output $\dot{y}_2=0$ with feedback $\dot{z}_2=2$. This will continue forever. The agent is not motivated to explore other $y$'s as $f(0)$ is already smaller than the expectation of $f(1)$. This is obviously not what we want. The greedy model fails. The agent ought to be inventive and try other outputs when given enough time.

The general reason for the failure of the greedy approach is that the information contained in the feedback $z_k$ depends on the output $y_k$. A FM agent can actively influence the knowledge it receives from the environment by the choice in $y_k$. It may be more advantageous to first collect certain knowledge about $f$ by an (in greedy sense) nonoptimal choice for $y_k$, rather than to minimize the $z_k$ expectation immediately. The nonminimality of $z_k$ might be overcompensated in the long run by exploiting this knowledge. In SP, the received information is always the current bit of the sequence, independent of what SP predicts for this bit. This is why a greedy strategy in the SP case is already optimal.

### 6.4.3 The General FM$\mu/\xi$ Model

To get a useful model we have to think more carefully about what we really want. Should the FM agent output a good minimum in the last output in a limited number of cycles $m$, or should the average of the $z_1,...,z_m$ values be minimal, or does it suffice that just one of the $z$ is as small as possible? Let us define the FM$\mu$ model as to minimize the $\mu$-averaged weighted sum $\alpha_1 z_1+...+\alpha_m z_m$ for some given $\alpha_k\geq 0$. Building the $\mu$ average by summation over the $z_i$ and minimizing w.r.t. the $y_i$ has to be performed in the correct chronological order. With a similar reasoning as in (6.1) to (6.4) we get

$$\dot{y}_k^{\text{FM}} = \arg\min_{y_k}\sum_{z_k}...\min_{y_m}\sum_{z_m}(\alpha_1 z_1+...+\alpha_m z_m)\cdot\mu(\dot{y}_1\dot{z}_1...\dot{y}_{k-1}\dot{z}_{k-1}y_k\underline{z}_k...y_m\underline{z}_m) \tag{6.26}$$

If we want the final output $\dot{y}_m$ to be optimal we should choose $\alpha_k=0$ for $k<m$ and $\alpha_m=1$ (final model FMF$\mu$). If we want to already have a good approximation during intermediate cycles, we should demand that the output of all cycles together be optimal in some average sense, so we should choose $\alpha_k=1$ for all $k$ (sum model FMS$\mu$). If we want to have something in between, for instance, increase the pressure to produce good outputs, we could choose the $\alpha_k=\mathrm{e}^{\gamma(k-m)}$ exponentially increasing for some $\gamma>0$ (exponential model FME$\mu$). For $\gamma\to\infty$ we get the FMF$\mu$; for $\gamma\to 0$ the FMS$\mu$ model. If we want to demand that the best of the outputs $y_1...y_k$ is optimal, we must replace the $\alpha$-weighted $z$-sum by $\min\{z_1,...,z_m\}$ (minimum model FMM$\mu$). We expect the behavior to be very similar to the FMF$\mu$ model, and do not consider it further (see Section 8.5.1 item 4).

By construction, the FM$\mu$ models guarantee optimal results in the usual sense that no other model knowing only $\mu$ can be expected to produce better results. The variety of FM variants is not a fault of the theory. They just reflect the fact that there is some interpretational freedom of what is meant by minimization within $m$ function calls. In most applications, probably FMF is appropriate. In the NPT application one might prefer the FMS model.

The interesting case (in AI) is when $\mu$ is unknown. For this case we define the FM$\xi$ model by replacing $\mu(f)$ with some $\xi(f)$, which should assign high probability to functions $f$ of low complexity. So we might define $\xi(f)=\sum_{q:\forall x[U(qx)=f(x)]}2^{-\ell(q)}$. The problem with this definition is that it is, in general, undecidable whether a TM $q$ is an implementation of a function $f$. $\xi(f)$ defined in this way is uncomputable, and not even approximable. As we only need a $\xi$ analogous to the l.h.s. of (6.25), the following definition is natural

$$\xi^{\text{FM}}(y_1\underline{z}_1...y_n\underline{z}_n) \;:=\; \sum_{q:q(y_i)=z_i\;\forall 1\leq i\leq n} 2^{-\ell(q)} \tag{6.27}$$

$\xi^{\text{FM}}$ is actually equivalent to inserting the uncomputable $\xi(f)$ into (6.25). $\xi^{\text{FM}}$ is an enumerable semimeasure and dominates all enumerable probability distributions of the form (6.25). We will not prove this here.

Alternatively, we could have constrained the sum in (6.27) by $q(y_1...y_n)= z_1...z_n$ analogous to (6.6), but these two definitions are not equivalent. Definition (6.27) ensures the symmetry[3] in its arguments, and $\xi^{\text{FM}}(...y\underline{z}...y\underline{z}'...)=0$ for $z\neq z'$. It incorporates all general knowledge we have about function minimization, whereas (6.6) does not. But this extra knowledge has only low

[3] See [Sol99] for a discussion on symmetric universal distributions on unordered data.

information content (complexity of $O(1)$), so we do not expect FM$\xi$ to perform much worse when using (6.6) instead of (6.27). But there is no reason to deviate from (6.27) at this point.

We can now define an "error" measure $E_m^{\text{FM}\mu}$ as (6.26) with $k=1$ and $\text{argmin}_{y_1}$ replaced by $\min_{y_1}$ and, additionally, $\mu$ replaced by $\xi$ for $E_m^{\text{FM}\xi}$. We expect $|E_m^{\text{FM}\xi}-E_m^{\text{FM}\mu}|$ to be bounded in a way that justifies the use of $\xi$ instead of $\mu$ for computable $\mu$, i.e. computable $f_0$ in the deterministic case. The arguments are the same as for the AI$\xi$ model.

### 6.4.4 Is the General Model Inventive?

In the following we will show that FM$\xi$ will never cease searching for minima, but will test an infinite set of different $y$'s for $m\to\infty$.

Let us assume that the agent tests only a finite number of $y_i\in\mathcal{A}\subset\mathcal{Y}$, $|\mathcal{A}|<\infty$. Let $t-1$ be the cycle in which the last new $y\in\mathcal{A}$ is selected (or some later cycle). Selecting $y$'s in cycles $k\geq t$ a second time, the feedback $z$ does not provide any new information, i.e. does not modify the probability $\xi^{\text{FM}}$. The agent can minimize $E_m^{\text{FM}\xi}$ by outputting in cycles $k\geq t$ the best $y\in\mathcal{A}$ found so far (in the case $\alpha_k=0$, the output does not matter). Let us fix $f$ for a moment. Then we have

$$E^a \;:=\; \alpha_1 z_1+...+\alpha_m z_m \;=\; \sum_{k=1}^{t-1}\alpha_k f(y_k)+f_1\cdot\sum_{k=t}^{m}\alpha_k, \qquad f_1:=\min_{1\leq k<t} f(y_k).$$

Let us now modify the agent and assume that it tests one additional $y_t\notin\mathcal{A}$ in cycle $t$, but no other $y\notin\mathcal{A}$. Again, it will keep to the best output for $k>t$, which is either the one of the previous agent or $y_t$.

$$E^b \;=\; \sum_{k=1}^{t}\alpha_k f(y_k)+\min\{f_1,f(y_t)\}\cdot\sum_{k=t+1}^{m}\alpha_k.$$

The difference can be represented in the form

$$E^a-E^b \;=\; \left(\sum_{k=t}^{m}\alpha_k\right)\cdot f^+-\alpha_t\cdot f^-, \qquad f^\pm \;:=\; \max\{0,\pm(f_1-f(y_t))\}\;\geq\;0.$$

As the true FM strategy is the one that minimizes $E$, assumption $a$ is ruled out if $E^a>E^b$. We then say that $b$ is favored over $a$, which does not mean that $b$ is the correct strategy, only that $a$ is not the true one. For probability distributed $f$, $b$ is favored over $a$ when

$$E^a-E^b \;=\; \left(\sum_{k=t}^{m}\alpha_k\right)\cdot\langle f^+\rangle-\alpha_t\cdot\langle f^-\rangle \;>\; 0 \quad\Leftrightarrow\quad \sum_{k=t}^{m}\alpha_k>\alpha_t\frac{\langle f^-\rangle}{\langle f^+\rangle},$$

where $\langle f^\pm\rangle$ is the $\xi$ expectation of $\pm(f_1-f(y_t))$ under the condition that $\pm f_1\geq\pm f(y_t)$ and under the constraints imposed in cycles $1...t-1$. As $\xi$ assigns a strictly positive probability to every nonempty event, $\langle f^+\rangle \neq 0$. Inserting $\alpha_k = e^{\gamma(k-m)}$, assumption $a$ is ruled out in model FME$\xi$ if

$$m-t > \frac{1}{\gamma}\ln\left[1+\frac{\langle f^-\rangle}{\langle f^+\rangle}(e^\gamma - 1)\right] - 1 \rightarrow \begin{cases} 0 & \text{for } \gamma\to\infty \text{ (FMF}\xi) \\ \langle f^-\rangle/\langle f^+\rangle - 1 & \text{for } \gamma\to 0 \text{ (FMS}\xi) \end{cases}$$

We see that if the condition is not satisfied for some $t$, it will remain wrong for all $t'>t$. So the FME$\xi$ agent will test each $y$ only once up to a point from which on it always outputs the best found $y$. Further, for $m\to\infty$ the condition always gets satisfied. As this is true for any finite $\mathcal{A}$, the assumption of a finite $\mathcal{A}$ is wrong. For $m\to\infty$ the agent tests an increasing number of different $y$'s, provided $\mathcal{Y}$ is infinite. The FMF$\xi$ model will never repeat any $y$ except in the last cycle $m$, where it chooses the best found $y$. The FMS$\xi$ model will test a new $y_t$ for fixed $m$, only if the expected value of $f(y_t)$ is not too large.

The above does not necessarily hold for other choices of $\alpha_k$. The above also holds for the FMF$\mu$ agent if $\langle f^+\rangle \neq 0$. $\langle f^+\rangle = 0$ if the agent can already exclude that $y_t$ is a better guess, so there is no reason to test it explicitly.

Nothing has been said about the quality of the guesses, but for the FM$\mu$ agent they are optimal by definition. If $K(\mu)$ for the true distribution $\mu$ is finite, we expect the FM$\xi$ agent to solve the 'exploration versus exploitation' problem in a universally optimal way, as $\xi$ converges to $\mu$.

### 6.4.5 Using the AI Models for Function Minimization

The AI$\mu$ model can be used for function minimization in the following way: The output $y_k$ of cycle $k$ is a guess for a minimum of $f$, like in the FM model. The reward $r_k$ should be high for small function values $z_k = f(y_k)$. The reward should also be weighted with $\alpha_k$ to reflect the same strategy as in the FM case. The choice of $r_k = -\alpha_k z_k$ is natural. Here, the feedback is not binary but $r_k \in \mathcal{R} \subset I\!R$, with $\mathcal{R}$ being a countable subset of $I\!R$, e.g. the computable reals or all rational numbers. The feedback $o_k$ should be the function value $f(y_k)$. So we set $o_k = z_k$. Note, that there is a redundancy if $\alpha_{()}$ is a computable function with no zeros, as $r_k = -\alpha_k o_k$. So, for small $K(\alpha_{()})$ like in the FMS model, one might set $x_k \equiv \epsilon$. If we keep $o_k$ the AI prior probability is

$$\mu^{\text{AI}}(y_1\underline{x}_1...y_n\underline{x}_n) = \begin{cases} \mu^{\text{FM}}(y_1\underline{z}_1...y_n\underline{z}_n) & \text{for } r_k = -\alpha_k z_k,\ o_k = z_k,\ x_k = r_k o_k \\ 0 & \text{else.} \end{cases} \tag{6.28}$$

Inserting this into (6.4) with $m_k = m$ we get

$$\dot y_k^{\text{AI}} = \arg\max_{y_k}\sum_{x_k}...\max_{y_m}\sum_{x_m}(r_k+ ... +r_m)\cdot\mu^{\text{AI}}(\dot y_1\dot x_1...y_k\underline{x}_k...y_m\underline{x}_m)$$

$$= \arg\min_{y_k}\sum_{z_k}...\min_{y_m}\sum_{z_m}(\alpha_k z_k+ ... +\alpha_m z_m)\cdot\mu^{\text{FM}}(\dot y_1\dot z_1...y_k\underline{z}_k...y_m\underline{z}_m) = \dot y_k^{\text{FM}}$$

where $\dot{y}_k^{\mathrm{FM}}$ was defined in (6.26). The proof of equivalence was so simple because the FM model already has a rather general structure, which is similar to the full AI$\mu$ model.

One might expect no problems when going from the already very general FM$\xi$ model to the universal AI$\xi$ model (with $m_k = m$), but there is a pitfall in the case of the FMF model. All rewards $r_k$ are zero in this case, except for the last one, which is $r_m$. Although there is a feedback $z_k$ in every cycle, the AI$\xi$ agent cannot learn from this feedback, as it is not told that in the final cycle $r_m$ will equal to $-z_m$. There is no problem in the FM$\xi$ model because in this case this knowledge is hardcoded into $\xi^{\mathrm{FM}}$. The AI$\xi$ model must first learn that it has to minimize a function, but it can only learn if there is a nontrivial credit assignment $r_k$. FMF works for repeated minimization of (different) functions, such as minimizing $N$ functions in $N \cdot m$ cycles. In this case there are $N$ nontrivial feedbacks, and AI$\xi$ has time to learn that there is a relation between $r_{k \cdot m}$ and $o_{k \cdot m}$ every $m^{th}$ cycle. This situation is similar to the case of (repeated) strategic games discussed in Section 6.3.

There is no problem in applying AI$\xi$ to FMS because the $r$ feedback provides enough information in this case. The only thing the AI$\xi$ model has to learn is to ignore the $o$ feedbacks since all information is already contained in $r$. Interestingly the same argument holds for the FME model if $K(\gamma)$ and $K(m)$ are small.[4] The AI$\xi$ model has additionally only to learn the relation $r_k = -\mathrm{e}^{\gamma(k-m)} o_k$. This task is simple, as every cycle provides one data point for a simple function to learn. This argument is no longer valid for $\gamma \to \infty$ since $K(\gamma) \to \infty$ in this case.

### 6.4.6 Remark on TSP

The traveling salesman problem (TSP) seems to be trivial in the AI$\mu$ model but nontrivial in the AI$\xi$ model, because (6.26) just implements an internal complete search, as $\mu(f) = \delta_{f f^{\mathrm{TSP}}}$ contains all necessary information. AI$\mu$ outputs, from the very beginning, the exact minimum of $f^{\mathrm{TSP}}$. This "solution" is, of course, unacceptable from a performance perspective. As long as we give no efficient approximation $\xi^c$ of $\xi$, we have not contributed anything to a solution of the TSP by using AI$\xi^c$. The same is true for any other problem where $f$ is computable and easily accessible. Therefore, TSP is not (yet) a good example because all we have done is to replace an NP complete problem with the uncomputable AI$\xi$ model or by a computable AI$\xi^c$ model, for which we have said nothing about computation time yet. It is simply overkill to reduce "easy" problems to AI$\xi$. TSP is a simple problem in this respect, until we consider the AI$\xi^c$ model seriously. For the other examples, where $f$ is inaccessible or complicated, an AI$\xi^c$ model would provide a true solution to the minimization problem, since an explicit definition of $f$ is not needed for AI$\xi$ and AI$\xi^c$. A computable version of AI$\xi$ will be defined in Section 7.2.

---

[4] Setting $\alpha_k = \mathrm{e}^{\gamma k}$ we see that the condition on $K(m)$ can be dropped.

## 6.5 Supervised Learning from Examples (EX)

The developed AI models provide a frame for reinforcement learning. The environment provides feedback $r$, informing the agent about the quality of its last (or earlier) output $y$; it assigns reward $r$ to output $y$. In this sense, reinforcement learning is explicitly integrated into the AI$\rho$ model. AI$\mu$ maximizes the true expected reward, whereas the AI$\xi$ model is a universal, environment-independent reinforcement learning algorithm.

There is another type of learning method: supervised learning by presentation of examples (EX). Many problems learned by this method are association problems of the following type. Given some examples $o \in R \subset \mathcal{O}$, the agent should reconstruct, from a partially given $o'$, the missing or corrupted parts, i.e. complete $o'$ to $o$ such that relation $R$ contains $o$. In many cases, $\mathcal{O}$ consists of pairs $(z,v)$, where $v$ is the possibly missing part.

### 6.5.1 Applications/Examples

Learning functions by presenting $(z,f(z))$ pairs and asking for the function value of $z$ by presenting $(z,?)$ falls into the category of supervised learning from examples, e.g. $f(z)$ may be the class label or category of $z$.

A basic example is learning properties of geometrical objects coded in some way. For instance, if there are 18 different objects characterized by their size (small or big), their colors (red, green, or blue) and their shapes (square, triangle, or circle), then $(object,property) \in R$ if the *object* possesses the *property*. Here, $R$ is a relation that is not the graph of a single-valued function.

When teaching a child by pointing to objects and saying "this is a tree" or "look how green" or "how beautiful", one establishes a relation of $(object,property)$ pairs in $R$. Pointing to a (possibly different) tree later and asking "what is this ?" corresponds to a partially given pair $(object,?)$, where the missing part "?" should be completed by the child saying "tree".

A final example we want to give is chess. We have seen that, in principle, chess can be learned by reinforcement learning. In the extreme case the environment only provides reward $r=1$ when the agent wins. The learning rate is completely inacceptable from a practical point of view, due to the very low amount of information feedback. A more practical method of teaching chess is to present example games in the form of sensible (*board-state,move*) sequences. They contain information about legal and good moves (but without any explanation). After several games have been presented, the teacher could ask the agent to make its own move by presenting (*board-state*,?) and then evaluate the answer of the agent.

### 6.5.2 Supervised Learning with the AI$\mu/\xi$ Model

Let us define the EX model as follows: The environment presents inputs $o_{k-1} = z_k v_k \equiv (z_k,v_k) \in R \cup (\mathcal{Z} \times \{?\}) \subset \mathcal{Z} \times (\mathcal{Y} \cup \{?\}) = \mathcal{O}$ to the agent in cycle $k-1$.

The agent is expected to output $y_k$ in the next cycle, which is evaluated with $r_k=1$ if $(z_k,y_k)\in R$ and 0 otherwise. To simplify the discussion, an output $y_k$ is expected and evaluated even when $v_k(\neq ?)$ is given. To complete the description of the environment, the probability distribution $\mu_R(\underline{o_1...o_n})$ of the examples and questions $o_i$ (depending on $R$) has to be given. Wrong examples should not occur, i.e. $\mu_R$ should be 0 if $o_i\notin R\cup(\mathcal{Z}\times\{?\})$ for some $1\leq i\leq n$. The relations $R$ might also be probability distributed with $\sigma(\underline{R})$. The example prior probability in this case is

$$\mu(\underline{o_1...o_n}) \;=\; \sum_R \mu_R(\underline{o_1...o_n})\cdot\sigma(\underline{R}). \tag{6.29}$$

The knowledge of the valuation $r_k$ on output $y_k$ restricts the possible relations $R$, consistent with $R(z_k,y_k)=r_k$, where $R(z,y):=1$ if $(z,y)\in R$ and 0 otherwise. The prior probability for the input sequence $x_1...x_n$ if the output sequence of AI$\mu$ is $y_1...y_n$, is therefore

$$\mu^{\text{AI}}(y_1\underline{x}_1...y_n\underline{x}_n) \;=\; \sum_{R:\forall 1<i\leq n[R(z_i,y_i)=r_i]} \mu_R(\underline{o_1...o_n})\cdot\sigma(\underline{R}),$$

where $x_i=r_io_i$ and $o_{i-1}=z_iv_i$ with $v_i\in\mathcal{Y}\cup\{?\}$. In the I/O sequence $y_1x_1y_2x_2...=y_1r_1z_2v_2y_2r_2z_3v_3...$ the $y_1r_1$ are dummies, after which regular behavior starts with example $(z_2,v_2)$.

The AI$\mu$ model is optimal by construction of $\mu^{\text{AI}}$. For computable prior $\mu_R$ and $\sigma$, we expect a near-optimal behavior of the universal AI$\xi$ model if $\mu_R$ additionally satisfies some separability property. In the following, we give some motivation why the AI$\xi$ model takes into account the supervisor information contained in the examples and why it learns faster than by reinforcement.

We keep $R$ fixed and assume $\mu_R(o_1...o_n)=\mu_R(o_1)\cdot...\cdot\mu_R(o_n)\neq 0\Leftrightarrow o_i\in R\cup(\mathcal{Z}\times\{?\})$ $\forall i$ to simplify the discussion. Short codes $q$ contribute most to $\xi^{\text{AI}}(y_1\underline{x}_1...y_n\underline{x}_n)$. As $o_1...o_n$ is distributed according to the computable probability distribution $\mu_R$, a short code of $o_1...o_n$ for large enough $n$ is a Huffman code w.r.t. the distribution $\mu_R$. So we expect $\mu_R$ and hence $R$ to be coded in the dominant contributions to $\xi^{\text{AI}}$ in some way, where the plausible assumption was made that the $y$ on the input tape do not matter. Much more than one bit per cycle will usually be learned, i.e. relation $R$ will be learned in $n\ll K(R)$ cycles by appropriate examples. This coding of $R$ in $q$ evolves independently of the feedbacks $r$. To maximize the feedback $r_k$, the agent has to learn to output a $y_k$ with $(z_k,y_k)\in R$. The agent has to invent a program extension $q'$ to $q$, which extracts $z_k$ from $o_{k-1}=(z_k,?)$ and searches for and outputs a $y_k$ with $(z_k,y_k)\in R$. As $R$ is already coded in $q$, $q'$ can reuse this coding of $R$ in $q$. The size of the extension $q'$ is, therefore, of order 1. To learn this $q'$, the agent requires feedback $r$ with information content $O(1)=K(q')$ only.

Let us compare this with reinforcement learning, where only $o_{k-1}=(z_k,?)$ pairs are presented. A coding of $R$ in a short code $q$ for $o_1...o_n$ is of no use and will therefore be absent. Only the rewards $r$ force the agent to learn $R$. $q'$ is therefore expected to be of size $K(R)$. The information content in the $r$'s must be of the order $K(R)$. In practice, there are often only very few $r_k = 1$ at the beginning of the learning phase, and the information content in $r_1...r_n$ is much less than $n$ bits. The required number of cycles to learn $R$ by reinforcement is, therefore, at least but in many cases much larger than $K(R)$.

Finally, consider a slightly easier setup, where after each prediction $y$, we inform the agent about the correct function value. This setup now coincides with a (partial) sequential classification task, which can be reduced to sequence prediction, as described in Section 3.7.3. Hence, the results of Chapter 3 apply, showing that the agent EX$\xi$, which is based on a mixture over $\mu$ of the restricted form (6.29), performs excellent. This is another indication that AI$\xi$ should also learn supervised well.

Although AI$\xi$ was never designed or told to learn supervised, it learns how to take advantage of the examples from the supervisor. $\mu_R$ and $R$ are learned from the examples; the rewards $r$ are not necessary for this process. The remaining task of learning how to learn supervised is then a simple task of complexity $O(1)$, for which the rewards $r$ are necessary.

## 6.6 Other Aspects of Intelligence

In AI, a variety of general ideas and methods have been developed. In the last sections, we saw how several problem classes can be formulated within AI$\xi$. As we claim universality of the AI$\xi$ model, we want to illuminate which of and how the other AI methods are incorporated in the AI$\xi$ model by looking at its structure. Some methods are directly included, while others are or should be emergent. We do not claim the following list to be complete.

*Probability theory* and *utility theory* are the heart of the AI$\mu/\xi$ models. The probability $\xi$ is a universal belief about the true environmental behavior $\mu$. The utility function is the total expected reward, called value, which should be maximized. Maximization of an expected utility function in a probabilistic environment is usually called *sequential decision theory*, and is explicitly integrated in full generality in our model. In a sense this includes probabilistic (a generalization of deterministic) *reasoning*, where the objects of reasoning are not true and false statements, but the prediction of the environmental behavior. *Reinforcement learning* is explicitly built in, due to the rewards. Supervised learning is an emergent phenomenon (Section 6.5). *Algorithmic information theory* leads us to use $\xi$ as a universal estimate for the prior probability $\mu$.

For horizon $>1$, the expectimax series in (6.4) and the process of selecting maximal values may be interpreted as abstract *planning*. The expectimax series is a form of *informed search*, in the case of AI$\mu$, and *heuristic search*, for AI$\xi$, where $\xi$ could be interpreted as a heuristic for $\mu$. The minimax strategy of *game playing* in case of AI$\mu$ is also subsumed. The AI$\xi$ model converges to the minimax strategy if the environment is a minimax player, but it can also take advantage of environmental players with limited rationality. *Problem solving* occurs (only) in the form of how to maximize the expected future reward.

*Knowledge* is accumulated by AI$\xi$ and is stored in some form not specified further on the work tape. Any kind of information in any representation on the inputs $y$ is exploited. The problem of *knowledge engineering* and *representation* appears in the form of how to train the AI$\xi$ model. More practical aspects, like *language or image processing*, have to be learned by AI$\xi$ from scratch.

Other theories, like *fuzzy logic, possibility theory, Dempster-Shafer theory*, etc. are partly outdated and partly reducible to Bayesian probability theory [Che85, Che88]. The interpretation and consequences of the evidence gap $g:= 1-\sum_{x_k}\xi(y\!x_{<k}y\!\underline{x}_k)>0$ in $\xi$ may be similar to those in Dempster-Shafer theory. Boolean logical reasoning about the external world plays, at best, an emergent role in the AI$\xi$ model.

Other methods that do not seem to be contained in the AI$\xi$ model might also be emergent phenomena. The AI$\xi$ model has to construct short codes of the environmental behavior, and the AI$\xi^{\tilde{t}\tilde{l}}$ (see next chapter) has to construct short action programs. If we would analyze and interpret these programs for realistic environments, we might find some of the unmentioned or unused or new AI methods at work in these programs. This is, however, pure speculation at this point. More important: when trying to make AI$\xi$ practically usable, some other AI methods, like neural nets or genetic algorithms, especially for I/O pre/postprocessing, may be useful.

The main thing we wanted to point out is that the AI$\xi$ model does not lack any important known property of intelligence or known AI methodology. What *is* missing, however, are computational aspects, which are addressed in the next chapter.

## 6.7 Problems

**6.1 (Self-Optimization)** [C35uo] Formally define the environmental classes $\mathcal{M}_{\text{EC}}$ for EC $\in$ {FMS, SGR, EX} similarly to EC $\in$ {SP, AI}. $\mathcal{M}_{\text{EC}}$ shall be the class of all (lower-semi)computable environments consistent with the problem setup EC, $\xi^{\text{EC}}:=\sum_{\nu\in\mathcal{M}_{\text{EC}}}2^{-K(\nu)}\nu$ the corresponding universal prior, and EC$\xi$ alias $p^\xi_{\text{EC}}$ the Bayes optimal policy. Show that function minimization (FMS), repeated strategic games (SGR), and supervised learning (EX) admit self-optimizing policies (cf. Problem 5.10), hence FMS$\xi$, SGR$\xi$ and EX$\xi$ are self-optimizing. Interpret the results and compare them to the properties of SP$\xi$ and AI$\xi$.

**6.2 (Prediction loss bounds for AI$\xi$)** [C40oi] In Section 6.2.2 we derived a bound (6.17) exponential in $K(\dot{z})$ on the number of prediction errors made by AI$\xi$ with horizon $h_k=1$ in deterministic passive environments. Try to generalize/improve this bound to (*a*) general loss functions, (*b*) bounds on $E_n^{\mathrm{AI}\xi} - E_n^{\mathrm{AI}\mu}$ for probabilistic passive environments, (*c*) the case $h_k > 1$, (*d*) bounds linear or polynomial in $K(\dot{z})$ as in the case of SP$\xi$ – or – find examples demonstrating the impossibility of such generalizations/improvements.

**6.3 (Posterization of prediction errors)** [C20u/C40o] Show $\xi^{\mathrm{AI}}(\underline{yx}_{1:n}) \stackrel{\times}{\geq} \xi^{\mathrm{SP}}(\underline{z}_{1:n})$, where $z_k = \delta_{y_k x_k}$, and that the other direction $\stackrel{\times}{\leq}$ is wrong. Use this result to "improve" the bound (6.17) to $E_n^{\mathrm{AI}\xi} \stackrel{\times}{\leq} [\xi^{\mathrm{SP}}(\dot{\underline{z}}_{1:n})]^{-1}$. Posterialize this to a bound $E_{kn}^{\mathrm{AI}\xi} \stackrel{\times}{\leq} \xi^{\mathrm{AI}}(\dot{\underline{y}}\dot{\underline{x}}_{<k})/\xi^{\mathrm{SP}}(\dot{\underline{z}}_{1:n})$ on the number of errors in cycles $k$ to $n$. Is it possible to improve the numerator to $\xi^{\mathrm{SP}}(\dot{\underline{z}}_{<k})$ and to bound the expression by $\approx 2^{K(\dot{z}_{k:n}|\dot{z}_{<k})}$ (cf. Problem 3.13)?

*Only math nerds would call $2^{500}$ finite* — Leonid Levin

*The biggest difference between time and space is that you can't reuse time* — Merrick Furst

*The only reason for time is so that everything doesn't happen at once* — Albert Einstein

*You insist that there is something a machine cannot do. If you will tell me precisely what it is that a machine cannot do, then I can always make a machine which will do just that!* — John von Neumann

John von Neumann (1903–1957)

# 7 Computational Aspects



Up to now we have shown the universal character of the AIXI model but have completely ignored computational aspects, which we make up for in this chapter.

We start in Section 7.1 by developing a general algorithm $M_{p^*}^\varepsilon$ that is capable of solving any well-defined problem $p^*$ as quickly as the fastest algorithm computing a solution to $p^*$, save for a factor of $1+\varepsilon$ and lower-order additive terms. $M_{p^*}^\varepsilon$ optimally distributes resources between the execution of provably correct $p^*$-solving programs and an enumeration of all proofs, including relevant proofs of program correctness and of time bounds on program runtimes. The solution is somewhat involved from an implementational aspect. An implementation would include first-order logic, the definition of a universal Turing machine within it and proof theory. $M_{p^*}^\varepsilon$ avoids Blum's speedup theorem by ignoring programs without correctness proof. $M_{p^*}^\varepsilon$ has broader applicability and can be faster than Levin's universal search, the fastest method for inverting functions save for a large multiplicative constant. Kolmogorov complexity is extended in two natural ways to measure function complexity. One of them is used to show that the most efficient program computing some function $f$ is also among the shortest programs provably computing $f$.

Based on a similar idea, we construct in Section 7.2 a computable version of the AIXI model. Let us assume that there exists some algorithm $p$ of size $l$ with computation time per interaction cycle $t$, which behaves in a sufficiently intelligent way (this assumption is the very basis of AI). The algorithm $p^*$ should run all algorithms of length $\leq l$ for $t$ time steps per cycle and select the best output among them. So we have an algorithm that runs in time $t \cdot 2^l$ and is at least as good as $p$, i.e. it also serves our needs apart from the very large but constant multiplicative factor in computation time. This idea of the 'typing monkeys', one of them eventually producing 'Shakespeare', is well known and widely used in theoretical computer science. The difficult part here is the selection of the algorithm with the best output. A further complication is that the selection process itself must have only limited computation time. We present a suitable modification of the AIXI model that solves these difficult problems. The assumptions behind this construction are discussed at the end. The reduction of the factor $2^l$ to $1+\varepsilon$ as for $M_{p^*}^\varepsilon$ is possible, but will not be presented here.

## 7.1 The Fastest & Shortest Algorithm for All Well-Defined Problems

### 7.1.1 Introduction & Main Result

Searching for fast algorithms to solve certain problems is a central and difficult task in computer science. Positive results usually come from explicit constructions of efficient algorithms for specific problem classes. A wide class of problems can be phrased in the following way. Given a formal specification of a problem depending on some parameter $x \in X$, we are interested in a fast algorithm computing solution $y \in Y$. This means that we are interested in a fast algorithm computing $f : X \to Y$, where $f$ is a formal (logical, mathematical, not necessarily algorithmic) specification of the problem. Ideally, we would like to have the fastest algorithm, maybe apart from some small constant factor in computation time. Unfortunately, Blum's speed-up theorem [Blu67, Blu71] shows that there are problems for which an (incomputable) sequence of speed-improving algorithms (of increasing size) exists, but no fastest algorithm.

In the approach presented here, we consider only those algorithms that *provably* solve a given problem and have a fast (i.e. quickly computable) time bound. Neither the programs themselves nor the proofs need to be known in advance. Under these constraints we construct the asymptotically fastest algorithm save a factor of $1+\varepsilon$ that solves any well-defined problem $f$.

**Theorem 7.1 (The fastest algorithm)** Let $p^*$ be a given algorithm computing $p^*(x)$ from $x$, or, more generally, a specification of a function. Let $p$ be any algorithm computing provably the same function as $p^*$ with computation time provably bounded by the function $t_p(x)$ for all $x$. $time_{t_p}(x)$ is the time needed to compute the time bound $t_p(x)$. Fix some $\varepsilon \in (0,\frac{1}{2})$. Then the algorithm $M_{p^*}^\varepsilon$ constructed in Section 7.1.5 computes $p^*(x)$ in time

$$time_{M_{p^*}^\varepsilon}(x) \;\leq\; (1+\varepsilon)\cdot t_p(x) + \tfrac{d_p}{\varepsilon}\cdot time_{t_p}(x) + \tfrac{c_p}{\varepsilon}$$

with constants $c_p$ and $d_p$ depending on $p$ but not on $x$. Neither $p$, $t_p$, nor the proofs need to be known in advance for the construction of $M_{p^*}^\varepsilon$.

Known time bounds for practical problems can often be computed quickly, i.e. $time_{t_p}(x)/time_p(x)$ often converges very quickly to zero. Furthermore, from a practical point of view, the provability restrictions are often rather weak. Hence, we have constructed for all those problems a solution that is asymptotically only a factor $1+\varepsilon$ slower than the (provably) fastest algorithm! There is no large multiplicative factor and the problems are not restricted to inversion problems, as in Levin's algorithm (see Section 7.1.2). What somewhat spoils the practical applicability of $M_{p^*}^\varepsilon$ is the large additive constant $c_p$, which will be estimated in Section 7.1.6.

An interesting and counter-intuitive consequence of Theorem 7.1, derived in Section 7.1.8, is that the fastest program that computes a certain function is also among the shortest programs that provably computes this function. Looking for larger programs saves at most a finite number of computation steps but cannot improve the time order.

In Section 7.1.2 we review Levin search and the universal search algorithms SIMPLE and SEARCH, described in [LV97]. We point out that SIMPLE has the same asymptotic time complexity as SEARCH not only w.r.t. the problem instance, but also w.r.t. to the problem class. In Section 7.1.3 we elucidate Theorem 7.1 and the applicability to an example problem (matrix multiplication) unsolvable by Levin search. Section 7.1.4 discusses the general applicability of $M_{p^*}^\varepsilon$. In Section 7.1.5 we give formal definitions of the expressions *time*, *proof*, *compute*, etc., which occur in Theorem 7.1, and define the fast algorithm $M_{p^*}^\varepsilon$. In Section 7.1.6 we analyze the algorithm $M_{p^*}^\varepsilon$, especially its computation time, prove Theorem 7.1, and give upper bounds for the constants $c_p$ and $d_p$. Subtleties regarding the underlying machine model are briefly discussed in Section 7.1.7. In Section 7.1.8 we show that the fastest program computing a certain function is also among the shortest programs provably computing this function. For this purpose, we extend the definition of the Kolmogorov complexity of a string and define two new natural measures for the complexity of functions and programs. Section 7.1.9 outlines generalizations of Theorem 7.1 to I/O streams and other time measures. Conclusions are given in Section 7.1.10.

### 7.1.2 Levin Search

Levin search is one of the few rather general speed-up algorithms. Within a typically large factor, it is the fastest algorithm for inverting a function $g:Y\to X$, if $g$ can be evaluated quickly [Lev73b, Lev84]. Given $x$, an inversion algorithm $p$ tries to find a $y\in Y$, called $g$-witness for $x$, with $g(y)=x$. Levin search just runs and verifies the result of *all* algorithms $p$ in parallel with relative computation time $2^{-\ell(p)}$; i.e. a time fraction $2^{-\ell(p)}$ is devoted to execute $p$, where $\ell(p)$ is the length of program $p$ (coded in binary). Verification is necessary since the output of *any* program can be *anything*. This is the reason why Levin search is only effective if a fast implementation of $g$ is available. Levin search halts if the first $g$-witness has been produced and verified. The total computation time to find a solution (if one exists) is bounded by $2^{\ell(p)}\cdot time_p^+(x)$, where $time_p^+(x)$ is the runtime of $p(x)$ *plus* the time to verify the correctness of the result $(g(p(x))=x)$ by a *known* implementation for $g$.

Li and Vitányi [LV97, p503] propose a very simple variant, called SIMPLE($g$), which runs all programs $p_1p_2p_3...$ one step at a time according to the following scheme: $p_1$ is run every second step, $p_2$ every second step in the remaining unused steps, $p_3$ every second step in the remaining unused steps, and so forth, i.e. according to the sequence of indices 121312141213121512.... If $p_k$ inverts $g$ on $x$ in $time_{p_k}(x)$ steps, then SIMPLE($g$) will do the same in *at most* $2^k time_{p_k}^+(x)+2^{k-1}$ steps. In order to improve the factor $2^k$, they define the algorithm SEARCH($g$), which runs all $p$ (of length less than $i$) for $2^i 2^{-\ell(p)}$ steps in phase $i=1,2,3,...$, until it has inverted $g$ on $x$. The computation time of SEARCH($g$) is bounded by $2^{K(k)+O(1)}time_{p_k}^+(x)$, where $K(k)\leq\ell(p_k)\leq 2\log_2 k$ is the Kolmogorov complexity of $k$ (Definition 2.9). They suggest that SIMPLE has worse asymptotic behavior w.r.t. $k$ than SEARCH, but actually this is not the case.

In fact, SIMPLE and SEARCH have the same asymptotics also in $k$, because SEARCH itself is an algorithm with some index $k_{\text{SEARCH}}=O(1)$. Hence, SIMPLE executes SEARCH every $2^{k_{\text{SEARCH}}}$ steps, and can at most be a constant (independent of $k$ and $x$) factor $2^{k_{\text{SEARCH}}}=O(1)$ slower than SEARCH. However, in practice, SEARCH should be favored, because constants also matter, and $2^{k_{\text{SEARCH}}} \approx 2^{2^{\ell(p_{k_{\text{SEARCH}}})}}$ is rather large.

Levin search can be modified to handle time-limited optimization problems as well [Sol86]. Many, but not all, problems are of inversion or optimization type. The matrix multiplication example (Section 7.1.3), the *decision* problem SAT [LV97, p503], and reinforcement learning (Section 7.2), for instance, are not of this form. Furthermore, the large factor $2^{\ell(p)}$ somewhat limits the applicability of Levin search.

Levin search in program space cannot be used directly in $M_{p^*}^\varepsilon$ for computing $p^*$ because we have to decide somehow whether a certain program solves our problem or computes something else. For this, we have to search through the space of proofs. In order to avoid the large time factor $2^{\ell(p)}$, we also have to search through the space of time-bounds. Only *one* (fast) program should be executed for a significant time interval. The algorithm $M_{p^*}^\varepsilon$ essentially consists of three interwoven algorithms: *sequential* program execution, sequential search through proof space, and Levin search through time-bound space. A tricky scheduling prevents performance degradation from computing slow $p$'s before *the* $p$ has been found.

### 7.1.3 Fast Matrix Multiplication

To illustrate Theorem 7.1, we consider the problem of multiplying two $n \times n$ matrices. If $p^*$ is the standard algorithm for multiplying two matrices[1] $x \in R^{n \cdot n} \times R^{n \cdot n}$ of size $\ell(x) \propto n^2$ over some ring $R$, then $t_{p^*}(x) := 2n^3$ upper bounds the true computation time $time_{p^*}(x) = n^2(2n-1)$. We know there exist algorithms $p'$ for matrix multiplication with $time_{p'}(x) \leq t_{p'}(x) := c \cdot n^\omega$ ($\omega = 2.81$ [Str69], $\omega = 2.50$ [CW82], $\omega = 2.38$ [CW90], ...). The time-bound function (cast to an integer) can, as in many cases, be computed very quickly, $time_{t_{p'}}(x) = O(\log^2 n)$. Hence, using Theorem 7.1, also $M_{p^*}^\varepsilon$ is fast, $time_{M_{p^*}^\varepsilon}(x) \leq (1+\varepsilon)c \cdot n^\omega + O(\log^2 n)$. Of course, $M_{p^*}^\varepsilon$ would be of no real use if $p'$ is already the fastest program, since $p'$ is known and could be used directly. We do not know however, whether there is an algorithm $p''$ with $time_{p''}(x) \leq d \cdot n^2 \log n$, for instance. But if it does exist, $time_{M_{p^*}^\varepsilon}(x) \leq (1+\varepsilon) d \cdot n^2 \log n + O(1)$ for all $x$ is guaranteed.

There is no contradiction to the result [CW82] that there is no fastest bilinear $\lambda$-algorithm (b$\lambda$A) for matrix multiplication. For every b$\lambda$A $p_i$ with computation time $n^{\omega_i}$ one can find another b$\lambda$A $p_{i+1}$ with computation time $n^{\omega_{i+1}}$ and $\omega_{i+1} < \omega_i$, but there is no b$\lambda$A with computation time $n^{\omega_0}$ and $\omega_0 = \inf_i\{\omega_i\}$. On the other hand, this says nothing about the existence of a non-b$\lambda$A $M$ with computation time of, for instance, $n^{\omega_0} \log n$, which is faster than all b$\lambda$A $p_i$. Indeed, a formal construction of such an algorithm is easy. The

[1] Instead of interpreting $R$ as the set of real numbers one might take the field $\mathbb{F}_2 = \{0,1\}$ to avoid subtleties arising from large numbers. Arithmetic operations are assumed to need one unit of time.

sequence $\{p_1,p_2,p_3,...\}$ is enumerable, i.e. there is an algorithm that creates the programs $p_1,p_2,p_3,...$, say in time $\tau_1,\tau_2,\tau_3,...$. We enumerate $p_1,p_2,p_3,...$ and start executing them in parallel as soon as they have been constructed and assign a fraction $\frac{1}{i(i+1)}$ of time to $p_i(x)$. The first $p_i$ that halts outputs the result. The total computation time of this (meta)algorithm $M$ is

$$time_M(x) \;=\; \min_i\{\tau_i + i(i+1)\cdot time_{p_i}(x)\} \;=\; O(time_{p_i}(x))\;\forall i.$$

Hence $M$ has better time complexity than any of the $p_i$. For instance, for $\omega_i = \omega_0 + O(i^{-2})$ and $\tau_i$ polynomial in $i$, it is easy to see that $time_M(x) = O(n^{\omega_0}\log n)$. The construction above works in general as long as the program sequence is enumerable. It fails for incomputable sequences, like in Blum's speed-up construction.

The matrix multiplication example was chosen for specific reasons. First, it is not an inversion or optimization problem directly suitable for Levin search. The computation time of Levin search is lower-bounded by the time to verify the solution with a known algorithm (which is currently $c\cdot n^{2.376...}$) multiplied with the (large) number of necessary verifications. Second, although matrix multiplication is a very important and time-consuming issue, $p'$ is not used in practice, since $c$ is so large that for all practically occurring $n$, the cubic algorithm is faster. The same is true for $c_p$ and $d_p$, but we must admit that although $c$ is large, the bounds we obtain for $c_p$ and $d_p$ are tremendous. On the other hand, even Levin search, which has a tremendous multiplicative factor, can be successfully applied [Sch97, SZW97, Sch03a, Sch04], when handled with care. The same should hold for Theorem 7.1, as will be discussed. We avoid the $O()$ notation as far as possible, as it can be severely misleading (e.g. $10^{42} = O(1)^{O(1)} = O(1)$). This chapter could be viewed as another $O()$ warning showing how important factors and even subdominant additive terms are.

### 7.1.4 Applicability of the Fast Algorithm $M_{p^*}^\varepsilon$

An obvious time bound for $p$ is the actual computation time itself. An obvious algorithm to compute $time_p(x)$ is to count the number of steps needed for computing $p(x)$. Hence, inserting $t_p = time_p$ into Theorem 7.1 and using $time_{time_p}(x) \le time_p(x)$, we see that the computation time of $M_{p^*}^\varepsilon$ is optimal within a multiplicative constant $(d_p+1+\varepsilon)$ and an additive constant $c_p$. This result is weaker than the one in Theorem 7.1, but no assumption concerning the computability of time bounds has to be made.

When do we trust that a fast algorithm solves a given problem? At least for well-specified problems, like satisfiability, solving a combinatoric puzzle, computing the digits of $\pi$, ..., we usually invent algorithms, prove that they solve the problem and, in many cases, also can prove good and quickly computable time bounds. In these cases, the provability assumptions in Theorem 7.1 are no real restriction. The same holds for approximate algorithms that guarantee a precision $\varepsilon$ within a known time bound (many numerical algorithms are of this kind). For exact or approximate programs provably computing or converging to the right answer (e.g. traveling salesman problem, and also many numerical programs), but for which no good and easy to compute time bound exists, $M_{p*}^{\varepsilon}$ is only optimal apart from a huge constant factor $1+\varepsilon+d_p$ in time, as discussed above. Universal reinforcement learning could be a problem of this kind. There is no known efficient algorithm for computing the optimal policy for sequential decision problems in non-Markov environments. The algorithm AIXI$tl$ developed in Section 7.2 is based on a similar idea as the $M_{p^*}^{\varepsilon}$. It creates an incremental policy for an agent in an unknown general (non-Markov) environment, which is superior to any other time $t$ and length $l$ bounded agent. The computation time of AIXI$tl$ is of the order $t\cdot 2^l$. For poorly specified problems, Theorem 7.1 does not help at all.

### 7.1.5 The Fast Algorithm $M_{p^*}^{\varepsilon}$

One ingredient of algorithm $M_{p^*}^{\varepsilon}$ is an enumeration of proofs of increasing length in some formal axiomatic system. If a proof actually proves that some $p$ is functionally equivalent $p^*$, and $p$ has time bound $t_p$, then $(p,t_p)$ is added to a list $L$. The program $p$ in $L$ with the currently smallest time bound $t_p(x)$ is executed. By construction, the result $p(x)$ is identical to $p^*(x)$. The trick to achieve the time bound stated in Theorem 7.1 is to schedule everything in a proper way, in order not to lose too much performance by computing slow $p$'s and $t_p$'s before *the* $p$ has been found.

To avoid confusion, we formally define $p$ and $t_p$ to be binary strings. That is, $p$ is neither a program nor a function, but can be informally interpreted as such. A formal definition of the interpretations of $p$ is given below. We say "p computes function f", when a universal reference Turing machine $U$ on input $p$ and $x$ computes $f(x)$ for all $x$. This is denoted by $U(p,x)=f(x)$. To be able to talk about proofs, we need a formal logic system $(\forall,\lambda,y_i,c_i,f_i,R_i,\rightarrow,\wedge,=,...)$, and axioms and inference rules. A proof is a sequence of formulas, where each formula is either an axiom or inferred from previous formulas in the sequence by applying the inference rules. See [Fit96, Sho67] or any other textbook on logic or proof theory. We only need to know that *provability*, *Turing machines*, and *computation time* can be formalized:

1. The set of correct proofs is enumerable.
2. A term $u$ can be defined such that the formula $[\forall y: u(p,y)=u(p^*,y)]$ is true if and only if $U(p,x)=U(p^*,x)$ for all $x$, i.e. if $p$ and $p^*$ describe the same function.
3. A term $tm$ can be defined such that the formula $[tm(p,x)=n]$ is true if and only if the computation time of $U$ on $(p,x)$ is $n$, i.e. if $n=time_p(x)$.

We say that $p$ is provably equivalent to $p^*$ if the formula $[\forall y: u(p,y)=u(p^*,y)]$ can be proven. $M_{p^*}^{\varepsilon}$ runs three algorithms $A$, $B$, and $C$ in parallel:

**Algorithm $M^{\varepsilon}_{p^*}(x)$**

Initialize the shared variables $L:=\{\}$, $t_{fast}:=\infty$, $p_{fast}:=p^*$.
Start algorithms $A$, $B$, and $C$ in parallel with relative computational resources $\varepsilon$, $\varepsilon$, and $1-2\varepsilon$, respectively.
That is, $C$ performs about $\frac{1}{\varepsilon}$ steps when $A$ and $B$ perform one step each.

**Algorithm $A$**

```
for i:=1,2,3,... do
   pick the i^th proof in the list of all proofs and
   isolate the last formula in the proof.
   if this formula is equal to [∀y:u(p,y)=u(p*,y)∧u(t,y)≥tm(p,y)]
   for some strings p and t,
   then add (p,t) to L.
next i
```

**Algorithm $B$**

```
for all (p,t)∈L
   run U on all (t,x) in parallel for all t with relative computational resources
   2^(-ℓ(p)-ℓ(t)).
   if U halts for some t and U(t,x)<t_fast,
   then t_fast:=U(t,x) and p_fast:=p and restart algorithm C.
continue (p,t)
```

**Algorithm $C$**

```
   run U on (p_fast,x). For each executed step decrease t_fast by 1.
   if U halts then print result U(p_fast,x) and abort computation of A, B
   and C.
```

Note that $A$ and $B$ only terminate when aborted by $C$. The discussion of the algorithms in the following subsections clarifies details and proves Theorem 7.1.

### 7.1.6 Time Analysis

Henceforth we return to the convenient abbreviations $p(x):=U(p,x)$ and $t_p(x):=U(t_p,x)$. Let $p'$ be some fixed algorithm that is provably equivalent to $p^*$, with computation time $time_{p'}$ provably bounded by $t_{p'}$. Let $\ell(proof(p'))$ be the length of the binary coding of the, for instance, shortest proof. Here, *computation time* refers to true overall computation time, whereas *computation steps* refer to instruction steps. Hence $steps = \alpha\cdot time$, if a percentage $\alpha$ of computation time is assigned to an algorithm.

$A$) To write down (not to invent!) a proof requires $O(\ell(proof))$ steps. A time $O(N_{axiom}\cdot\ell(F_i))$ is needed to check whether a formula $F_i$ in the proof $F_1F_2...F_n$ is an axiom, where $N_{axiom}$ is the number of axioms or axiom schemes, which is finite. Variable substitution (binding) can be performed in linear time. For a suitable finite set of axiom schemes, the only necessary inference rule is modus ponens. If $F_i$ is not an axiom, one searches for a formula $F_j$, $j<i$ of the form $F_k \rightarrow F_i$ and then for the formula $F_k$, $k<i$. This takes time $O(\ell(proof))$. There are $n \leq O(\ell(proof))$ formulas $F_i$ to check in this way. Whether the sequence of formulas constitutes a valid proof can, hence, be checked in $O(\ell(proof)^2)$ steps. There are less than $2^{l+1}$ proofs of (binary) length $\leq l$. Algorithm $A$ receives a fraction $\varepsilon$ of relative computation time. Hence, for a proof of $(p',t_{p'})$ to occur, and for $(p',t_{p'})$ to be added to $L$, at most time $T_A \leq \frac{1}{\varepsilon}\cdot 2^{\ell(proof(p'))+1}\cdot O(\ell(proof(p'))^2)$ is needed. Note that the same program $p$ can and will be accompanied by different time bounds $t_p$; for instance $(p,time_p)$ will occur.

$B$) The time assignment of algorithm $B$ to the $t_p$'s only works if the Kraft inequality $\sum_{(p,t_p)\in L} 2^{-\ell(p)-\ell(t_p)} \leq 1$ is satisfied [Kra49]. This can be ensured by using prefix-free (e.g. Shannon-Fano) codes [Sha48, LV97]. The number of steps to calculate $t_{p'}(x)$ is, by definition, $time_{t_{p'}}(x)$. The relative computation time available for computing $t_{p'}(x)$ is $\varepsilon\cdot 2^{-\ell(p')-\ell(t_{p'})}$. Hence, $t_{p'}(x)$ is computed and $t_{fast} \leq t_{p'}(x)$ is checked after time $T_B \leq T_A + \frac{1}{\varepsilon}\cdot 2^{\ell(p')+\ell(t_{p'})}\cdot time_{t_{p'}}(x)$. We have to add $T_A$, since $B$ has to wait, in the worst case, time $T_A$ before it can start executing $t_{p'}(x)$.

$C$) If algorithm $C$ halts, its construction guarantees that the output is correct. In the following, we show that $C$ always halts, and give a bound for the computation time.

(*i*) Assume that algorithm $C$ stops before $B$ performed the check $t_{p'}(x) < t_{fast}$, because a different $p$ already computed $p(x)$. In this case $T_C \leq T_B$.

(*ii*) Assume that $B$ performs the check $t_{p'}(x) < t_{fast}$ and the check succeeds. Runtime $T_B$ has passed until this point. $C$ is restarted and computes $p_{fast}(x) = p'(x)$ in time $t_{fast} := t'_p$, or faster, if during the computation, $p_{fast}$ gets replaced by an even faster algorithm constructed by $A$ and $B$ ($t_{fast}$ is a decreasing variable). Since a fraction $1-2\varepsilon$ of relative computation time is assigned to $C$, it halts after time $T_C \leq T_B + \frac{1}{1-2\varepsilon} t_{p'}(x)$.

(*iii*) At any point in time the remaining time until $C$ halts is bounded by $\frac{1}{1-2\varepsilon} t_{fast}$, since $t_{fast}$ is never increasing. Hence, if the check $t_{p'}(x) < t_{fast}$ fails, $T_C \leq T_B + \frac{1}{1-2\varepsilon} t_{fast} \leq T_B + \frac{1}{1-2\varepsilon} t_{p'}(x)$.

The maximum of the cases (*i*) to (*iii*) bounds the computation time of $C$ and, hence, of $M_{p^*}^{\varepsilon}$ by

$$time_{M_{p^*}^{\varepsilon}}(x) = T_C \;\leq\; T_B + \tfrac{1}{1-2\varepsilon} t_p(x) \;\leq\; (1+3\varepsilon)\cdot t_p(x) + \tfrac{d_p}{3\varepsilon}\cdot time_{t_p}(x) + \tfrac{c_p}{3\varepsilon},$$

$$d_p = 3\cdot 2^{\ell(p)+\ell(t_p)}, \quad c_p = 3\cdot 2^{\ell(proof(p))+1}\cdot O(\ell(proof(p)^2)),$$

where we have dropped the prime from $p$ and used $\frac{1}{1-2\varepsilon} \leq 1+3\varepsilon$ for $\varepsilon \leq \frac{1}{6}$. We have also suppressed the dependency of $c_p$ and $d_p$ on $p^*$ ($proof(p)$ depends on $p^*$ too), since we considered $p^*$ to be a fixed given algorithm. Rescaling $\varepsilon \rightsquigarrow \varepsilon/3$ leads to the bound in Theorem 7.1.

### 7.1.7 Assumptions on the Machine Model

In the time analysis above we have assumed that program simulation with abort possibility and scheduling parallel algorithms can be performed in realtime, i.e. without loss of performance. Parallel computation can be avoided by sequentially performing time slices of $N$ operations and then switching to the next task. Algorithms $A$ and $C$, and every $(p,t) \in L$ in algorithm $B$ constitute a task. If switching between time slices needs constant time $s$ and we choose $N \sim \frac{1}{s\varepsilon}$, then time slicing increases computation time by a factor $1+\varepsilon$. Also, in order to avoid a possible slowdown of $p$ in algorithm $C$ caused by decrementing $t_{fast}$, one should decrement $t_{fast}$ by $N$ every $N^{th}$ time step, possibly synchronously to the task switching. Counting can be performed in time $O(1)$ [SV88].

A thorough construction of a realtime machine $U$ goes beyond the scope of this book. The above discussion should be a motivation that universal realtime machines $U$ are something reasonable. Note that we use the same universal Turing machine $U$ with the same underlying Turing machine model (number of heads, symbols, ...) for measuring computation time of all programs (strings) $p$, including $M_{p^*}^\varepsilon$. This prevents us from applying the linear speedup theorem (which is cheating somewhat anyway), but allows the possibility of designing a $U$ that allows realtime simulation with abort possibility. Theorem 7.1 should also hold for suitable Kolmogorov-Uspenskii [KU63] and Pointer machines [Sch80].

### 7.1.8 Algorithmic Complexity and the Shortest Algorithm

Data compression is a very important issue in computer science. Saving space or channel capacity are obvious applications. In Chapter 2 we saw that a less obvious, but not far-fetched, application is that of inductive inference in various forms (hypothesis testing, forecasting, classification, ...). A free interpretation of Occam's razor is that the shortest theory consistent with past data is the most likely to be correct. This was put into a rigorous scheme by [Sol64] and proved to be optimal in Chapter 3. Kolmogorov complexity is a universal notion of the information content of a string [Kol65, Cha66, ZL70]. It is defined as the length of the shortest program computing string $x$:

$$K_U(x) \;:=\; \min_p\{\ell(p) : U(p) = x\} \;=\; K(x) + O(1),$$

where $U$ is some universal Turing machine. It can be shown that $K_U(x)$ varies, at most, by an additive constant independent of $x$ by varying the machine $U$. Hence, *the* Kolmogorov complexity $K(x)$ is universal in the sense that it is uniquely defined up to an additive constant. $K(x)$ can be approximated from above (is co-enumerable), but is not finitely computable. Refer to Chapter 2 for details on Kolmogorov complexity and its application to prediction.

Recently, Schmidhuber [Sch00, Sch02a] has generalized Kolmogorov complexity in various ways to the limits of computability and beyond. In the following, we also need a generalization, but of a different kind. We need a short description of a function, rather than of a string. The following definition of the complexity of a function $f$,

$$K'(f) := \min_p \{\ell(p) : U(p,x) = f(x)\,\forall x\}$$

seems natural, but suffers from not even being approximable (see Definition 2.12). There exists no algorithm converging to $K'(f)$, because it is in general undecidable whether a program $p$ is equivalent to (some formal definition of) a function $f$. Even if we have a program $p^*$ computing $f$, $K'(p^*)$ is not approximable. Using $K(p^*)$ is not a suitable alternative, since $K(p^*)$ might be considerably larger than $K'(p^*)$, as in the former case all information conveyed by $p^*$ will be kept, even that which is functionally irrelevant (e.g. dead code). An alternative is to restrict ourselves to provably equivalent programs. The length of the shortest one is

$$K''(p^*) \;:=\; \min_p \{\ell(p) : \text{a proof of } [\forall y\!:\!u(p,y) = u(p^*,y)] \text{ exists}\}.$$

It can be approximated from above, since the set of all programs provably equivalent to $p^*$ is enumerable.

Having obtained, after some time, a very short description $p'$ of $p^*$ for some purpose (e.g. for defining a prior probability for some inductive inference scheme), it is usually also necessary to obtain values for some arguments. We are now concerned with the computation time of $p'$. Could we get slower and slower algorithms by compressing $p^*$ more and more? Interestingly, this is not the case. Inventing complex (long) programs is *not* necessary to construct asymptotically fast algorithms, under the stated provability assumptions, in contrast to Blum's theorem [Blu67, Blu71]. The following theorem roughly says that there is a *single* program that is the fastest *and* the shortest program.

**Theorem 7.2 (The fastest & shortest algorithm)** Let $p^*$ be a given algorithm or formal specification of a function. There exists a program $\tilde p$, equivalent to $p^*$, for which the following holds

$$\begin{aligned} i)\quad & \ell(\tilde p) && \le\; K''(p^*) + O(1), \\ ii)\quad & time_{\tilde p}(x) && \le\; (1+\varepsilon)\cdot t_p(x) + \tfrac{d_p}{\varepsilon}\cdot time_{t_p}(x) + \tfrac{c_p}{\varepsilon}, \end{aligned}$$

where $p$ is any program provably equivalent to $p^*$ with computation time provably less than $t_p(x)$. The constants $c_p$ and $d_p$ depend on $p$ but not on $x$.

To prove the theorem, we just insert the shortest algorithm $p'$ provably equivalent to $p^*$ into $M$, that is, $\tilde{p}:=M_{p'}^\varepsilon$. As only $O(1)$ instructions are needed to build $M_{p'}^\varepsilon$ from $p'$, $M_{p'}^\varepsilon$ has size $\ell(p')+O(1)=K''(p^*)+O(1)$. The computation time of $M_{p'}^\varepsilon$ is the same as of $M_{p^*}^\varepsilon$ apart from "slightly" different constants $c_p$ and $d_p$.

The following subtlety was pointed out by Peter van Emde Boas. Neither $M_{p^*}^\varepsilon$ nor $\tilde{p}$ is *provably* equivalent to $p^*$. The construction of $M_{p^*}^\varepsilon$ in Section 7.1.5 shows equivalence of $M_{p^*}^\varepsilon$ (and of $\tilde{p}$) to $p^*$, but it is a meta-proof that cannot be formalized within the considered proof system. A formal proof of the correctness of $M_{p^*}^\varepsilon$ would prove the consistency of the proof system, which is impossible by Gödel's second incompleteness theorem [Göd31]. See [Har79] for details in a related context.

### 7.1.9 Generalizations

If $p^*$ has to be evaluated repeatedly, algorithm $A$ can be modified to remember its current state and continue operation for the next input ($A$ is independent of $x$!). The large offset time $c_p$ is only needed on the first invocation.

$M_{p^*}^\varepsilon$ can be modified to handle I/O streams, definable by a Turing machine with monotone input and output tapes (and bidirectional work tapes) receiving an input stream and producing an output stream. The currently read prefix of the input stream is $x$. $time_p(x)$ is the time used for reading $x$. $M_{p^*}^\varepsilon$ caches the input and output streams, so that algorithm $C$ can repeatedly read/write the streams for each new $p$. The true input/output tapes are used for requesting/producing a new symbol. Algorithm $B$ is reset after 1,2,4,8,... steps (not after reading the next symbol of $x$!) to appropriately take into account increased prefixes $x$. Algorithm $A$ just continues. The bound of Theorem 7.1 holds for this case too, with slightly increased $d_p$.

The construction above also works if time is defined as a function of the current output rather than the current input $x$. This measure is, for example, used for the time-complexity of calculating the $n^{th}$ digit of a computable real (e.g. $\pi$), where there is no input, but only an output stream.

### 7.1.10 Summary & Outlook

We presented an algorithm $M_{p^*}^\varepsilon$ that accelerates the computation of a program $p^*$. $M_{p^*}^\varepsilon$ combines ($A$) sequential search through proof space, ($B$) Levin search through time-bound space, and ($C$) *sequential* program execution, using a somewhat tricky scheduling. Under certain provability constraints, $M_{p^*}^\varepsilon$ is the asymptotically fastest algorithm for computing $p^*$ apart from a factor $1+\varepsilon$ in computation time. Blum's theorem shows that the provability constraints are essential. We showed that the conditions on Theorem 7.1 are often, but not always, satisfied for practical problems. For complex approximation problems, for instance, where no good and quickly computable time bound exists, $M_{p^*}^\varepsilon$ is still optimal, but in this case, only apart from a large multiplicative factor. We briefly outlined how $M_{p^*}^\varepsilon$ can be modified to handle I/O streams and other time measures. An interesting and counterintuitive consequence of Theorem 7.1 was that the fastest program computing a certain function is also among the shortest programs provably computing this function. Looking for larger programs saves at most a finite number of computation steps, but cannot improve the time order. To quantify this statement, we extended the definition of Kolmogorov complexity and defined two new natural measures for the complexity of a function. The large constants $c_p$ and $d_p$ seem to spoil a direct implementation of $M_{p^*}^\varepsilon$. On the other hand, Levin search has been successfully adapted/generalized and applied to solve rather difficult machine learning problems [Sch97, SZW97, Sch03a, Sch04], even though it suffers from a large multiplicative factor of similar origin. The use of more elaborate theorem provers, rather than brute-force enumeration of all proofs, could lead to smaller constants and bring $M_p^*$ closer to practical applications, possibly restricted to subclasses of problems [RV01]. A more fascinating (and more speculative) way may be the utilization of so-called transparent or holographic proofs [BFLS91]. The correctness of these proofs can be checked by only reading a logarithmic number of their bits. This would mean that exponentially many proofs are checked simultaneously, reducing the constants $c_p$ and $d_p$ to their logarithm. I would like to conclude with a general question. Will the ultimate search for asymptotically fastest programs typically lead to fast or slow programs for arguments of practical size? Levin search, matrix multiplication and the algorithm $M_{p^*}^\varepsilon$ seem to support the latter, but this might be due to our inability to do better.

## 7.2 Time-Bounded AIXI Model

### 7.2.1 Introduction

Until now, we have not bothered with the non-computability of the universal probability distribution $\xi:=\xi_U$. As all universal models in this book are based on $\xi$, they are not effective in this form. In this section, we outline how the previous models and results can be modified/generalized to the time-bounded case. Indeed, the situation is not as bad as it could be. $\xi$ is enumerable and $\dot{y}_k$ is still approximable, i.e. there exists an algorithm that will produce a sequence of outputs eventually converging to the exact output $\dot{y}_k$, but we can never be sure whether we have already reached it. Besides this, the convergence is extremely slow, so this type of asymptotic computability is of no direct (practical) use, but will nevertheless be important later.

Let $\tilde{p}$ be a program that calculates within a reasonable time $\tilde{t}$ per cycle a reasonable intelligent output, i.e. $\tilde{p}(\dot{x}_{<k})=\dot{y}_{1:k}$. This sort of computability assumption, that a general-purpose computer of sufficient power is able to behave in an intelligent way, is the very basis of AI, justifying the hope to be able to construct agents which eventually reach and outperform human

intelligence. For a contrary viewpoint see [Luc61, Pen89, Pen94]. It is not necessary to discuss here what is meant by 'reasonable time/intelligence' and 'sufficient power'. What we are interested in, in this section, is whether there is a computable version AIXI$\tilde{t}$ of the AIXI agent that is superior or equal to any $p$ with computation time per cycle of at most $\tilde{t}$. By 'superior', we mean 'more intelligent', so what we need is an order relation, like (5.14) for intelligence.

The best result we could think of would be an AIXI$\tilde{t}$ with computation time $\leq\tilde{t}$ at least as intelligent as any $p$ with computation time $\leq\tilde{t}$. If AI is possible at all, we would have reached the final goal, the construction of the most intelligent algorithm with computation time $\leq\tilde{t}$. Just as there is no universal measure in the set of computable measures (within time $\tilde{t}$), neither may such an AIXI$\tilde{t}$ exist.

What we can realistically hope to construct is an AIXI$\tilde{t}$ agent of computation time $c\cdot\tilde{t}$ per cycle for some constant $c$. The idea is to run all programs $p$ of length $\leq\tilde{l}:=\ell(\tilde{p})$ and time $\leq\tilde{t}$ per cycle and pick the best output. The total computation time is $c\cdot\tilde{t}$ with $c=2^{\tilde{l}}$. This sort of idea of 'typing monkeys' with one of them eventually writing Shakespeare, has been applied in various forms and contexts in theoretical computer science. The realization of this *best vote* idea, in our case, is not straightforward and will be outlined in this section. A related idea is that of basing the decision on the majority of algorithms. This 'democratic vote' idea was used in [LW94, Vov92] for sequence prediction and is referred to as 'weighted majority' (see Section 3.7.4).

### 7.2.2 Time-Limited Probability Distributions

In the literature one can find time-limited versions of Kolmogorov complexity [Dal73, Dal77, Ko86] and time-limited universal semimeasures [LV91, LV97, Sch02b]. In the following, we utilize and adapt the latter and see how far we get. One way to define a time-limited universal chronological semimeasure is similar to the unbounded case (5.5) as a mixture, but restricted to enumerable chronological semimeasures computable within time $\tilde{t}$ and of size at most $\tilde{l}$.

$$\xi^{\tilde{t}\tilde{l}}(y\!x_{1:n}) \;:=\; \sum_{\rho\,:\,\ell(\rho)\leq\tilde{l}\,\wedge\,t(\rho)\leq\tilde{t}} 2^{-\ell(\rho)}\rho(y\!x_{1:n}). \tag{7.3}$$

Let us assume that the true environmental prior probability $\mu^{\mathrm{AI}}$ is equal to or sufficiently accurately approximated by a $\rho$ with $\ell(\rho)\leq\tilde{l}$ and $t(\rho)\leq\tilde{t}$ with $\tilde{t}$ and $\tilde{l}$ of reasonable size. There are several AI problems that fall into this class. In function minimization of Section 6.4, the computation of $f$ and $\mu^{\mathrm{FM}}$ are often feasible. In many cases, the sequences of Section 6.2 that should be predicted can be easily calculated when $\mu^{\mathrm{SP}}$ is known. In a classifier problem, the probability distribution $\mu^{\mathrm{CF}}$, according to which examples are presented, is, in many cases, also elementary. But not all AI problems are of this "easy" type. For the strategic games of Section 6.3, the environment itself is usually a highly complex strategic player with a $\mu^{\mathrm{SG}}$ that is difficult to calculate, although one might argue that the environmental player may have limited capabilities too. But it is easy to think of a difficult-to-calculate physical (probabilistic) environment like the chemistry of biomolecules.

The number of interesting applications makes this restricted class of AI problems, with time- and space-bounded environment $\mu^{\tilde{t}\tilde{l}}$, worthy of study. Superscripts to a probability distribution except for $\xi^{\tilde{t}\tilde{l}}$ indicate their length and maximal computation time. $\xi^{\tilde{t}\tilde{l}}$ defined in (7.3), with a yet to be determined computation time, multiplicatively dominates all $\mu^{\tilde{t}\tilde{l}}$ of this type. Hence, an AI$\xi^{\tilde{t}\tilde{l}}$ model, where we use $\xi^{\tilde{t}\tilde{l}}$ as prior probability, is universal relative to all AI$\mu^{\tilde{t}\tilde{l}}$ models in the same way as AIXI is universal to AI$\mu$ for all enumerable chronological semimeasures $\mu$. The $\text{argmax}_{y_k}$ in (5.3) selects a $y_k$ for which $\xi^{\tilde{t}\tilde{l}}$ has the highest expected utility $V_{km_k}$, where $\xi^{\tilde{t}\tilde{l}}$ is the weighted average over the $\rho^{\tilde{t}\tilde{l}}$; output $\dot{y}_k^{\text{AI}\xi^{\tilde{t}\tilde{l}}}$ is determined by a weighted majority. We expect AI$\xi^{\tilde{t}\tilde{l}}$ to outperform all (bounded) AI$\rho^{\tilde{t}\tilde{l}}$, analogous to the unrestricted case.

In the following we analyze the computability properties of $\xi^{\tilde{t}\tilde{l}}$ and AI$\xi^{\tilde{t}\tilde{l}}$, i.e. of $\dot{y}_k^{\text{AI}\xi^{\tilde{t}\tilde{l}}}$. To compute $\xi^{\tilde{t}\tilde{l}}$ according to the definition (7.3) we have to enumerate all chronological enumerable semimeasures $\rho^{\tilde{t}\tilde{l}}$ of length $\leq\tilde{l}$ and computation time $\leq\tilde{t}$. This can be done similarly to the unbounded case (5.42)–(5.44). All $2^{\tilde{l}}$ enumerable functions of length $\leq\tilde{l}$, computable within time $\tilde{t}$ have to be converted to chronological probability distributions. For this, one has to evaluate each function for $|\mathcal{X}|\cdot k$ different arguments. Hence, $\xi^{\tilde{t}\tilde{l}}$ is computable within time[2] $t(\xi^{\tilde{t}\tilde{l}}(y\!x_{1:k}))=O(|\mathcal{X}|\cdot k\cdot 2^{\tilde{l}}\cdot\tilde{t})$. The computation time of $\dot{y}_k^{\text{AI}\xi^{\tilde{t}\tilde{l}}}$ depends on the size of $\mathcal{X}$, $\mathcal{Y}$ and $m_k$. $\xi^{\tilde{t}\tilde{l}}$ has to be evaluated $|\mathcal{Y}|^{h_k}|\mathcal{X}|^{h_k}$ times in (5.3). It is possible to optimize the algorithm and perform the computation within time

$$t(\dot{y}_k^{\text{AI}\xi^{\tilde{t}\tilde{l}}}) \;=\; O(|\mathcal{Y}|^{h_k}|\mathcal{X}|^{h_k}\cdot 2^{\tilde{l}}\cdot\tilde{t}) \tag{7.4}$$

per cycle. If we assume that the computation time of $\mu^{\tilde{t}\tilde{l}}$ is exactly $\tilde{t}$ for all arguments, the brute-force time $\bar{t}$ for calculating the sums and maxs in (4.17) is $\bar{t}(\dot{y}_k^{\text{AI}\mu^{\tilde{t}\tilde{l}}})\geq|\mathcal{Y}|^{h_k}|\mathcal{X}|^{h_k}\cdot\tilde{t}$. Combining this with (7.4), we get

$$t(\dot{y}_k^{\text{AI}\xi^{\tilde{t}\tilde{l}}}) \;=\; O(2^{\tilde{l}}\cdot\bar{t}(\dot{y}_k^{\text{AI}\mu^{\tilde{t}\tilde{l}}})).$$

This result has the proposed structure that there is a universal AI$\xi^{\tilde{t}\tilde{l}}$ agent with computation time $2^{\tilde{l}}$ times the computation time of a special AI$\mu^{\tilde{t}\tilde{l}}$ agent.

Unfortunately, the class of AI$\mu^{\tilde{t}\tilde{l}}$ systems with brute-force evaluation of $\dot{y}_k$, based on (4.17) is completely uninteresting from a practical point of view. For example in the context of chess, the above result says that the AI$\xi^{\tilde{t}\tilde{l}}$ is

[2] We assume that a (Turing) machine can be simulated by another in linear time.

superior within time $2^{\tilde{l}}\cdot\tilde{t}$ to any brute-force minimax strategy of computation time $\tilde{t}$. Even if the factor of $2^{\tilde{l}}$ in computation time would not matter, the $\text{AI}\xi^{\tilde{t}\tilde{l}}$ agent is nevertheless practically useless, as a brute-force minimax chess player with reasonable time $\tilde{t}$ is a very poor player.

Note that in the case of binary sequence prediction ($h_k=1$, $|\mathcal{Y}|=|\mathcal{X}|=2$) the computation time of $\rho$ coincides with that of $\dot{y}_k^{\text{AI}\rho}$ within a factor of 2. The class $\text{AI}\rho^{\tilde{t}\tilde{l}}$ includes *all* non-incremental sequence prediction algorithms of size $\leq\tilde{l}$ and computation time $\leq\tilde{t}/2$. By non-incremental, we mean that no information of previous cycles is taken into account for speeding up the computation of $\dot{y}_k$ of the current cycle.

The shortcomings (mentioned and unmentioned ones) of this approach are cured in the next subsections, by deviating from the standard way of defining a time-bounded $\xi$ as a sum over functions or programs.

### 7.2.3 The Idea of the Best Vote Algorithm

A general agent is a chronological program $p(x_{<k})=y_{1:k}$. This form, introduced in Section 4.1, is general enough to include any AI system (and also less intelligent systems). In the following, we are interested in programs $p$ of length $\leq\tilde{l}$ and computation time $\leq\tilde{t}$ per cycle. One important point in the time-limited setting is that $p$ should be incremental, i.e. when computing $y_k$ in cycle $k$, the information of the previous cycles stored on the work tape can be reused. Indeed, there is probably no practically interesting, non-incremental AI system at all.

In the following, we construct a policy $p^*$, or more precisely, policies $p_k^*$, for every cycle $k$ that outperform all time- and length-limited AI systems $p$. In cycle $k$, $p_k^*$ runs all $2^{\tilde{l}}$ programs $p$ and selects the one with the best output $y_k$. This is a 'best vote' type of algorithm, as compared to the 'weighted majority' type algorithm of the last subsection. The ideal measure for the quality of the output would be the $\xi$-expected future reward

$$V_{km}^{p\xi}(\dot{y}\dot{x}_{<k}) \;:=\; \sum_{q\in\dot{Q}_k} 2^{-\ell(q)}V_{km}^{pq}, \qquad V_{km}^{pq} \;:=\; r(x_k^{pq})+...+r(x_m^{pq}). \tag{7.5}$$

The program $p$ that maximizes $V_{km_k}^{p\xi}$ should be selected. We have dropped the normalization $\mathcal{N}$ unlike in (5.13), as it is independent of $p$ and does not change the order relation in which we are solely interested here. Furthermore, without normalization, $V_{km}^{*\xi}(\dot{y}\dot{x}_{<k}):=\max_{p\in\dot{P}}V_{km}^{p\xi}(\dot{y}\dot{x}_{<k})$ is enumerable, which will be important later.

### 7.2.4 Extended Chronological Programs

In the functional form of the AIXI model it was convenient to maximize $V_{km_k}$ over all $p\in\dot{P}_k$, i.e. all $p$ consistent with the current history $\dot{y}\dot{x}_{<k}$. This was not

a restriction, because for every possibly inconsistent program $p$ there exists a program $p' \in \dot{P}_k$ consistent with the current history and identical to $p$ for all future cycles $\geq k$. For the time-limited best vote algorithm $p^*$ it would be too restrictive to demand $p \in \dot{P}_k$. To prove universality, one has to compare *all* $2^{\tilde{l}}$ algorithms in every cycle, not just the consistent ones. An inconsistent algorithm may become the best one in later cycles. For inconsistent programs we have to include the $\dot{y}_k$ into the input, i.e. $p(\dot{y}\dot{x}_{<k}) = y^p_{1:k}$ with $\dot{y}_i \neq y^p_i$ possible. For $p \in \dot{P}_k$ this was not necessary, as $p$ knows the output $\dot{y}_k \equiv y^p_k$ in this case. The $r^{pq}_i$ in the definition of $V_{km}$ are the rewards emerging in the I/O sequence, starting with $\dot{y}\dot{x}_{<k}$ (emerging from $p^*$) and then continued by applying $p$ and $q$ with $\dot{y}_i := y^p_i$ for $i \geq k$.

Another problem is that we need $V_{km_k}$ to select the best policy, but unfortunately $V_{km_k}$ is uncomputable. Indeed, the structure of the definition of $V_{km_k}$ is very similar to that of $\dot{y}_k$, hence a brute-force approach to approximate $V_{km_k}$ requires too much computation time as for $\dot{y}_k$. We solve this problem in a similar way, by supplementing each $p$ with a program that estimates $V_{km_k}$ by $w^p_k$ within time $\tilde{t}$. We combine the calculation of $y^p_k$ and $w^p_k$ and extend the notion of a chronological program once again to

$$p(\dot{y}\dot{x}_{<k}) \;=\; w^p_1 y^p_1 ... w^p_k y^p_k \tag{7.6}$$

with chronological order $w^p_1 y^p_1 \dot{y}_1 \dot{x}_1 w^p_2 y^p_2 \dot{y}_2 \dot{x}_2 ...$.

### 7.2.5 Valid Approximations

Policy $p$ might suggest any output $y^p_k$, but it is not allowed to rate it with an arbitrarily high $w^p_k$ if we want $w^p_k$ to be a reliable criterion for selecting the best $p$. We demand that no policy is allowed to claim that it is better than it actually is. We define a (logical) predicate VA($p$) called *valid approximation*, which is true if and only if $p$ always satisfies $w^p_k \leq V^{p\xi}_{km_k}$, i.e. never overrates itself.

$$\text{VA}(p) \;\equiv\; [\forall k \forall w^p_1 y^p_1 \dot{y}_1 \dot{x}_1 ... w^p_k y^p_k : p(\dot{y}\dot{x}_{<k}) = w^p_1 y^p_1 ... w^p_k y^p_k \Rightarrow w^p_k \leq V^{p\xi}_{km_k}(\dot{y}\dot{x}_{<k})] \tag{7.7}$$

In the following, we restrict our attention to programs $p$ for which VA($p$) can be proven in some formal axiomatic system. A very important point is that $V^{*\xi}_{km_k}$ is enumerable. This ensures the existence of sequences of programs $p_1, p_2, p_3, ...$ for which VA($p_i$) can be proven and $\lim_{i\to\infty} w^{p_i}_k = V^{*\xi}_{km_k}$ for all $k$ and all I/O sequences. $p_i$ may be defined as the naive (nonhalting) approximation scheme (by enumeration) of $V^{*\xi}_{km_k}$, but terminated after $i$ time steps and using the approximation obtained so far for $w^{p_i}_k$ together with the corresponding output $y^{p_i}_k$. The convergence $w^{p_i}_k \overset{i\to\infty}{\longrightarrow} V^{*\xi}_{km_k}$ ensures that $V^{*\xi}_{km_k}$, which we claimed to be the universally optimal value, can be approximated by $p$ with provable VA($p$) arbitrarily well, when given enough time. The approximation is not uniform in $k$, but this does not matter as the selected $p$ is allowed to change from cycle to cycle.

Another possibility would be to consider only those $p$ that check $w_k^p \leq V_{km_k}^{p\xi}$ online in every cycle, instead of the pre-check VA($p$), either by constructing a proof (on the work tape) for this special case, or $w_k^p \leq V_{km_k}^{p\xi}$ is already evident by the construction of $w_k^p$. In cases where $p$ cannot guarantee $w_k^p \leq V_{km_k}^{p\xi}$ it sets $w_k = 0$ and, hence, trivially satisfies $w_k^p \leq V_{km_k}^{p\xi}$. On the other hand, for these $p$ it is also no problem to prove VA($p$), as one has simply to analyze the internal structure of $p$ and recognize that $p$ shows the validity internally itself, cycle by cycle, which is easy by assumption on $p$. The cycle-by-cycle check is therefore a special case of the pre-proof of VA($p$).

### 7.2.6 Effective Intelligence Order Relation

In Section 5.1 we introduced an intelligence order relation $\succeq$ on AI systems, based on the value $V_{km_k}^{p\xi}$. In the following we need an order relation $\succeq^c$ based on the claimed value $w_k^p$, which might be interpreted as an approximation to $\succeq$.

**Definition 7.8 (Effective intelligence order relation)** We call $p$ *effectively more or equally intelligent* than $p'$ if

$$p \succeq^c p' \;:\Leftrightarrow\; \forall k \forall \dot{y}\dot{x}_{<k} \exists w_{1:n} w'_{1:n} :$$
$$p(\dot{y}\dot{x}_{<k}) = w_1 * ... w_k * \;\wedge\; p'(\dot{y}\dot{x}_{<k}) = w'_1 * ... w'_k * \;\wedge\; w_k \geq w'_k,$$

i.e. if $p$ always claims higher value estimate $w$ than $p'$.

Relation $\succeq^c$ is a co-enumerable partial order relation on extended chronological programs. Restricted to valid approximations it orders the policies w.r.t. the quality of their outputs *and* their ability to justify their outputs with high $w_k$.

### 7.2.7 The Universal Time-Bounded AIXI*tl* Agent

In the following, we describe the algorithm $p^*$ underlying the universal time-bounded AIXI$\tilde{t}\tilde{l}$ agent. It is essentially based on the selection of the best algorithms $p_k^*$ out of the time $\tilde{t}$ and length $\tilde{l}$ bounded $p$, for which there exists a proof of VA($p$) with length $\leq l_P$.

1. Create all binary strings of length $l_P$ and interpret each as a coding of a mathematical proof in the same formal logic system in which VA($\cdot$) was formulated. Take those strings that are proofs of VA($p$) for some $p$ and keep the corresponding programs $p$.
2. Eliminate all $p$ of length $>\tilde{l}$.

3. Modify the behavior of all remaining $p$ in each cycle $k$ as follows: Nothing is changed if $p$ outputs some $w_k^p y_k^p$ within $\tilde{t}$ time steps. Otherwise stop $p$ and write $w_k = 0$ and some arbitrary $y_k$ to the output tape of $p$. Let $P$ be the set of all those modified programs.
4. Start first cycle: $k := 1$.
5. Run every $p \in P$ on extended input $\dot{y}\dot{x}_{<k}$, where all outputs are redirected to some auxiliary tape: $p(\dot{y}\dot{x}_{<k}) = w_1^p y_1^p ... w_k^p y_k^p$. This step is performed incrementally by adding $\dot{y}\dot{x}_{k-1}$ for $k > 1$ to the input tape and continuing the computation of the previous cycle.
6. Select the program $p$ with highest claimed reward $w_k^p$: $p_k^* := \text{argmax}_p w_k^p$.
7. Write $\dot{y}_k := y_k^{p_k^*}$ to the output tape.
8. Receive input $\dot{x}_k$ from the environment.
9. Begin next cycle: $k := k+1$, goto step 5.

It is easy to see that the following theorem holds.

**Theorem 7.9 (Optimality of AIXI*tl*)** Let $p$ be any extended chronological (incremental) program like (7.6) of length $\ell(p) \leq \tilde{l}$ and computation time per cycle $t(p) \leq \tilde{t}$, for which there exists a proof of VA($p$) defined in (7.7) of length $\leq l_P$. The algorithm $p^*$ constructed in the last paragraph, which depends on $\tilde{l}$, $\tilde{t}$ and $l_P$ but not on $p$, is effectively more or equally intelligent, according to $\succeq^c$ (see Definition 7.8) than any such $p$. The size of $p^*$ is $\ell(p^*) = O(\log(\tilde{l} \cdot \tilde{t} \cdot l_P))$, the setup time is $t_{setup}(p^*) = O(l_P^2 \cdot 2^{l_P})$ and the computation time per cycle is $t_{cycle}(p^*) = O(2^{\tilde{l}} \cdot \tilde{t})$.

Roughly speaking, the theorem says that if there exists a computable solution to some or all AI problems at all, the explicitly constructed algorithm $p^*$ is such a solution. Although this theorem is quite general, there are some limitations and open questions that we discuss in the following.

The construction of the algorithm $p^*$ needs the specification of a formal logic system $(\forall, \lambda, y_i, c_i, f_i, R_i, \rightarrow, \wedge, =, ...)$, and axioms, and inference rules. A proof is a sequence of formulas, where each formula is either an axiom or inferred from previous formulas in the sequence by applying the inference rules. Details were presented in Section 7.1.5. We only need to know that *provability* and *Turing machines* can be formalized. The setup time in the theorem is just the time needed to verify the $2^{l_P}$ proofs, each needing time $O(l_P^2)$.

### 7.2.8 Limitations and Open Questions

- Formally, the total computation time of $p^*$ for cycles $1...k$ increases linearly with $k$, i.e. is of order $O(k)$ with a coefficient $2^{\tilde{l}} \cdot \tilde{t}$. The unreasonably large factor $2^{\tilde{l}}$ is a well-known drawback in best/democratic vote models and will be taken without further comments, whereas the factor $\tilde{t}$ can be assumed to be of reasonable size. If we do not take the limit $k \to \infty$ but consider reasonable $k$, the practical significance of the time bound on $p^*$ is somewhat limited, due to the additional additive constant $O(l_P^2 \cdot 2^{l_P})$. It is much larger than $k \cdot 2^{\tilde{l}} \cdot \tilde{t}$ as typically $l_P \gg \ell(\text{VA}(p)) \geq \ell(p) \equiv \tilde{l}$.

- $p^*$ is superior only to those $p$ that justify their outputs by (large $w_k^p$). It might be possible that there are $p$ which produce good outputs $y_k^p$ within reasonable time, but it takes an unreasonably long time to justify their outputs by sufficiently high $w_k^p$. We do not think that (from a certain complexity level onwards) there are policies where the process of constructing a good output is completely separated from some sort of justification process. But this justification might not be translatable (at least within reasonable time) into a reasonable estimate of $V_{km_k}^{p\xi}$.
- The (inconsistent) programs $p$ must be able to continue strategies started by other policies. It might happen that a policy $p$ steers the environment to a direction for which $p$ is specialized. A "foreign" policy might be able to displace $p$ only between loosely connected episodes. There is probably no problem for factorizable $\mu$. Think of a chess game, where it is usually very difficult to continue the game or strategy of a different player. When the game is over, it is usually advantageous to replace a player by a better one for the next game. There might also be no problem for sufficiently separable $\mu$.
- There might be (efficient) valid approximations $p$ for which VA($p$) is true but not provable, or for which only a very long ($>l_P$) proof exists.

### 7.2.9 Remarks

- The idea of suggesting outputs and justifying them by proving reward bounds implements one aspect of human thinking. There are several possible reactions to an input. Each reaction possibly has far-reaching consequences. Within a limited time one tries to estimate the consequences as well as possible. Finally, each reaction is valuated, and the best one is selected. What is inferior to human thinking is that the estimates $w_k^p$ must be rigorously proved and the proofs are constructed by blind exhaustive search, further, that *all* behaviors $p$ of length $\leq \tilde{l}$ are checked. It is inferior "only" in the sense of necessary computation time but not in the sense of the quality of the outputs.
- In practical applications there are often cases with short and slow programs $p_s$ performing some task $T$, e.g. the computation of the digits of $\pi$, for which there exist long but quick programs $p_l$ too. If it is not too difficult to prove that this long program is equivalent to the short one, then it is possible to prove $K^{t(p_l)}(T) \stackrel{+}{\leq} \ell(p_s)$ with $K^t$ being the time-bounded Kolmogorov complexity. Similarly, the method of proving bounds $w_k$ for

$V_{km_k}$ can give high lower bounds without explicitly executing these short and slow programs, which mainly contribute to $V_{km_k}$.

- Dovetailing all length- and time-limited programs is a well-known elementary idea (e.g. typing monkeys). The crucial part, which was developed here, is the selection criterion for the most intelligent agent.
- The construction of AIXI$\widetilde{tl}$ and the enumerability of $V_{km_k}$ ensure arbitrary close approximations of $V_{km_k}$, hence we expect that the behavior of AIXI$\widetilde{tl}$ converges to the behavior of AIXI in the limit $\tilde{t},\tilde{l},l_P \to \infty$, in some sense.
- Depending on what you know or assume that a program $p$ of size $\tilde{l}$ and computation time per cycle $\tilde{t}$ is able to achieve, the computable AIXI$\widetilde{tl}$ model will have the same capabilities. For the strongest assumption of the existence of a Turing machine that outperforms human intelligence, AIXI$\widetilde{tl}$ will do too, within the same time frame, up to an (unfortunately very large) constant factor.

*In spite of its incomputability, Algorithmic Probability can serve as a kind of 'Gold Standard' for induction systems* — Ray Solomonoff

*It's hard to make predictions, especially about the future* — Niels Bohr

*We have the mathematical theory of decision making under uncertainty. What the mathematical theory is worth, it is hard to say. It does have the advantage, though, of providing definite rules* — Richard Bellman

Ray Solomonoff

# 8 Discussion



This chapter critically reviews what has been achieved in the book and discusses some otherwise unmentioned topics of general interest. We summarize the major results and compare performance and generality of AIXI($tl$) to those of other approaches to AI. We remark on various topics, including concurrent actions and perceptions, the choice of the I/O spaces, treatment of encrypted information, and peculiarities of mortal embodied agents. We also make some personal comments and speculations on the present status and the future of the research fields of AI and machine learning themselves. We continue with an outlook on further research. Since many ideas have already been presented in the problems and conclusions sections of the various chapters, we concentrate on nontechnical open questions of general importance, including optimality, down-scaling, implementation, approximation, elegance, extra knowledge, and training of/for AIXI($tl$). Furthermore, we collect and state all explicit or implicit assumptions, problems and limitations of AIXI($tl$). We briefly discuss some relevant philosophical issues: the free will versus determinism paradox, the existence of objective probabilities, and the Turing test. We also include some (personal) remarks on non-computable physics, the number of wisdom $\Omega$, and consciousness. As it should be, the book concludes with conclusions.

## 8.1 What has been Achieved

### 8.1.1 Results

The major theme of the book was to develop a mathematical foundation of artificial intelligence. This is not an easy task since intelligence has many (often ill-defined) faces. More specifically, our goal was to develop a theory for rational agents acting optimally in any environment. Thereby we touched various scientific areas, including reinforcement learning, algorithmic information theory, Kolmogorov complexity, computational complexity theory, information theory and statistics, Solomonoff induction, Levin search, sequential decision theory, adaptive control theory, and many more. The conceptual ingredients of AIXI may be depicted in as follows:

$$\begin{array}{ccc} \text{Decision Theory} & = & \text{Probability + Utility Theory} \\ + & & + \\ \text{Universal Induction} & = & \text{Occam + Epicurus + Bayes} \\ \| & & \| \\ \multicolumn{3}{c}{\text{Universal Artificial Intelligence without Parameters}} \end{array}$$

The major achievements were the following:

- We presented the philosophical and mathematical foundations of universal induction: Occam's razor principle, Epicurus' principle of multiple explanations, subjective versus objective probabilities, Cox's axioms for beliefs, Kolmogorov's axioms of probability, conditional probability and Bayes' rule, Turing machines, Kolmogorov complexity, culminating in universal Solomonoff induction (Chapter 2).

- We derived various convergence results, (tight) loss bounds, and Pareto optimality for predictors based on Bayes mixture priors. We gave an Occam's razor argument that using Solomonoff's prior leads to a universally optimal prediction scheme (Chapter 3).
- We presented sequential decision theory in a very general form in which actions and perceptions may depend on arbitrary past events. The development was more of a formal exercise and optimality of the AI$\mu$ model for known environment $\mu$ is obvious by construction (Chapter 4).
- We unified sequential decision theory and Solomonoff's theory of universal induction, both optimal in their own domain. The resulting parameter-free AIXI model constitutes an agent for which we gave strong arguments that it behaves optimally in any environment. That is, it copes with exploration versus exploitation, large state spaces, generalization and function approximation, non-stationary and partially observable environments, and so on (Chapter 5).
- We defined a universal intelligence order relation $\prec$ regarding which AIXI is also the most intelligent agent and argued this order relation to be reasonable (Section 5.1.4).
- We discussed the difficulties in extending the optimality results from the prediction case to AIXI. Along these lines, we suggested various potentially relevant environmental (separability) concepts (Sections 5.2 and 5.3).
- We discussed the choice of the horizon and came to the conclusion that a reward discounting (like near-harmonic) which leads to an effective horizon that increases in proportion to the current age of the agent is best (Section 5.7).
- For restricted environmental classes and Bayes mixtures $\xi$ we showed that AI$\xi$ is self-optimizing and Pareto optimal (Section 5.4 and 5.6).
- We showed how AIXI is suitable for dealing with a number of important problem classes, including sequence prediction, strategic games, function minimization, and supervised learning (Chapter 6).
- Based on the mathematical (incomputable) AIXI model we developed a computable model, AIXI$tl$, with optimal order of computation time, apart from a large multiplicative constant (Section 7.2).
- We developed a general-purpose algorithm — the asymptotically fastest (and shortest) algorithm for all well-defined problems. We got rid of the large multiplicative constant as in Levin search and AIXI$tl$, at the expense of an (unfortunately even larger) additive constant (Section 7.1).

All in all, the results show that artificial intelligence can be framed by an elegant mathematical theory. Some progress has also been made toward an elegant *computational* theory of intelligence.

### 8.1.2 Comparison to Other Approaches

A different way to measure the achievements of this book is to compare AIXI($tl$) to other AI approaches. In Table 8.1 we compare various learning algorithms that are rather general in purpose, have an agent-like setup, are popular or are otherwise interesting or promising. We subjectively rate the different approaches w.r.t. various performance and generality criteria. We use a gray scale between YES and NO, since the evaluation is often arguable, especially if there are many algorithm variants. For most table cells one can imagine a variant and an application for which rating YES would be justified, and one for which rating NO would be justified. Hence, the presented ratings refer to typical algorithm variants and typical (intended) applications. It is beyond the scope of this book to describe all these approaches and to justify each rating in detail.

**Table 8.1 (Properties of learning algorithms)** The table compares various important properties of learning algorithms limited to their typical domains. The evaluation is often subjective and arguable, especially because there are many variants, so we introduce the gray scale YES → yes → yes/no → no/yes → no → NO (see Section 8.1.2 for further explanation).

| Algorithm / Properties | time efficient | data efficient | exploration | convergence | global optimum | generalization | POMDP | learning | active |
|---|---|---|---|---|---|---|---|---|---|
| Value/Policy iteration | yes/no | yes | – | YES | YES | NO | NO | NO | yes |
| TD with finite $\mathcal{S}$ | yes/no | NO | NO | YES | YES | NO | NO | YES | YES |
| TD linear func.approx. | yes/no | NO | NO | yes | yes/no | YES | NO | YES | YES |
| TD general func.approx. | no/yes | NO | NO | no/yes | NO | YES | NO | YES | YES |
| Direct Policy Search | no/yes | YES | NO | no/yes | NO | YES | no | YES | YES |
| Logic Planners | yes/no | YES | yes | YES | YES | no | no | YES | yes |
| RL with Split Trees | yes | YES | no | YES | NO | yes | YES | YES | YES |
| Pred.w. Expert Advice | yes/no | YES | – | YES | yes/no | yes | NO | YES | NO |
| Adaptive LS | no/yes | no | no | yes | yes/no | yes | YES | YES | YES |
| OOPS | yes/no | no | – | yes | yes/no | YES | YES | YES | YES |
| Market/Economy RL | yes/no | no | NO | no | no/yes | yes | yes/no | YES | YES |
| SPXI | no | YES | – | YES | YES | YES | NO | YES | NO |
| AIXI | NO | YES | YES | yes | YES | YES | YES | YES | YES |
| AIXI$tl$ | no/yes | YES | YES | YES | yes | YES | YES | YES | YES |
| Human | yes | yes | yes | no/yes | NO | YES | YES | YES | YES |

We consider the following *properties* in the different columns: An algorithm is *time efficient* if it runs on a present-day (2004) computer in acceptable time for "interesting" applications. An algorithm is *data efficient* if it exploits (learns from) the information contained in the received data in a theoretically near-optimal fashion. An algorithm gets a *yes* in the *exploration* column only if it addresses the exploration versus exploitation problem in a fundamental near-optimal way. (Many algorithms are greedy with some simple random exploration added). *Convergence* of an algorithm may be just to *some* policy, to a local optimum, or to a/the *global optimum*. An important issue is whether learning algorithms are capable of *generalizing* from previous experience to *similar* situations. We also indicate whether an algorithm is capable of or designed for dealing with non-Markovian environments, e.g. POMDPs. All selected algorithms are capable of *learning* by experience, except value and policy iteration, which need an exact description of the environment in advance. The last column distinguishes between passive predictors and *active* agents.

The first group of algorithms in Table 8.1 contains "classical" reinforcement learning algorithms. See [SB98, BT96] for a description of value and policy iteration, and temporal difference (TD) learning with finite $\mathcal{S}$tate space versus linear/general function approximation. See [BB01, KHS01a] for an introduction to direct gradient-based reinforcement learning. The second group in Table 8.1 contains various other learning algorithms: logic planners [RN95, Part IV], split trees [Rin94, McC95], adaptive Levin search [SZW97], optimal ordered problem solver (OOPS) [Sch03a, Sch04], prediction with expert advice (PEA) [CB97, HP04], market/economy-based reinforcement learning [Bau99, KHS01b]. The third group in Table 8.1 lists the main models of this book: sequence prediction based on Solomonoff's prior (SPXI) and the AIXI($tl$) model(s). The last line lists the capabilities of human agents. Schmidhuber's recent self-referential and self-improving Gödel machine [Sch03b] (not in the table) is a promising idea to overcome the huge constants in AIXI$tl$, but it is unclear for now whether self-improvements can take place if the only environment axiom is $\mu \in \mathcal{M}_U$, the utility axiom is to maximize $V_\mu^p$, and the initial software is AIXI($tl$). Overall, it can be said that the models in the first two groups are applicable in limited domains with feasible computation time, whereas the models of the last group are completely general, but computationally not feasible without further approximations.

## 8.2 General Remarks

This section remarks on some otherwise unmentioned topics of general interest. The logically disconnected subsections discuss concurrent actions and perceptions, the choice of the I/O spaces, (universal) prior knowledge, treatment of encrypted information, and peculiarities of mortal embodied agents.

### 8.2.1 Miscellaneous

**Game theory.** In game theory [OR94] one often wants to model the situation of simultaneous actions, whereas the AI$\xi$ models have serial I/O. Simultaneity can be simulated by withholding the environment from the current agent's output $y_k$, until $x_k$ has been received by the agent. Formally, this means that $\mu(y\!x_{<k}y\!\underline{x}_k)$ is independent of the last output $y_k$. The AI$\xi$ agent is already of simultaneous type in an abstract view if the behavior $p$ is interpreted as the action. In this sense, AI$\xi$ is the policy $p^\xi$ that maximizes the utility function (value), under the assumption that the environment acts according to $\xi$. The situation is different from game theory, as the environment $\xi$ is not a second 'player' that tries to optimize his own utility (see Section 6.3).

**Input/output spaces.** In various examples we have chosen differently specialized input and output spaces $\mathcal{X}$ and $\mathcal{Y}$. It should be clear that, in principle, this is unnecessary, as large enough spaces $\mathcal{X}$ and $\mathcal{Y}$ (e.g. the set of strings of length $2^{32}$) serve every need and can always be Turing-reduced to the specific presentation needed internally by the AIXI agent itself. But it is clear that, using a generic interface, such as camera and monitor for learning tic-tac-toe, for example, adds the task of learning vision and drawing.

### 8.2.2 Prior Knowledge

In many problems in practice we have extra information about the problem at hand, which could and should be used to guide the forecasting. If the prior knowledge is of the form that it includes only environments of certain structures, e.g. MDPs, one can use the appropriate Bayes mixture over these environments. If there is reason to believe that certain environments are less or more likely than Occam's razor tells us, then this could be coded in the weights $w_\nu$. Unfortunately, this procedure is often intractable in practice, since one has only a (possibly vague) description of prior facts, which are hard to translate into classes $\mathcal{M}$ and/or weights $w_\nu$. Fortunately, there is a simple way of incorporating all prior knowledge $D$ in an easy and optimal way. The trick is to get rid of all prior knowledge by prefixing the observation sequence $x_1x_2...$ by *some* binary coding $d_{1:l}$ of $D$. Using then Solomonoff's prior $M$ on $d_{1:l}x_{1:n}$ for prediction on cycles $l+1$ to $n+l$ one gets loss bounds (to logarithmic accuracy) in terms of $K(\mu|D)$. If $D$ contains information about $\mu$ it will reduce the Kolmogorov complexity of $\mu$, if not we cannot expect $D$ to improve prediction accuracy. This also solves the often mentioned concern of how to make good predictions for short sequences (sparse data) of length $n=O(1)$. As long as $n+l$ is larger than the typical compiler constants, predictions based on $\xi_U$ are good. It *seems* that in science one often faces problems with data of information content, say 200 bits only, and none or very little prior knowledge, say only 100 bits is available. For instance, having thrown a biased coin 200 times and describing our prior knowledge as "i.i.d. with uniform second-order prior over bias $\theta$" *seems* not to contribute more than 100 bits on prior information. Laplace' law of succession [Lap12] leads to reasonable estimates of $\theta$ and predictions of further tosses, whereas $\xi_U$ is far from $\mu$ in cycle 300 for typical $U$. But it is an illusion that we only have 100 bits of prior knowledge. We spent at least 12 years in school, before having heard about uniform priors and Laplace' rule. Our whole scientific knowledge serves as prior knowledge. If we take for $D$ a representative collection of scientific books (+ some language books), $l$ is much larger than the typical compiler constants and $\xi_U(x_n|d_{1:l}x_{<n})$ will be very close to the true bias $\theta$! This also holds true for more complex examples. *We* can make non-arbitrary predictions given a sequence of $l+n-1$ bits only if $\xi_U$ leads to the same prediction for all "reasonably complex" universal Turing machines $U$.

### 8.2.3 Universal Prior Knowledge

There are people who believe universal AI is not possible, that one *has* to incorporate some/sufficient prior knowledge. I disagree in a sense described in Section 8.4. A different approach is to exclude only those environments that we are sure not to be realized. This approach is worth considering, but has the following problems:

1) Physical knowledge is never 100% sure. For instance, 100 years ago everybody would have assumed a flat three-dimensional universe. I'm not too concerned about this, since today's physical theories are very accurate and reliable, at least the parts that seem to be *relevant* for (in a very broad sense) human-sized and equipped agents. Instead of eliminating universes that seem to be excluded by our observations and theories, one may only reduce the prior belief in odd universes, but this does not help in substantially increasing the prior belief of the likely universes.

2) More seriously, $\mu$ does not describe the total universe, but only a small fraction, from the subjective perspective of the agent. It is (somewhat/much?) harder to characterize the set of possible universes $\mathcal{M}$ from the subjective agent perspective.

3) One may take into account only general properties of the universe like locality, continuity, or the existence of (manipulable) objects with properties and relations in a manifold. The major problem is that, although the universe seems to be a local continuous MDP (ignoring quantum effects), $\mu$ is neither an MDP nor local. What the agent directly observes (with his sensors, like a camera) is not the complete MDP state and often appears nonlocal. So probably very little *really* exploitable can be said about $\mu$.

Of course, the scientific approach is to simply *assume* some properties (whether true in real life or not) and analyze the performance of the resulting models.

### 8.2.4 How AIXI($tl$) Deals with Encrypted Information

Consider the task of decrypting a message that was encrypted by a public key encrypter like RSA. A message $m$ is encrypted using a product $n$ of two large primes $p_1$ and $p_2$, resulting in encrypted message $c=\text{RSA}(m|n)$. RSA is a simple algorithm of size $O(1)$. If AIXI is given the public key $n$ and encrypted message $c$, in order to reconstruct the original message $m$ it only has to "learn" the function $\text{RSA}^{-1}(c|n) := \overline{\text{RSA}}(c|p_1,p_2) = m$. $\text{RSA}^{-1}$ can itself be described in length $O(1)$, since $\overline{\text{RSA}}$ is $O(1)$ and $p_1$ and $p_2$ can be reconstructed from $n$. Only very little information is needed to learn $O(1)$ bits. In this sense decryption is easy for AIXI (like TSP, see Section 6.4.6). The problem is that while $\overline{\text{RSA}}$ is efficient, $\text{RSA}^{-1}$ is an extremely slow algorithm, since it has to find the prime factors from the public key. But note, in AIXI we are not talking about computation time, we are only talking about information efficiency (learning in the least number of interaction cycles). One of the key insights in this book that allowed for an elegant theory of AI was this separation of data efficiency from computation time efficiency. Of course, in the real world computation time matters, so we introduced AIXI$tl$. AIXI$tl$ can do every job as well as the best length $l$ and time $t$ bounded agent, apart from time factor $2^l$ and a huge offset time. No practical offset time is sufficient to find the factors of $n$, but in theory, enough offset time allows also AIXI$tl$ to (once-and-for-all) find the factorization, and then, decryption is easy of course.

### 8.2.5 Mortal Embodied Agents

The examples we gave in this book, particularly those in Chapter 6, were mainly bodiless agents: predictors, gamblers, optimizers, learners. There are some peculiarities with reinforcement-learning autonomous embodied robots in real environments.

We can still reward the robot according to how well it solves the task we want it to do. A minimal requirement is that the robot's hardware functions properly. If the robot starts to malfunction its capabilities degrade, resulting in lower reward. So, in an attempt to maximize reward, the robot will also maintain itself. The problem is that some parts will malfunction rather quickly when no appropriate actions are performed, e.g. flat batteries, if not recharged in time. Even worse, the robot may work perfectly until the battery is nearly empty, and then suddenly stop its operation (death), resulting in zero reward from then on. There is too little time to learn how to maintain itself before it's too late. An autonomous embodied robot cannot start from scratch but must have some rudimentary built-in capabilities (which may not be that rudimentary at all) that allow it to at least survive. This is similar to the problem discussed in Section 6.4.5 of using AIXI in the FMF setting with too late a reward. Using FMF$\xi$ instead, corresponds to incorporating some rudimentary capability. Animals survive due to reflexes, innate behavior, an internal reward attached to the condition of their organs, and a guarding environment during childhood. Different species emphasize different aspects. Reflexes and innate behaviors are stressed in lower animals versus years of safe childhood for humans. The same variety of solutions is available for constructing autonomous robots (which we will not detail here).

Another problem connected, but possibly not limited to embodied agents, especially if they are rewarded by humans, is the following: Sufficiently intelligent agents may increase their rewards by psychologically manipulating their human "teachers", or by threatening them. This is a general sociological problem that successful AI will cause, which has nothing specifically to do with AIXI. Every intelligence superior to humans is capable of manipulating the latter. In the absence of manipulable humans, e.g. where the reward structure serves a survival function, AIXI may directly hack into its reward feedback. Since this is unlikely to increase its long-term survival, AIXI will probably resist this kind of manipulation (just as most humans don't take hard drugs, because of their long-term catastrophic consequences).

## 8.3 Personal Remarks

It is hard to predict the future, as it is to predict the development of research areas like AI. Nevertheless, I would like to risk a try. To be more specific, in the following I suggest a framework for machine learning research. It is a mixture of how I expect and would like the field to look in the near future.

### 8.3.1 On the Foundations of Machine Learning

Instead of addressing machine learning directly, let us first consider a different research area such as algorithm and complexity theory. The goal of algorithm theory is to find and analyze fast algorithms, while the goal of complexity theory is to show lower bounds on the time needed to solve certain problem classes. All concepts are rigorously defined: algorithm, Turing machine, problem class, computation time, and so on. Most disciplines generally start with an informal way of attacking a subject. With time the discipline becomes more and more formalized, often up to a point where it is completely rigorous. Examples are number theory, set theory, proof theory, probability theory, infinitesimal calculus, and quantum field theory. Each theory experienced a time in which it was dealt with in an informal way, but after a while it was made rigorous, is now completely axiomatized, and is rarely questioned.[1] Of course, not all disciplines are axiomatized yet or are axiomatizable at all (e.g. biology), and new research areas emerge, starting in an informal condition, but the point is that once a field has emerged, the path is toward increasing rigor.

What can be said about machine learning? In machine learning one tries to build and understand systems that learn from past data, make good predictions, are able to generalize, act intelligently, etc. Many of these and other terms are only vaguely defined or have many alternative definitions. As discussed in Chapter 2 and elsewhere, from a formal point of view, all learning tasks can be unified in the framework of sequence prediction. We propose Occam's razor, quantified in terms of Kolmogorov complexity or Solomonoff's

[1] Quantum field theory may be argued not to be in a completely mathematically satisfactory condition, yet.

prior, combined with the chain rule for conditional probabilities, and possibly sequential decision theory as a rigorous mathematical/axiomatic definition of machine learning. More precisely, Solomonoff's induction scheme should be "used" for sequence prediction tasks, and, when combined with sequential decision theory, for making sequential decisions. The results of this book and also the results of the more applied MML, MDL, and SRM principles support the power of Occam's razor. As long as there is no convincing evidence against Occam's razor, and what is even more important, as long as there is no alternate suggestion of how to define machine learning rigorously, it is worth assuming Occam's razor and studying its consequences. Indeed, we showed in Chapter 3 that the performance of Solomonoff's universal induction scheme, as compared to any other prediction scheme in any environment, is so good that one may be tempted to take the results as proof of Occam's razor. Whereas the theorems were proven with mathematical rigor, one has to be careful about their interpretation and the underlying assumptions, especially in Theorem 3.70, which was more a self-consistency or bootstrap result.

We expect that in the future machine learning will, by default, be based on Occam's razor. Real-world machine learning tasks will with overwhelming majority be solved by developing algorithms that approximate Kolmogorov complexity or Solomonoff's prior (e.g. MML, MDL, SRM, and more specific ones, like SVM, LZW, neural/Bayes nets with complexity penalty, ...). Machine learning theory will derive results on convergence speed and approximation quality of the various approximation schemes. Only a minority will investigate nonstandard ML by modifying or replacing Occam's razor "axiom" in the hope of finding something better.

### 8.3.2 In a World Without Occam

Finally, I would like to remark on an analogy to Peano's axioms for the natural numbers and especially the induction axiom. Remove this axiom and replace it with a vague concept that resembles this axiom.[2] It is then not unreasonable to call this concept *induction principle* since it infers properties valid for *all* natural numbers from a local $n \to n+1$ property. Imagine if arithmetic were still in this situation. Most modern mathematical theorems would evaporate, and with them nearly all modern technology. Fortunately, we have this induction axiom, but as a formal rule it is now purely deductive!

I believe the same is true for Occam's razor. Without the vague concept of Occam's razor, science and, hence, machine learning would probably be not existent at all. Informal Occam's razor is directly or indirectly the basis of all scientific induction. The establishment of a formal version of Occam's razor would give machine learning in particular, and maybe even science in general, a significant boost. I anticipate this by looking at the practical success

[2] I guess that there was a time in history when arithmetic was exactly in this condition.

of MML, MDL, SRM, and SVM, and the theoretical impact of Kolmogorov complexity and Solomonoff induction, which are all formalizations of Occam's razor.

## 8.4 Outlook & Open Questions

Many ideas for further studies were already stated in the various chapters of the book, especially in the problems and conclusions sections. This outlook only contains nontechnical open questions regarding AIXI($tl$) of general importance.

**Value bounds.** Rigorous proofs for non-asymptotic value bounds for AI$\xi$ are the major theoretical challenge — general ones, as well as tighter bounds for special environments $\mu$, e.g. for rapidly mixing MDPs. For AIXI other performance criteria have to be found and proved. Although not necessary from a practical point of view, the study of continuous classes $\mathcal{M}$, restricted policy classes, and/or infinite $\mathcal{Y}$, $\mathcal{X}$ and $m$ may lead to useful insights.

**Scaling AIXI down.** A direct implementation of the AIXI$tl$ model is, at best, possible for small-scale (toy) environments due to the large factor $2^l$ in computation time. But there are other applications of the AIXI theory. We saw in several examples how to integrate problem classes into the AIXI model. Conversely, one can downscale the AI$\xi$ model by using more restricted forms of $\xi$. This could be done in the same way as the theory of universal induction was downscaled with many insights to the minimum description length principle [LV92a, Ris89] or to the domain of finite automata [FMG92]. The AIXI model might similarly serve as a supermodel or as the very definition of (universal unbiased) intelligence, from which specialized models could be derived.

**Implementation and approximation.** With a reasonable computation time, the AIXI model would be a solution of AI (see the next point if you disagree). The AIXI$tl$ model was the first step, but the elimination of the factor $2^l$ without introducing a large additive constant like in $M_{p*}^\varepsilon$ and without giving up universality will almost certainly be a very difficult task. One could try to select programs $p$ and prove VA($p$) in a more clever way than by mere enumeration, to improve performance without destroying universality. All kinds of ideas like genetic algorithms, advanced theorem provers, and many more could be incorporated. It remains to be seen whether these handwaving suggestions can be substantiated. The Gödel machine is a promising recent approach.

**Computability.** We seem to have transferred the AI problem just to a different level, to proving VA($p$). This shift has some advantages (and also some disadvantages) but does not present a practical solution. Nevertheless, we want to stress that we have reduced the AI problem to (mere) computational questions. Even the most general other systems the author is aware of depend on some (more than complexity) assumptions about the environment, or it is far from clear whether they are, indeed, universally optimal. Although computational questions are themselves highly complicated, this reduction is a nontrivial result. A formal theory of something, even if not computable, is often a great step toward solving a problem and also has merits of its own, and AI should not be different in this respect (see previous item).

**Elegance.** Many researchers in AI believe that intelligence is something complicated and cannot be condensed into a few formulas. It is more a combining of enough *methods* and much explicit *knowledge* in the right way. From a theoretical point of view we disagree, as the AIXI model is simple and seems to serve all needs. From a practical point of view we agree, to the following extent: To reduce the computational burden one should provide special-purpose algorithms (*methods*) from the very beginning, probably many of them related to reduce the complexity of the input and output spaces $\mathcal{X}$ and $\mathcal{Y}$ by appropriate pre/postprocessing *methods*.

**Extra knowledge.** There is no need to incorporate extra *knowledge* from the very beginning. It can be presented in the first few cycles in *any* format. As long as the algorithm to interpret the data is of size $O(1)$, the AIXI agent will "understand" the data after a few cycles (see Sections 8.2.2 and 6.5). If the environment $\mu$ is complicated but extra knowledge $z$ makes $K(\mu|z)$ small, one can show that the bound (5.9)–(5.10) reduces roughly to $\ln 2 \cdot K(\mu|z)$ when $x_1 \equiv z$, i.e. when $z$ is presented in the first cycle. The special-purpose algorithms could be presented in $x_1$ too, but it would be cheating to say that no special-purpose algorithms were implemented in AIXI. The boundary between implementation and training is unsharp in the AIXI model.

**Training.** We have not said much about the training process itself, as it is not specific to the AIXI model and has been discussed in literature in various forms and disciplines [Sol86, Sch03a, Sch04]. By a training process we mean a sequence of simple-to-complex tasks to solve, with the simpler ones helping in learning the more complex ones. A serious discussion would be out of place. To repeat a truism, it is, of course, important to present enough knowledge $o_k$ and evaluate the agent output $y_k$ with $r_k$ in a reasonable way. To maximize the information content in the reward, one should start with simple tasks and give positive reward to approximately the better half of the outputs $y_k$.

## 8.5 Assumptions, Problems, Limitations

Just as every approach to AI (or any other field) makes assumptions and has its problems and limitations, so does AIXI($tl$). It is time to take a critical look at all explicit or implicit assumptions, problems and limitations.

### 8.5.1 Assumptions

- The central assumption of this book is Occam's razor. Since Occam's razor seems to be at the heart of science and intelligent behavior in any case, it is not a restrictive assumption, but nevertheless a profound one. Occam's razor actually only serves as a motivation in this book; the actual assumption we use is different, see next item.

- The environment is sampled from a computable probability distribution with a reasonable program size on a natural Turing machine. Assumption 2.5 ensures that AIXI($tl$) is essentially independent of whatever universal Turing machine is chosen.

- We assumed the existence of objective randomness/probabilities respecting Kolmogorov's probability Axioms 2.14. As remarked in Section 2.3, this assumption is not essential, since we can restrict the setting to deterministic environments $\mu$. Using Bayes mixtures as subjective probabilities also did not involve any assumptions, since they were justified decision-theoretically.

- In reinforcement learning, the total reward is defined as the *sum* of rewards $r_1+...+r_m$ over cycles, and so did we. In finance, where money can be reinvested, a product is common, but this can be converted to a sum by taking the logarithm. More generally, assume the goal is to maximize some function of the rewards $R_k := R(r_1,...,r_k)$, e.g. their maximum ($R=\max$, see Section 6.4.3). If we reward AIXI (5.3) in each cycle $k$ not with $r_k$ but instead with $r'_k := R_k - R_{k-1}$ (with $R_0 := 0$), we see that AIXI tries to maximize expected $r'_1+...+r'_m \equiv R_m$, as desired. The original rewards may need to be retained as observations ($o'_k := o_k r_k$). So, restricting to reward sums is not a real limitation.

- For probabilistic environments we defined the value of a policy which shall be maximized in the standard way as the *expected* reward sum. More generally, one may define for each policy a probability distribution for the total reward. The question is how to compare these distributions. Besides the popular mean, one may want to compare medians or quantiles. The most frequent argument for departing from the mean is to achieve more robust policies, e.g. policies that, with high probability, have a high lower bound on their reward sum. We believe that robustness is never a primary goal in itself. The reason for wanting robustness is that one dislikes low rewards more than the rewards themselves express. A natural solution is to take the expectation of $f$-transformed rewards, where $f$ is a monotone increasing concave function (like log). Function $f$ penalizes small rewards and leads to more robustness.

- We assumed finite action/perception spaces $\mathcal{Y}/\mathcal{X}$, which are sufficient for all practical purposes. From a theoretical point of view infinite spaces may be attractive in certain situations. Countable $\mathcal{X}$ should cause no problems;

in case of countable $\mathcal{Y}$ only $\varepsilon$-optimal policies may exist. For continuous $\mathcal{X}$ and $\mathcal{Y}$ one has to somehow generalize the notion of Kolmogorov complexity and Solomonoff's prior. Continuous environmental classes $\mathcal{M}$ were briefly discussed in Section 3.7.2.

- We assumed bounded nonnegative rewards $r_k \in [0, r_{max}]$. Nonnegativity is not essential, but boundedness is essential for ensuring existence of values. Again, from a practical point of view this should not be restrictive.
- We assumed finite horizon or near-harmonic discounting to ensure the existence of values. We provided motivations for the choice of the latter, but we are not sure whether it represents a final answer.

After all this, one should not forget that all other known approaches to AI implicitly or explicitly make (many) more assumptions.

### 8.5.2 Problems

- Assume AIXI is used in a multi-agent setup interacting with other agents. For simplicity we only discuss the case of a single other agent in a competitive setup, i.e. a two-person zero-sum game situation. We can entangle agents $A$ and $B$ by $o_k(A) = y_k(B)$, $o_{k+1}(B) = y_k(A)$. The rewards $r_k(A)$ and $r_k(B)$ are provided externally by the rules of the game. The situation where $A$ is AIXI and $B$ is a perfect minimax player was analyzed in Section 6.3. In multi-agent systems one is mostly interested in a symmetric setup, i.e. $B$ is also an AIXI. Whereas both AIXIs *may* be able to learn the game and improve their strategies (to optimal minimax), this setup violates one of our basic assumptions. Since AIXI is incomputable, AIXI($B$) does not constitute a computable environment for AIXI($A$). More generally, starting with any class of environments $\mathcal{M}$, then $\mu \widehat{=} \text{AI}\xi_{\mathcal{M}}$ seems not to belong to class $\mathcal{M}$ for most (all?) choices of $\mathcal{M}$. Various results of the book can no longer be applied, since $\mu \notin \mathcal{M}$ when coupling two AI$\xi$s. Many questions arise: Are there interesting environmental classes for which $\text{AI}\xi_{\mathcal{M}} \in \mathcal{M}$ or $\text{AI}\xi tl_{\mathcal{M}} \in \mathcal{M}$? Do AIXI($A/B$) converge to optimal minimax players? Do AIXIs perform well in general multi-agent setups?

### 8.5.3 Limitations

- Although AIXI may be regarded as a formal definition or a mathematical solution of AI, it is not a practical solution due to its incomputability. AIXI*tl* is a step in the direction of a computable theory of AI, but is also not practically feasible. Whether AIXI can be scaled down in a systematic way to yield practical AI algorithms or whether it will only serve as a guiding principle in attacking difficult AI problems remains to be seen.

## 8.6 Philosophical Issues

Many arguments against strong and weak AI have been proposed: Lucas' and Penrose's arguments based on Gödel's incompleteness theorem [Luc61, Pen89, Pen94], Searle's Chinese room argument [Sea80], the lookup table argument [Chu86], Moravec's brain prosthesis experiment [Mor88], the free will argument and, of course, various religious reasons. All of them have loopholes and can be refuted, but this is not the place to repeat this discussion. We only discuss the free will paradox. There are also objections to the existence of objective probabilities. We present a possibly new one below. We also briefly comment on the famous Turing test, which also fits under the heading of this section. Finally, we speculate on the *big* questions of AI in general and the AIXI model in particular, related to non-computable physics, the number of wisdom $\Omega$, and consciousness.

### 8.6.1 Turing Test

The Turing test [Tur50] was designed to decide whether an AI system is intelligent. We should concede a machine true intelligence if it passes the Turing test, but to deny intelligence in case of failure may be too harsh. The true problem in using the Turing test (e.g. instead of AIXI) as a *definition* of intelligent systems is another. The test involves a human interrogator and, hence, cannot be formalized mathematically, therefore it does also not allow the development of a computational theory of intelligence.

### 8.6.2 On the Existence of Objective Probabilities

Throughout the book we have assumed the existence of objective probabilities respecting Kolmogorov's probability Axioms 2.14. As remarked in Section 2.3 we could have restricted the development to classes $\mathcal{M}$ of deterministic environments, thus avoiding objective probabilities.[3] In the following we give an argument which makes the belief in objective probabilities look somewhat "unscientific". The assumption that an event occurs with some objective probability expresses the opinion that the occurrence of an individual stochastic event has no explanation, i.e. is inherently impossible to predict for sure. One central goal of science is to *explain* things. Often we do not have an explanation (yet) that is acceptable, but to say that "something can principally not be explained" means to stop even *trying* to find an explanation. From a distance, tossing a coin looks objectively random, but looking at it closer the outcome is just subjectively unknown due to most observers' lack of knowledge of the initial conditions and external influences on the coin during its throw. When

[3] Using Bayes mixtures as subjective probabilities did not need any (e.g. Cox's) axioms for justification, but received a decision-theoretic justification.

knowing the exact initial conditions and the exact equations of motion, classical physics is predictable (this includes chaotic systems). Physicists claim that quantum mechanics is truly random, and there is indeed quite some evidence to suggest this, but experiments cannot exclude the possibility that quantum events are only pseudo-random [Sch02b]. It seems safer and more honest to say that with our current technology and understanding we can only determine (subjective) outcome probabilities. If a sufficiently large community of people arrive at the same subjective probabilities from their prior knowledge, one may want to call these probabilities objective. For instance, for most people (those with no special equipment and education) a fair coin comes up head in 50% of the cases. And for *all* people so far, if they measure the spin of one photon in a para-positronium decay, it is up in 50% of the cases. On one hand, we have to abandon objective probabilities because their assumption seems unscientific, but on the other hand, their assumption is very convenient. Without objective probabilities there would be no (objective) unbiased coins, dice, MDPs, radioactive decays, etc. Maybe one should admit a gray scale of more or less subjective probabilities.

### 8.6.3 Free Will versus Determinism

For illustrational purpose we replace determinism with computability.

**The paradox.** If the brain of a human were computable we could predict the action of the human with a computer. If we tell the human his action in advance, he is forced to perform this action and hence loses his free will. Assuming humans have free will refutes the computability assumption of the brain.

This paradox between computability and free will is sometimes used as an argument against the possibility of AI. However, it vanishes by a more careful reasoning. That a part of the universe is computable, is defined as follows:

**Assumption 1.** Given a box (part of the universe) in state $s$ at time $t$ we can compute the next (or some more distant future) state $s'$ at time $t'>t$ *if* there is no interaction between the box and the rest of the universe during time $t...t'$.

Without this independence assumption in time interval $[t,t']$ the possibility of correct prediction cannot be guaranteed.

**Assumption 2.** Assume that the brain is computable. It receives input $x$ at time $t$ and computes action $y$ at time $t'$. During the thinking period $[t,t']$ it is completely separated from the environment.

After input $x$, the brain $B$ is in a state $s$ and Assumption 1 applies, i.e. we can compute, say with algorithm $p:\mathcal{X}\to\mathcal{Y}$, the brain's decision $y$. We cannot inform the brain in period $[t,t']$ of this decision without violating Assumption 2. We are free to hand out $y$ in a closed envelope to $B$. After $B$ has made its decision, it is allowed to open the letter and realizes that its decision was predictable. Such an experiment will have enormous psychological, social, and legal consequences, and looks paradoxical, but does not lead to any contradictions!

Assume we allow intermittent interaction and inform the brain about $y$, then the brain $B'$ maps input $(x,y)$ to an action $y'$, which is possibly different from $y$. There is no contradiction, since $p$ maps $\mathcal{X}$ to $\mathcal{Y}$, whereas $B'$ maps $\mathcal{X}\times\mathcal{Y}\to\mathcal{Y}$, so these functions have nothing to do with each other.

Consider a variant of the paradox, where $B$ itself reliably predicts/precomputes its own action, and then "decides" to deviate from its own prediction. More formally, the assumption is that a part $B_2$ of brain $B\equiv B_1$ can simulate $B_1$. By assumption $B_2$ is functionally identical to $B_1$. Only for illustrational purposes we make the further assumption that $B_2$ also operates identically to $B_1$ in the sense that $B_2$ itself contains a part, say $B_3$, which simulates $B_2$, etc. We have an infinite recursion. The first question is not *what* the output of $B$ is and whether it is finitely computable, but whether this infinite recursion has a value *at all*. What we need is a fixed point. Insert a decision $y$ (as a possible candidate for $B_2$) into brain $B$ and test whether $B$ computes the same decision $y$. If it does, then $y$ is a fixed point of the recursion. If such a fixed point exists (and is unique) we may define the value of the infinite recursion as this fixed-point value. Finally, we would have to check whether this fixed point can be found by a finite algorithm. It is well known that not every recursion $y=B(y)$ has a fixed point. The paradox in our case is just that we implicitly assumed the existence of a (unique) fixed point. The paradox is resolved by noting that this fixed point simply does not exist. Sometimes fixed points can be found by iteration. One starts with some value $y_1$ for $y$ and iterates $y_2=B(y_1)$, ..., $y_n=B(y_{n-1})$. If the limit $y_\infty$ exists, then it is a fixed point.

Assume our function $B$ acts with $y=1$ if $B_2$ predicts $y=0$, and vice versa. In this case $y_n=1-y_{n-1}$ oscillates, and $y_\infty$ does not exist. In the case of a binary decision $y\in\mathcal{Y}\in\{0,1\}$ this proves that a fixed point does not exist. So this paradox is about nonexistent fixed points. A self-contradictory brain simply does not exist (when starting from Assumptions 1 and 2). We may lower our demands to probabilistic predictions. Mathematically we lift the mapping $B:\mathcal{Y}\to\mathcal{Y}$ to a linear mapping $B':[0,1]^\mathcal{Y}\to[0,1]^\mathcal{Y}$ between (probability) vectors over $\mathcal{Y}$. $B'$ always has a fixed point. For our example we get $B'(p)=1-p$ with fixed point $p=\frac{1}{2}$, meaning that $B_2$ has no idea of what $B$ is going to do. Whatever, it is not possible to set up a brain with a part predicting its own behavior reliably in every situation. The same analysis holds for (an infinite regression of) *external* predictors, telling the human his action in advance.

Note that neither the paradox, nor the solution has anything specific to do with *computable functions*. We could have formulated the paradox in terms of general mathematical functions (mappings).

### 8.6.4 The Big Questions

**On non-computable physics & brains.** There are two possible objections to AI in general and, therefore, to AIXI in particular. Non-computable physics (which is not too odd) could make Turing-computable AI impossible. As at least the world that is relevant for humans seems mainly to be computable, we do not believe that it is necessary to integrate non-computable devices into an AI system. The (clever and nearly convincing) Gödel argument by Penrose [Pen89, Pen94], refining Lucas [Luc61], that non-computational physics *must* exist and *is* relevant to the brain, has (in our opinion convincing) loopholes.

**Evolution & the number of wisdom.** A more serious problem is the evolutionary information gathering process. It has been shown that the 'number of wisdom' $\Omega$ contains a very compact tabulation of $2^n$ undecidable problems in its first $n$ binary digits [Cha75, CHKW98]. $\Omega$ is only enumerable with computation time increasing more rapidly with $n$ than any recursive function. The enormous computational power of evolution could have developed and coded something like $\Omega$ into our genes, which significantly guides human reasoning [Cha91]. In short: Intelligence could be something complicated, and evolution toward it from an even cleverly designed algorithm of size $O(1)$ could be too slow. As evolution has already taken place, we could add the information from our genes or brain structure to any/our AI system, but this would mean that the important part is still missing, and that it is principally impossible to derive an efficient algorithm from a simple formal definition of AI.

**Consciousness.** For what is probably the *biggest question*, that of *consciousness*, we want to give a physical analogy. Quantum (field) theory is the most accurate and universal physical theory ever invented. Although already developed in the 1930s, the *big* question, regarding the interpretation of the wave function collapse, is still open. Although this is extremely interesting from a philosophical point of view, it is completely irrelevant from a practical point of view.[4] We believe the same to be valid for *consciousness* in the field of artificial intelligence: philosophically highly interesting but practically unimportant. Whether consciousness *will* be explained some day is another question.

## 8.7 Conclusions

All tasks that require intelligence to be solved can naturally be formulated as a maximization of some expected utility in the framework of agents. We presented a functional (4.7) and an iterative (4.17) formulation of such a decision-theoretic agent in Chapter 4, which is general enough to cover all AI problem

[4] In the Theory of Everything, the collapse might become of "practical" importance and must or will be solved.

classes, as demonstrated by several examples. The main remaining problem is the unknown prior probability distribution $\mu$ of the environment(s). Conventional learning algorithms are unsuitable, because they can neither handle large (unstructured) state spaces, nor do they converge in the theoretically minimal number of cycles, nor can they handle non-stationary environments appropriately. On the other hand, Solomonoff's universal prior $M \stackrel{\times}{=} \xi_U$ (2.26), based on ideas from algorithmic information theory, solves the problem of the unknown prior distribution for induction problems, as was demonstrated in Chapters 2 and 3. No explicit learning procedure is necessary, as $\xi_U$ automatically converges to $\mu$. We unified the theory of universal sequence prediction with the decision-theoretic agent by replacing the unknown true distribution $\mu$ by an appropriately generalized universal semimeasure $\xi$ in Chapter 5. We gave various arguments that the resulting AIXI model is the most intelligent, parameter-free and environmental/application-independent model possible. We defined an intelligence order relation (5.14) to give a rigorous meaning to this claim. Furthermore, possible solutions to the horizon problem were discussed. In Chapter 6 we outlined how the AIXI model solves various problem classes. These included sequence prediction, strategic games, function minimization and, especially, learning to learn supervised. The list could easily be extended to other problem classes like classification, function inversion and many others. The major drawback of the AIXI model is that it is uncomputable, or more precisely, only asymptotically computable, which makes an implementation impossible. To overcome this problem, we constructed a modified model AIXI$tl$, which is still effectively more intelligent than any other time $t$ and length $l$ bounded algorithm (Section 7.2). The computation time of AIXI$tl$ is of the order $t \cdot 2^l$. A way of overcoming the large multiplicative $2^l$ constant was presented at the expense of an (unfortunately even larger) additive constant (Section 7.1). Possible further research was discussed. The main directions could be to prove general and special reward bounds, use AIXI as a supermodel and explore its relation to other specialized models and finally improve performance with or without giving up universality.

# Bibliography


[ACBFS02] P. Auer, N. Cesa-Bianchi, Y. Freund, and R. E. Schapire. The non-stochastic multiarmed bandit problem. *SIAM Journal on Computing*, 32(1):48–77, 2002.

[ACBG02] P. Auer, N. Cesa-Bianchi, and C. Gentile. Adaptive and self-confident on-line learning algorithms. *Journal of Computer and System Sciences*, 64(1):48–75, 2002.

[Ačz66] J. Ačzel. *Lectures on Functional Equations and Their Applications*. Academic Press, New York, 1966.

[AS83] D. Angluin and C. H. Smith. Inductive inference: Theory and methods. *ACM Computing Surveys*, 15(3):237–269, 1983.

[Bar00] A. R. Barron. Limits of information, Markov chains, and projection. In *Proc. IEEE International Symposium on Information Theory (ISIT)*, pages 25–25, Sorrento, Italy, 2000.

[Bau99] E. B. Baum. Toward a model of intelligence as an economy of agents. *Machine Learning*, 35(2):155–185, 1999.

[Bay63] T. Bayes. An essay towards solving a problem in the doctrine of chances. *Philosophical Transactions of the Royal Society*, 53:376–398, 1763. [Reprinted in *Biometrika*, 45, 243–315, 1958].

[BB01] J. Baxter and P. L. Bartlett. Infinite-horizon policy-gradient estimation. *Journal of Artificial Intelligence Research*, 15:319–350, 2001.

[BD62] D. Blackwell and L. Dubins. Merging of opinions with increasing information. *Annals of Mathematical Statistics*, 33:882–887, 1962.

[BEHW87] A. Blumer, A. Ehrenfeucht, D. Haussler, and M. K. Warmuth. Occam's razor. *Information Processing Letters*, 24(6):377–380, 1987.

[BEHW89] A. Blumer, A. Ehrenfeucht, D. Haussler, and M. K. Warmuth. Learnability and the Vapnik-Chervonenkis dimension. *Journal of the ACM*, 36(4):929–965, 1989.

[Bel57] R. E. Bellman. *Dynamic Programming*. Princeton University Press, Princeton, NJ, 1957.

[Ben98] C. H. Bennett et al. Information distance. *IEEE Transactions on Information Theory*, 44, 1998.

[Ber13] J. Bernoulli. *Ars Conjectandi*. Thurnisiorum, Basel, 1713. [Reprinted in: *Die Werke von Jakob Bernoulli*, pages 106–286, volume 3, Birkhäuser, Basel, 1975, and in: *A Source Book in Mathematics*, pages 85–90, Dover, New York, 1959. English translation of part IV (with limit theorem) by Bing Sung, Harvard Univ. Dept. of Statistics, Technical Report #2, 1966].

[Ber95a] D. P. Bertsekas. *Dynamic Programming and Optimal Control, volume I*. Athena Scientific, Belmont, MA, 1995.

[Ber95b] D. P. Bertsekas. *Dynamic Programming and Optimal Control, volume II*. Athena Scientific, Belmont, MA, 1995.

[BEYL04] Y. Baram, R. El-Yaniv, and K. Lutz. Online choice of active learning algorithms. *Journal of Machine Learning Research*, 5:255–291, 2004.

[BF85] D. A. Berry and B. Fristedt. *Bandit Problems: Sequential Allocation of Experiments.* Chapman and Hall, London, 1985.

[BFLS91] L. Babai, L. Fortnow, L. A. Levin, and M. Szegedy. Checking computations in polylogarithmic time. *STOC: 23rd ACM Symp. on Theory of Computation*, 23:21–31, 1991.

[BG79] C. H. Bennett and M. Gardner. The random number Omega bids fair to hold the mysteries of the universe. *Scientific American*, 241:20–34, 1979.

[BGHK92] F. Bacchus, A. Grove, J. Y. Halpern, and D. Koller. From statistics to beliefs. In *Proc. 10th National Conf. on Artificial Intelligence (AAAI-92)*, pages 602–608, San Jose, CA, 1992. AAAI Press.

[Blu67] M. Blum. A machine-independent theory of the complexity of recursive functions. *Journal of the ACM*, 14(2):322–336, 1967.

[Blu71] M. Blum. On effective procedures for speeding up algorithms. *Journal of the ACM*, 18(2):290–305, 1971.

[BM98] A. A. Borovkov and A. Moullagaliev. *Mathematical Statistics.* Gordon & Breach, 1998.

[BS84] B. G. Buchanan and E. H. Shortliffe. *Rule-Based Expert Systems: The MYCIN Experiments of the Stanford Heuristic Programming Project.* Addison Wesley, Reading, MA, 1984.

[BSA83] A. G. Barto, R. S. Sutton, and C. W. Anderson. Neuronlike adaptive elements that can solve difficult learning control problems. *IEEE Transactions on Systems, Man, and Cybernetics*, SMC-13:834–846, 1983.

[BT96] D. P. Bertsekas and J. N. Tsitsiklis. *Neuro-Dynamic Programming.* Athena Scientific, Belmont, MA, 1996.

[Cal02] C. Calude. *Information and Randomness.* Springer, Berlin, 2nd edition, 2002.

[Can74] G. Cantor. Über eine Eigenschaft des Inbegriffs aller reellen algebraischen Zahlen. *Journal für reine und angewandte Mathematik*, 77:258–262, 1874. [English translation: On a property of the set of real algebraic numbers. In *A Source Book in the Foundations of Mathematics*, volume 2, pages 839–843, Clarendon, Oxford].

[Car63] G. Cardano. Liber de ludo aleae, 1565/1663. Published in 1663 but completed already around 1565.

[Car48] R. Carnap. On the application of inductive logic. *Philosophy and Phenomenological Research*, 8:133–148, 1948.

[Car50] R. Carnap. *Logical Foundations of Probability.* University of Chicago Press, Chicago, 1950.

[CB90] B. S. Clarke and A. R. Barron. Information-theoretic asymptotics of Bayes methods. *IEEE Transactions on Information Theory*, 36:453–471, 1990.

[CB97] N. Cesa-Bianchi et al. How to use expert advice. *Journal of the ACM*, 44(3):427–485, 1997.

[CBL01] N. Cesa-Bianchi and G. Lugosi. Worst-case bounds for the logarithmic loss of predictors. *Machine Learning*, 43(3):247–264, 2001.

[Cha66] G. J. Chaitin. On the length of programs for computing finite binary sequences. *Journal of the ACM*, 13(4):547–569, 1966.

[Cha69] G. J. Chaitin. On the length of programs for computing finite binary sequences: Statistical considerations. *Journal of the ACM*, 16(1):145–159, 1969.

[Cha75] G. J. Chaitin. A theory of program size formally identical to information theory. *Journal of the ACM*, 22(3):329–340, 1975.

[Cha91] G. J. Chaitin. Algorithmic information and evolution. In *Perspectives on Biological Complexity*, pages 51–60. IUBS Press, 1991.

[Che85] P. Cheeseman. In defense of probability. In *Proc. 9th International Joint Conf. on Artificial Intelligence*, pages 1002–1009, Los Altos, CA, 1985. Morgan Kaufmann.

[Che88] P. Cheeseman. An inquiry into computer understanding. *Computational Intelligence*, 4(1):58–66, 1988.

[CHKW98] C. S. Calude, P. H. Hertling, B. Khoussainov, and Y. Wang. Recursively enumerable reals and Chaitin $\Omega$ numbers. In *15th Annual Symposium on Theoretical Aspects of Computer Science*, volume 1373 of *LNCS*, pages 596–606, Paris, 1998. Springer, Berlin.

[Chu40] A. Church. On the concept of a random sequence. *Bulletin of the American Mathematical Society*, 46:130–135, 1940.

[Chu86] P. S. Churchland. *Neurophilosophy: Toward a Unified Science of the Mind-Brain.* MIT Press, Cambridge, MA, 1986.

[Con97] M. Conte et al. Genetic programming estimates of Kolmogorov complexity. In *Proc. 17th International Conf. on Genetic Algorithms*, pages 743–750, East Lansing, MI, 1997. Morgan Kaufmann, San Francisco, CA.

[Cov74] T. M. Cover. Universal gambling schemes and the complexity measures of Kolmogorov and Chaitin. Technical Report 12, Statistics Department, Stanford University, Stanford, CA, 1974.

[Cox46] R. T. Cox. Probability, frequency, and reasonable expectation. *American Journal of Physics*, 14(1):1–13, 1946.

[Csi67] I. Csiszár. Information-type measures of difference of probability distributions and indirect observations. *Studia Scientiarum Mathematicarum Hungarica*, 2:299–318, 1967.

[CT91] T. M. Cover and J. A. Thomas. *Elements of Information Theory.* Wiley Series in Telecommunications. Wiley, New York, 1991.

[CV03] R. Cilibrasi and P. M. B. Vitányi. Clustering by compression. Technical report, CWI, Amsterdam, 2003. http://arXiv.org/abs/cs/0312044.

[CW82] D. Coppersmith and S. Winograd. On the asymptotic complexity of matrix multiplication. *SIAM Journal on Computing*, 11(3):472–492, 1982.

[CW90] D. Coppersmith and S. Winograd. Matrix multiplication via arithmetic progressions. *Journal of Symbolic Computation*, 9(3):251–280, 1990.

[Dal73] R. P. Daley. Minimal-program complexity of sequences with restricted resources. *Information and Control*, 23(4):301–312, 1973.

[Dal77] R. P. Daley. On the inference of optimal descriptions. *Theoretical Computer Science*, 4(3):301–319, 1977.

[Dau90] J. W. Dauben. *Georg Cantor: His Mathematics and Philosophy of the Infinite.* Princeton University Press, Princeton, NJ, 1990.

[Daw84] A. P. Dawid. Statistical theory. The prequential approach. *Journal of the Royal Statistical Society*, Series A 147:278–292, 1984.

[Dem68] A. P. Dempster. A generalization of Bayesian inference. *Journal of the Royal Statistical Society*, Series B 30:205–247, 1968.

[Doo53] J. L. Doob. *Stochastic Processes.* Wiley, New York, 1953.

[EF98] G. W. Erickson and J. A. Fossa. *Dictionary of Paradox.* University Press of America, Lanham, MD, 1998.

[Fel68] W. Feller. *An Introduction to Probability Theory and its Applications.* Wiley, New York, 3rd edition, 1968.

[Fer67] T. S. Ferguson. *Mathematical Statistics: A Decision Theoretic Approach.* Academic Press, New York, 3rd edition, 1967.

[Fin37] B. de Finetti. Le prévision: ses lois logiques, ses sources subjectives. *Ann. Inst. Poincaré,* 7:1–68, 1937. [English translation: Foresight: Its logical laws, its subjective sources. In *Studies in Subjective Probability.* Krieger, New York, pages 55–118, 1980].

[Fin73] T. L. Fine. *Theories of Probability.* Academic Press, New York, 1973.

[Fis22] R. A. Fisher. On the mathematical foundations of theoretical statistics. *Philosophical Transactions of the Royal Society of London,* Series A 222:309–368, 1922.

[Fit96] M. C. Fitting. *First-Order Logic and Automated Theorem Proving.* Graduate Texts in Computer Science. Springer, Berlin, 2nd edition, 1996.

[FMG92] M. Feder, N. Merhav, and M. Gutman. Universal prediction of individual sequences. *IEEE Transactions on Information Theory,* 38:1258–1270, 1992.

[FS97] Y. Freund and R. E. Schapire. A decision-theoretic generalization of on-line learning and an application to boosting. *Journal of Computer and System Sciences,* 55(1):119–139, 1997.

[FT91] D. Fudenberg and J. Tirole. *Game Theory.* MIT Press, Cambridge, MA, 1991.

[Gác74] P. Gács. On the symmetry of algorithmic information. *Soviet Mathematics Doklady,* 15:1477–1480, 1974.

[Gác83] P. Gács. On the relation between descriptional complexity and algorithmic probability. *Theoretical Computer Science,* 22:71–93, 1983.

[Gal68] R. G. Gallager. *Information Theory and Reliable Communication.* Wiley, New York, 1968.

[GCSR95] A. Gelman, J. B. Carlin, H. S. Stern, and D. B. Rubin. *Bayesian Data Analysis.* Chapman & Hall / CRC, 1995.

[GH02] P. D. Grünwald and J. Y. Halpern. Updating probabilities. In *Proc. 18th Conf. on Uncertainty in Artificial Intelligence (UAI-2002),* pages 187–196. Morgan Kaufmann, San Francisco, CA, 2002.

[Gin87] M. L. Ginsberg, editor. *Readings in Nonmonotonic Reasoning.* Morgan Kaufmann, Los Altos, CA, 1987.

[Git89] J. C. Gittins. *Multi-Armed Bandit Allocation Indices.* Wiley, New York, 1989.

[GJ74] J. C. Gittins and D. M. Jones. A dynamic allocation index for the sequential design of experiments. In *Progress in Statistics,* pages 241–266. North-Holland, Amsterdam, 1974.

[Göd31] K. Gödel. Über formal unentscheidbare Sätze der Principia Mathematica und verwandter Systeme I. *Monatshefte für Matematik und Physik,* 38:173–198, 1931. [English translation by E. Mendelsohn: On undecidable propositions of formal mathematical systems. In *The Undecidable,* pages 39–71, Raven Press, New York, 1965].

[Grü98] P. D. Grünwald. *The Minimum Discription Length Principle and Reasoning under Uncertainty.* PhD thesis, Universiteit van Amsterdam, 1998.

[GTV01] P. Gács, J. Tromp, and P. M. B. Vitányi. Algorithmic statistics. *IEEE Transactions on Information Theory*, 47(6):2443–2463, 2001.

[Hac75] I. Hacking. *The Emergence of Probability.* Cambridge University Press, Cambridge, MA, 1975.

[Hal90] A. Hald. *A History of Probability and Statistics and Their Applications Before 1750.* Wiley, New York, 1990.

[Hal99] J. Y. Halpern. A counterexample to theorems of Cox and Fine. *Journal of Artificial Intelligence Research*, 10:67–85, 1999.

[Har79] J. Hartmanis. Relations between diagonalization, proof systems, and complexity gaps. *Theoretical Computer Science*, 8(2):239–253, 1979.

[Hec88] D. E. Heckerman. An axiomatic framework for belief updates. In *Uncertainty in Artificial Intelligence 2*, volume 5 of *Machine Intelligence and Pattern Recognition*, pages 11–22. North-Holland, Amsterdam, 1988.

[HHL86] E. J. Horvitz, D. E. Heckerman, and C. P. Langlotz. A framework for comparing alternative formalisms for plausible reasoning. In *Proc. 5th National Conf. on Artificial Intelligence (AAAI-86)*, volume 1, pages 210–214, Philadelphia, PA, 1986. Morgan Kaufmann.

[HKW98] D. Haussler, J. Kivinen, and M. K. Warmuth. Sequential prediction of individual sequences under general loss functions. *IEEE Transactions on Information Theory*, 44(5):1906–1925, 1998.

[HM04] M. Hutter and An. A. Muchnik. Universal convergence of semimeasures on individual random sequences. In *Proc. 15th International Conf. on Algorithmic Learning Theory (ALT-2004)*, volume 3244 of *LNAI*, Padova, 2004. Springer, Berlin.

[HMU01] J. E. Hopcroft, R. Motwani, and J. D. Ullman. *Introduction to Automata Theory, Language, and Computation.* Addison-Wesley, 2nd edition, 2001.

[HP04] M. Hutter and J. Poland. Prediction with expert advice by following the perturbed leader for general weights. In *Proc. 15th International Conf. on Algorithmic Learning Theory (ALT-2004)*, volume 3244 of *LNAI*, Padova, 2004. Springer, Berlin.

[Hug89] R. I. G. Hughes. *Structure and Interpretation of Quantum Mechanics.* Harvard University Press, Cambridge, MA, 1989.

[Hum39] D. Hume. *A Treatise of Human Nature, Book I.* [Edited version by L. A. Selby-Bigge and P. H. Nidditch, Oxford University Press, 1978], 1739.

[Hut00] M. Hutter. A theory of universal artificial intelligence based on algorithmic complexity. Technical Report cs.AI/0004001, München, 62 pages, 2000. http://arxiv.org/abs/cs.AI/0004001.

[Hut01a] M. Hutter. Convergence and error bounds for universal prediction of nonbinary sequences. In *Proc. 12th European Conf. on Machine Learning (ECML-2001)*, volume 2167 of *LNAI*, pages 239–250, Freiburg, 2001. Springer, Berlin.

[Hut01b] M. Hutter. General loss bounds for universal sequence prediction. In *Proc. 18th International Conf. on Machine Learning (ICML-2001)*, pages 210–217, Williamstown, MA, 2001. Morgan Kaufmann.

[Hut01c] M. Hutter. New error bounds for Solomonoff prediction. *Journal of Computer and System Sciences*, 62(4):653–667, 2001.

[Hut01d] M. Hutter. Towards a universal theory of artificial intelligence based on algorithmic probability and sequential decisions. In *Proc. 12th European*

*Conf. on Machine Learning (ECML-2001)*, volume 2167 of *LNAI*, pages 226–238, Freiburg, 2001. Springer, Berlin.

[Hut01e] M. Hutter. Universal sequential decisions in unknown environments. In *Proc. 5th European Workshop on Reinforcement Learning (EWRL-5)*, volume 27, pages 25–26. Onderwijsinsituut CKI, Utrecht Univ., 2001.

[Hut02a] M. Hutter. The fastest and shortest algorithm for all well-defined problems. *International Journal of Foundations of Computer Science*, 13(3):431–443, 2002.

[Hut02b] M. Hutter. Self-optimizing and Pareto-optimal policies in general environments based on Bayes-mixtures. In *Proc. 15th Annual Conf. on Computational Learning Theory (COLT 2002)*, volume 2375 of *LNAI*, pages 364–379, Sydney, 2002. Springer, Berlin.

[Hut03a] M. Hutter. Convergence and loss bounds for Bayesian sequence prediction. *IEEE Transactions on Information Theory*, 49(8):2061–2067, 2003.

[Hut03b] M. Hutter. On the existence and convergence of computable universal priors. In *Proc. 14th International Conf. on Algorithmic Learning Theory (ALT-2003)*, volume 2842 of *LNAI*, pages 298–312, Sapporo, 2003. Springer, Berlin.

[Hut03c] M. Hutter. Optimality of universal Bayesian prediction for general loss and alphabet. *Journal of Machine Learning Research*, 4:971–1000, 2003.

[Hut03d] M. Hutter. Sequence prediction based on monotone complexity. In *Proc. 16th Annual Conf. on Learning Theory (COLT-2003)*, volume 2777 of *LNAI*, pages 506–521, Washington, DC, 2003. Springer, Berlin.

[Hut04a] M. Hutter. Sequential predictions based on algorithmic complexity. *Journal of Computer and System Sciences*, 2004. to appear.

[Hut04b] M. Hutter. *Universal Artificial Intelligence: Sequential Decisions based on Algorithmic Probability.* Springer, Berlin, 2004.

[Jay78] E. T. Jaynes. Where do we stand on maximum entropy? In *The Maximum Entropy Formalism*, pages 15–118. MIT Press, Cambridge, MA, 1978.

[Jay03] E. T. Jaynes. *Probability Theory: The Logic of Science.* Cambridge University Press, Cambridge, MA, 2003.

[Jef83] R. C. Jeffrey. *The Logic of Decision.* University of Chicago Press, Chicago, IL, 2nd edition, 1983.

[Key21] J. M. Keynes. *A Treatise on Probability.* Macmillan, London, 1921.

[KHS01a] I. Kwee, M. Hutter, and J. Schmidhuber. Gradient-based reinforcement planning in policy-search methods. In *Proc. 5th European Workshop on Reinforcement Learning (EWRL-5)*, volume 27, pages 27–29. Onderwijsinsituut CKI, Utrecht Univ., 2001.

[KHS01b] I. Kwee, M. Hutter, and J. Schmidhuber. Market-based reinforcement learning in partially observable worlds. In *Proc. International Conf. on Artificial Neural Networks (ICANN-2001)*, volume 2130 of *LNCS*, pages 865–873, Vienna, 2001. Springer, Berlin.

[KLC98] L. P. Kaelbling, M. L. Littman, and A. R. Cassandra. Planning and acting in partially observable stochastic domains. *Artificial Intelligence*, 101:99–134, 1998.

[Kle36] S. Kleene. General recursive functions of natural numbers. *Mathematische Annalen*, 112:727–742, 1936.

[KLM96] L. P. Kaelbling, M. L. Littman, and A. W. Moore. Reinforcement learning: a survey. *Journal of Artificial Intelligence Research*, 4:237–285, 1996.

[Knu73] D. E. Knuth. *The Art of Computer Programming, volume I: Fundamental Algorithms.* Addison-Wesley, Reading, MA, 1973.

[Ko86] K.-I. Ko. On the notion of infinite pseudorandom sequences. *Theoretical Computer Science*, 48(1):9–33, 1986.

[Kol33] A. N. Kolmogorov. *Grundlagen der Wahrscheinlichkeitsrechnung.* Springer, Berlin, 1933. [English translation: *Foundations of the Theory of Probability.* Chelsea, New York, 2nd edition, 1956].

[Kol63] A. N. Kolmogorov. On tables of random numbers. *Sankhya, the Indian Journal of Statistics*, Series A 25, 1963.

[Kol65] A. N. Kolmogorov. Three approaches to the quantitative definition of information. *Problems of Information and Transmission*, 1(1):1–7, 1965.

[Kol83] A. N. Kolmogorov. Combinatorial foundations of information theory and the calculus of probabilities. *Russian Mathematical Surveys*, 38(4):27–36, 1983.

[Kra49] L. G. Kraft. A device for quantizing, grouping and coding amplitude modified pulses. Master's thesis, Electrical Engineering Department, Massachusetts Institute of Technology, Cambridge, MA, 1949.

[KS98] M. J. Kearns and S. Singh. Near-optimal reinforcement learning in polynomial time. In *Proc. 15th International Conf. on Machine Learning*, pages 260–268. Morgan Kaufmann, San Francisco, CA, 1998.

[KU63] A. N. Kolmogorov and V. A. Uspenskii. On the definition of an algorithm. *American Mathematical Society Translations*, 29:216–245, 1963. [Translated from Russian Original Uspekhi Matematicheskikh Nauk 13(4):3–28, 1958].

[KU87] A. N. Kolmogorov and V. A. Uspenskii. Algorithms and randomness. *Theory of Probability and its Applications*, 3(32):389–412, 1987.

[KV86] P. R. Kumar and P. P. Varaiya. *Stochastic Systems: Estimation, Identification, and Adaptive Control.* Prentice Hall, Englewood Cliffs, NJ, 1986.

[KW99] J. Kivinen and M. K. Warmuth. Averaging expert predictions. In *Proc. 4th European Conf. on Computational Learning Theory (Eurocolt-99)*, volume 1572 of *LNAI*, pages 153–167. Springer, Berlin, 1999.

[Kyb77] H. E. Kyburg. Randomness and the right reference class. *The Journal of Philosophy*, 74(9):501–521, 1977.

[Kyb83] H. E. Kyburg. The reference class. *Philosophy of Science*, 50:374–397, 1983.

[Lam87] M. van Lambalgen. *Random Sequences.* PhD thesis, University of Amsterdam, 1987.

[Lap12] P. Laplace. *Théorie analytique des probabilités.* Courcier, Paris, 1812. [English translation by F. W. Truscott and F. L. Emory: *A Philosophical Essay on Probabilities.* Dover, 1952].

[Lev73a] L. A. Levin. On the notion of a random sequence. *Soviet Mathematics Doklady*, 14(5):1413–1416, 1973.

[Lev73b] L. A. Levin. Universal sequential search problems. *Problems of Information Transmission*, 9:265–266, 1973.

[Lev74] L. A. Levin. Laws of information conservation (non-growth) and aspects of the foundation of probability theory. *Problems of Information Transmission*, 10(3):206–210, 1974.

[Lev84] L. A. Levin. Randomness conservation inequalities: Information and independence in mathematical theories. *Information and Control*, 61:15–37, 1984.

[Li 03] M. Li et al. The similarity metric. In *Proc. 14th Annual ACM-SIAM Symposium on Discrete Algorithms (SODA-03)*, pages 863–872. ACM Press, New York, 2003.

[Lov69a] D. W. Loveland. On minimal-program complexity measures. In *Proc. 1st ACM Symposium on Theory of Computing*, pages 61–78. ACM Press, New York, 1969.

[Lov69b] D. W. Loveland. A variant of the Kolmogorov concept of complexity. *Information and Control*, 15(6):510–526, 1969.

[Luc61] J. R. Lucas. Minds, machines, and Gödel. *Philosophy*, 36:112–127, 1961.

[LV77] L. A. Levin and V. V. V'yugin. Invariant properties of informational bulks. In *Proc. 6th Symposium on Mathematical Foundations of Computer Science*, volume 53 of *LNCS*, pages 359–364. Springer, Berlin, 1977.

[LV91] M. Li and P. M. B. Vitányi. Learning simple concepts under simple distributions. *SIAM Journal on Computing*, 20(5):911–935, 1991.

[LV92a] M. Li and P. M. B. Vitányi. Inductive reasoning and Kolmogorov complexity. *Journal of Computer and System Sciences*, 44:343–384, 1992.

[LV92b] M. Li and P. M. B. Vitányi. Philosophical issues in Kolmogorov complexity (invited lecture). In *Proceedings on Automata, Languages and Programming (ICALP-92)*, pages 1–15. Springer, Berlin, 1992.

[LV97] M. Li and P. M. B. Vitányi. *An Introduction to Kolmogorov Complexity and its Applications*. Springer, Berlin, 2nd edition, 1997.

[LW89] N. Littlestone and M. K. Warmuth. The weighted majority algorithm. In *30th Annual Symposium on Foundations of Computer Science*, pages 256–261, Research Triangle Park, NC, 1989. IEEE.

[LW94] N. Littlestone and M. K. Warmuth. The weighted majority algorithm. *Information and Computation*, 108(2):212–261, 1994.

[MA93] A. W. Moore and C. G. Atkeson. Prioritized sweeping: Reinforcement learning with less data and less time. *Machine Learning*, 13:103–130, 1993.

[McC80] J. McCarthy. Circumscription—A form of non-monotonic reasoning. *Artificial Intelligence*, 13(1–2):27–39, 1980.

[McC95] A. K. McCallum. Instance-based utile distinctions for reinforcement learning with hidden state. In *Proc. 12th International Conf. on Machine Learning*, pages 387–395, Tahoe, CA, 1995. Morgan Kaufmann.

[MD80] D. McDermott and J. Doyle. Nonmonotonic logic 1. *Artificial Intelligence*, 13:41–72, 1980.

[MF98] N. Merhav and M. Feder. Universal prediction. *IEEE Transactions on Information Theory*, 44(6):2124–2147, 1998.

[Mic66] D. Michie. Game-playing and game-learning automata. In *Advances in Programming and Non-Numerical Computation*, pages 183–200. Pergamon, New York, 1966.

[Mis19] R. von Mises. Grundlagen der Wahrscheinlichkeitsrechnung. *Mathematische Zeitschrift*, 5:52–99, 1919. Correction, *Ibid.*, volume 6, 1920, [English translation in: *Probability, Statistics, and Truth*, Macmillan, 1939].

[Mis28] R. von Mises. *Wahrscheinlichkeit, Statistik und Wahrheit*. Springer, Berlin, 1928. [English translation: *Probability, Statistics, and Truth*, Allen and Unwin, London, 1957].

[ML66] P. Martin-Löf. The definition of random sequences. *Information and Control*, 9(6):602–619, 1966.

[Mor88] H. Moravec. *Mind Children: The Future of Robot and Human Intelligence.* Harvard University Press, Cambridge, MA, 1988.

[Mos65] F. Mosteller. *Fifty Challenging Problems in Probability with Solutions.* Addison-Wesley, Reading, MA, 1965.

[NM44] J. Von Neumann and O. Morgenstern. *Theory of Games and Economic Behavior.* Princeton University Press, Princeton, NJ, 1944.

[Odi89] P. Odifreddi. *Classical Recursion Theory, volume 1.* North–Holland, Amsterdam, 1989.

[Odi99] P. Odifreddi. *Classical Recursion Theory, volume 2.* Elsevier, Amsterdam, 1999.

[OR94] M. J. Osborne and A. Rubenstein. *A Course in Game Theory.* The MIT Press, Cambridge, MA, 1994.

[Par95] J. B. Paris. *The Uncertain Reasoner's Companion: A Mathematical Perspective.* Cambridge University Press, Cambridge, 1995.

[Pas54] B. Pascal. Letters to Fermat, 1654.

[Pen89] R. Penrose. *The Emperor's New Mind.* Oxford University Press, 1989.

[Pen94] R. Penrose. *Shadows of the Mind, A Search for the Missing Science of Consciousness.* Oxford University Press, 1994.

[PF97] X. Pintado and E. Fuentes. A forecasting algorithm based on information theory. In *Objects at Large*, page 209. Université de Genève, 1997.

[PH04a] J. Poland and M. Hutter. Convergence of discrete MDL for sequential prediction. In *Proc. 17th Annual Conf. on Learning Theory (COLT-2004)*, volume 3120 of *LNAI*, pages 300–314, Banff, 2004. Springer, Berlin.

[PH04b] J. Poland and M. Hutter. On the convergence speed of MDL predictions for Bernoulli sequences. In *Proc. 15th International Conf. on Algorithmic Learning Theory (ALT-2004)*, volume 3244 of *LNAI*, Padova, 2004. Springer, Berlin.

[Pin64] M. S. Pinsker. *Information and Information Stability of Random Variables and Processes.* Holden-Day, San-Francisco, CA, 1964. [Russian original, Izd. Akad. Nauk, 1960].

[Pop34] K. R. Popper. *Logik der Forschung.* Springer, Berlin, 1934. [English translation: *The Logic of Scientific Discovery* Basic Books, New York, 1959, and Hutchinson, London, revised edition, 1968].

[Pos44] E. L. Post. Recursively enumerable sets of positive integers and their decision problems. *Bulletin of the American Mathematical Society*, 50:284–316, 1944.

[Put63] H. Putnam. 'Degree of confirmation' and inductive logic. In *The Philosophy of Rudolf Carnap.* Open Court, La Salle, IL, 1963.

[Ram31] F. P. Ramsey. Truth and probability. In *The Foundations of Mathematics: Collected Papers of Frank P. Ramsey*, pages 156–198. Routledge and Kegan Paul, London, 1931.

[Rei49] H. Reichenbach. *The Theory of Probability: An Inquiry into the Logical and Mathematical Foundations of the Calculus of Probability.* University of California Press, Berkeley, CA, 2nd edition, 1949.

[Rei80] R. Reiter. A logic for default reasoning. *Artificial Intelligence*, 13:81–132, 1980.

[Res01] N. Rescher. *Paradoxes: Their Roots, Range, and Resolution.* Open Court, Lanham, MD, 2001.

[Rin94] M. Ring. *Continual Learning in Reinforcement Environments.* PhD thesis, University of Texas, Austin, 1994.

[Ris78] J. J. Rissanen. Modeling by shortest data description. *Automatica*, 14:465–471, 1978.

[Ris89] J. J. Rissanen. *Stochastic Complexity in Statistical Inquiry.* World Scientific, Singapore, 1989.

[Ris96] J. J. Rissanen. Fisher information and stochastic complexity. *IEEE Trans on Information Theory*, 42(1):40–47, 1996.

[RN95] S. J. Russell and P. Norvig. *Artificial Intelligence. A Modern Approach.* Prentice-Hall, Englewood Cliffs, NJ, 1995.

[Rob52] H. Robbins. Some aspects of the sequential design of experiments. *Bulletin of the American Mathematical Society*, 58:527–535, 1952.

[Rog67] H. Rogers. *Theory of Recursive Functions and Effective Computability.* McGraw-Hill, New York, 1967.

[RV01] A. Robinson and A. Voronkov, editors. *Handbook of Automated Reasoning.* Elsevier Science, 2001.

[Sam59] A. L. Samuel. Some studies in machine learning using the game of checkers. *IBM Journal on Research and Development*, 3:210–229, 1959.

[Sav54] L. J. Savage. *The Foundations of Statistics.* Wiley, New York, 1954.

[SB98] R. S. Sutton and A. G. Barto. *Reinforcement Learning: An Introduction.* MIT Press, Cambridge, MA, 1998.

[Sch71] C. P. Schnorr. *Zufälligkeit und Wahrscheinlichkeit.* Springer, Berlin, 1971.

[Sch73] C. P. Schnorr. Process complexity and effective random tests. *Journal of Computer and System Sciences*, 7(4):376–388, 1973.

[Sch80] A. Schönhage. Storage modification machines. *SIAM Journal on Computing*, 9(3):490–508, 1980.

[Sch95] J. Schmidhuber. Discovering solutions with low Kolmogorov complexity and high generalization capability. In *Proc. 12th International Conf. on Machine Learning*, pages 488–496. Morgan Kaufmann, 1995.

[Sch97] J. Schmidhuber. Discovering neural nets with low Kolmogorov complexity and high generalization capability. *Neural Networks*, 10(5):857–873, 1997.

[Sch99] M. Schmidt. Time-bounded Kolmogorov complexity may help in search for extra terrestrial intelligence (SETI). *Bulletin of the European Association for Theoretical Computer Science*, 67:176–180, 1999.

[Sch00] J. Schmidhuber. Algorithmic theories of everything. Report IDSIA-20-00, quant-ph/0011122, IDSIA, Manno (Lugano), Switzerland, 2000.

[Sch02a] J. Schmidhuber. Hierarchies of generalized Kolmogorov complexities and nonenumerable universal measures computable in the limit. *International Journal of Foundations of Computer Science*, 13(4):587–612, 2002.

[Sch02b] J. Schmidhuber. The speed prior: A new simplicity measure yielding near-optimal computable predictions. In *Proc. 15th Conf. on Computational Learning Theory (COLT-2002)*, volume 2375 of *LNAI*, pages 216–228, Sydney, 2002. Springer, Berlin.

[Sch03a] J. Schmidhuber. Bias-optimal incremental problem solving. In *Advances in Neural Information Processing Systems 15*, pages 1571–1578. MIT Press, Cambridge, MA, 2003.

[Sch03b] J. Schmidhuber. Gödel machines: Self-referential universal problem solvers making provably optimal self-improvements. Report IDSIA-17-03, IDSIA, Manno (Lugano), Switzerland, 2003.

[Sch04] J. Schmidhuber. Optimal ordered problem solver. *Machine Learning*, 54(3):211–254, 2004.

[Sea80] J. Searle. Minds, brains, and programs. *Behavioral & Brain Sciences*, 3:417–458, 1980.

[SH02] J. Schmidhuber and M. Hutter. Universal learning algorithms and optimal search. *NIPS 2001 Workshop*, 2002. http://www.idsia.ch/~marcus/idsia/nipsws.htm.

[Sha48] C. E. Shannon. A mathematical theory of communication. *Bell System Technical Journal*, 27:379–423, 623–656, 1948.

[Sha76] G. Shafer. *A Mathematical Theory of Evidence*. Princeton University Press, Princeton, NJ, 1976.

[Sha85] G. Shafer. Conditional probability. *International Statistical Review*, 53(3):261–277, 1985.

[Sho67] J. R. Shoenfield. *Mathematical Logic*. Addison-Wesley, Reading, MA, 1967.

[Sho76] E. H. Shortliffe. *Computer-Based Medical Consultations: MYCIN*. Elsevier/North-Holland, Amsterdam, 1976.

[Slo04] N. J. A. Sloane. The on-line encyclopedia of integer sequences. *AT&T*, 2004. http://www.research.att.com/~njas/sequences/.

[Sol64] R. J. Solomonoff. A formal theory of inductive inference: Parts 1 and 2. *Information and Control*, 7:1–22 and 224–254, 1964.

[Sol78] R. J. Solomonoff. Complexity-based induction systems: Comparisons and convergence theorems. *IEEE Transaction on Information Theory*, IT-24:422–432, 1978.

[Sol86] R. J. Solomonoff. The application of algorithmic probability to problems in artificial intelligence. In *Uncertainty in Artificial Intelligence*, pages 473–491. Elsevier Science/North-Holland, Amsterdam, 1986.

[Sol97] R. J. Solomonoff. The discovery of algorithmic probability. *Journal of Computer and System Sciences*, 55(1):73–88, 1997.

[Sol99] R. J. Solomonoff. Two kinds of probabilistic induction. *Computer Journal*, 42(4):256–259, 1999.

[Sto01] D. Stork. Foundations of Occam's razor and parsimony in learning. *NIPS 2001 Workshop*, 2001. http://www.rii.ricoh.com/~stork/OccamWorkshop.html.

[Str69] V. Strassen. Gaussian elimination is not optimal. *Numerische Mathematik*, 13:354–356, 1969.

[Sut88] R. S. Sutton. Learning to predict by the methods of temporal differences. *Machine Learning*, 3:9–44, 1988.

[SV88] J. I. Seiferas and P. M. B. Vitányi. Counting is easy. *Journal of the ACM*, 35(4):985–1000, 1988.

[Szé86] G. J. Székely. *Paradoxes in Probability Theory and Mathematical Statistics*. Reidel, Dordrecht, 1986.

[SZW97] J. Schmidhuber, J. Zhao, and M. A. Wiering. Shifting inductive bias with success-story algorithm, adaptive Levin search, and incremental self-improvement. *Machine Learning*, 28:105–130, 1997.

[Tes94] G. Tesauro. "TD"-Gammon, a self-teaching backgammon program, achieves master-level play. *Neural Computation*, 6(2):215–219, 1994.

[Tri69] M. Tribus. *Rational Descriptions, Decisions and Designs*. Pergamon, New York, 1969.

[Tur36] A. M. Turing. On computable numbers, with an application to the Entscheidungsproblem. *Proc. London Mathematical Society*, 2(42):230–265, 1936.

[Tur50] A. M. Turing. Computing machinery and intelligence. *Mind*, 1950.

[USS90] V. A. Uspenskii, A. L. Semenov, and A. K. Shen. Can an individual sequence of zeros and ones be random? *Russian Mathematical Surveys*, 45, 1990.

[Val84] L. G. Valiant. A theory of the learnable. *Communications of the ACM*, 27(11):1134–1142, 1984.

[Vap99] V. N. Vapnik. *The Nature of Statistical Learning Theory*. Springer, Berlin, 2nd edition, 1999.

[VL00] P. M. B. Vitányi and M. Li. Minimum description length induction, Bayesianism, and Kolmogorov complexity. *IEEE Transactions on Information Theory*, 46(2):446–464, 2000.

[Vog60] W. Vogel. An asymptotic minimax theorem for the two-armed bandit problem. *Annals of Mathematical Statistics*, 31:444–451, 1960.

[Vov87] V. G. Vovk. On a randomness criterion. *Soviet Mathematics Doklady*, 35(3):656–660, 1987.

[Vov92] V. G. Vovk. Universal forecasting algorithms. *Information and Computation*, 96(2):245–277, 1992.

[Vov01] V. G. Vovk. Competitive on-line statistics. *International Statistical Review*, 69:213–248, 2001.

[VV02] N. Vereshchagin and P. M. B. Vitányi. Kolmogorov's structure functions with an application to the foundations of model selection. In *Proc. 43rd Symposium on Foundations of Computer Science*, pages 751–760, Vancouver, 2002.

[VW98] V. G. Vovk and C. Watkins. Universal portfolio selection. In *Proc. 11th Conf. on Computational Learning Theory (COLT-98)*, pages 12–23. ACM Press, New York, 1998.

[Wal37] A. Wald. Die Widerspruchsfreiheit des Kollektivbegriffs in der Wahrscheinlichkeitsrechnung. In *Ergebnisse eines Mathematischen Kolloquiums*, volume 8, pages 38–72, 1937.

[Wal91] P. Walley. *Statistical Reasoning with Imprecise Probabilities*. Chapman and Hall, London, 1991.

[Wan96] Y. Wang. *Randomness and Complexity*. PhD thesis, Universität Heidelberg, 1996.

[Wat89] C. Watkins. *Learning from Delayed Rewards*. PhD thesis, King's College, Oxford, 1989.

[WB68] C. S. Wallace and D. M. Boulton. An information measure for classification. *Computer Journal*, 11(2):185–194, 1968.

[WD92] C. Watkins and P. Dayan. Q-learning. *Machine Learning*, 8:279–292, 1992.

[WM97] D. H. Wolpert and W. G. Macready. No free lunch theorems for optimization. *IEEE Transactions on Evolutionary Computation*, 1(1):67–82, 1997.

[WS96] M. A. Wiering and J. Schmidhuber. Solving POMDPs with Levin search and EIRA. In *Proc. 13th International Conf. on Machine Learning*, pages 534–542, Bari, Italy, 1996.

[WS98] M. A. Wiering and J. Schmidhuber. Fast online "Q"($\lambda$). *Machine Learning*, 33(1):105–116, 1998.

[Yam98] K. Yamanishi. A decision-theoretic extension of stochastic complexity and its applications to learning. *IEEE Transactions on Information Theory*, 44:1424–1439, 1998.

[YEYS04] R. Yaroshinsky, R. El-Yaniv, and S. Seiden. How to better use expert advice. *Machine Learning*, 55(3):271–309, 2004.

[Zad65] L. A. Zadeh. Fuzzy sets. *Information and Control*, 8:338–353, 1965.

[Zad78] L. A. Zadeh. Fuzzy sets as a basis for a theory of possibility. *Fuzzy Sets and Systems*, 1:3–28, 1978.

[Zim91] H.-J. Zimmermann. *Fuzzy Set Theory–And Its Applications.* Kluwer, Dordrecht, 2nd edition, 1991.

[ZL70] A. K. Zvonkin and L. A. Levin. The complexity of finite objects and the development of the concepts of information and randomness by means of the theory of algorithms. *Russian Mathematical Surveys*, 25(6):83–124, 1970.

# Index

## Q



## R



## S

## Monographs in Theoretical Computer Science • An EATCS Series

K. Jensen
**Coloured Petri Nets**
*Basic Concepts,* Analysis Methods and Practical Use, Vol. 1
2nd ed.

K. Jensen
**Coloured Petri Nets**
Basic Concepts, *Analysis Methods* and Practical Use, Vol. 2

K. Jensen
**Coloured Petri Nets**
Basic Concepts, Analysis Methods and *Practical Use,* Vol. 3

A. Nait Abdallah
**The Logic of Partial Information**

Z. Fülöp, H. Vogler
**Syntax-Directed Semantics**
Formal Models Based on Tree Transducers

A. de Luca, S. Varricchio
**Finiteness and Regularity in Semigroups and Formal Languages**

E. Best, R. Devillers, M. Koutny
**Petri Net Algebra**

S.P. Demri, E.S. Orłowska
**Incomplete Information: Structure, Inference, Complexity**

J.C.M. Baeten, C.A. Middelburg
**Process Algebra with Timing**

L.A. Hemaspaandra, L. Torenvliet
**Theory of Semi-Feasible Algorithms**

E. Fink, D. Wood
**Restricted-Orientation Convexity**

Zhou Chaochen, M. R. Hansen
**Duration Calculus**
A Formal Approach to Real-Time Systems

M. Große-Rhode
**Semantic Integration of Heterogeneous Software Specifications**

## Texts in Theoretical Computer Science • An EATCS Series

J. L. Balcázar, J. Díaz, J. Gabarró
**Structural Complexity I**

M. Garzon
**Models of Massive Parallelism**
Analysis of Cellular Automata
and Neural Networks

J. Hromkovič
**Communication Complexity
and Parallel Computing**

A. Leitsch
**The Resolution Calculus**

G. Păun, G. Rozenberg, A. Salomaa
**DNA Computing**
New Computing Paradigms

A. Salomaa
**Public-Key Cryptography**
2nd ed.

K. Sikkel
**Parsing Schemata**
A Framework for Specification
and Analysis of Parsing Algorithms

H. Vollmer
**Introduction to Circuit Complexity**
A Uniform Approach

W. Fokkink
**Introduction to Process Algebra**

K. Weihrauch
**Computable Analysis**
An Introduction

J. Hromkovič
**Algorithmics for Hard Problems**
Introduction to Combinatorial
Optimization, Randomization,
Approximation, and Heuristics
2nd ed.

S. Jukna
**Extremal Combinatorics**
With Applications
in Computer Science

P. Clote, E. Kranakis
**Boolean Functions
and Computation Models**

L. A. Hemaspaandra, M. Ogihara
**The Complexity Theory Companion**

C.S. Calude
**Information and Randomness.**
An Algorithmic Perspective
2nd ed.

J. Hromkovič
**Theoretical Computer Science**
Introduction to Automata,
Computability, Complexity,
Algorithmics, Randomization,
Communication and Cryptography

A. Schneider
**Verification of Reactive Systems**
Formal Methods and Algorithms

S. Ronchi Della Rocca, L. Paolini
**The Parametric Lambda Calculus**
A Metamodel for Computation

Y. Bertot, P. Castéran
**Interactive Theorem Proving
and Program Development**
Coq'Art: The Calculus
of Inductive Constructions

L. Libkin
**Elements of Finite Model Theory**

M. Hutter
**Universal Artificial Intelligence**
Sequential Decisions
Based on Algorithmic Probability